MANUFACTURING SCIENCE AND TECHNOLOGY

MANUFACTURING SCIENCE AND TECHNOLOGY

ANUFACTURING PROCESSES AND MACHINE TOOLS

(SECOND EDITION)

K VARA PRASADA RAO
BE, MTech. (Production IIT, Kharagpur)
FIE, FIIPE, LMISTE, LMISNT, LMFPSI
Former Principal
Loyala Institute of Technology and Management
Sattenapalli
&
Professor and Head of Mechanical Engineering
Narasaraopet Engineering College, Narasaraopet, (AP)

Publishing Globally

NEW AGE INTERNATIONAL (P) LIMITED, PUBLISHERS
LONDON • NEW DELHI • NAIROBI
Bangalore • Chennai • Cochin • Guwahati • Hyderabad • Kolkata • Lucknow • Mumbai
Visit us at www.newagepublishers.com

Published by New Age International (P) Ltd., Publishers
Second Edition: 2009
Reprint: 2017

GLOBAL OFFICES

- **New Delhi** **NEW AGE INTERNATIONAL (P) LIMITED, PUBLISHERS**
7/30 A, Daryaganj, New Delhi-110002, (INDIA)
Tel.: (011) 23253771, 23253472, **Telefax:** 23267437, 43551305
E-mail: contactus@newagepublishers.com • Visit us at www.newagepublishers.com
- **London** **NEW AGE INTERNATIONAL (UK) LTD.**
27 Old Gloucester Street, London, WC1N 3AX, UK
E-mail: info@newacademicscience.co.uk • Visit us at www.newacademicscience.co.uk
- **Nairobi** **NEW AGE GOLDEN (EAST AFRICA) LTD.**
Ground Floor, Westlands Arcade, Chiromo Road (Next to Naivas Supermarket)
Westlands, Nairobi, KENYA, **Tel.:** 00-254-713848772, 00-254-725700286
E-mail: kenya@newagepublishers.com

BRANCHES

- **Bangalore** 37/10, 8th Cross (Near Hanuman Temple), Azad Nagar, Chamarajpet, Bangalore- 560 018
Tel.: (080) 26756823, **Telefax:** 26756820, **E-mail: bangalore@newagepublishers.com**
- **Chennai** 26, Damodaran Street, T. Nagar, Chennai-600 017, **Tel.:** (044) 24353401, **Telefax:** 24351463
E-mail: chennai@newagepublishers.com
- **Cochin** CC-39/1016, Carrier Station Road, Ernakulam South, Cochin-682 016, **Tel.:** (0484) 2377303, **Telefax:** 4051304
E-mail: cochin@newagepublishers.com
- **Guwahati** Hemsen Complex, Mohd. Shah Road, Paltan Bazar, Near Starline Hotel, Guwahati-781 008,
Tel.: (0361) 2513881, **Telefax:** 2543669, **E-mail: guwahati@newagepublishers.com**
- **Hyderabad** 105, 1st Floor, Madhiray Kaveri Tower, 3-2-19, Azam Jahi Road, Near Kumar Theater, Nimboliadda
Kachiguda, Hyderabad-500 027, **Tel.:** (040) 24652456, **Telefax:** 24652457
E-mail: hyderabad@newagepublishers.com
- **Kolkata** RDB Chambers (Formerly Lotus Cinema) 106A, 1st Floor, S N Banerjee Road, Kolkata-700 014
Tel.: (033) 22273773, **Telefax:** 22275247, **E-mail: kolkata@newagepublishers.com**
- **Lucknow** 16-A, Jopling Road, Lucknow-226 001, **Tel.:** (0522) 2209578, 4045297, **Telefax:** 2204098
E-mail: lucknow@newagepublishers.com
- **Mumbai** 142C, Victor House, Ground Floor, N.M. Joshi Marg, Lower Parel, Mumbai-400 013
Tel.: (022) 24927869, **Telefax:** 24915415, **E-mail: mumbai@newagepublishers.com**
- **New Delhi** 22, Golden House, Daryaganj, New Delhi-110 002, **Tel.:** (011) 23262368, 23262370, **Telefax:** 43551305
E-mail: sales@newagepublishers.com

ISBN: 978-81-224-2675-5
C-16-11-9783

Printed in India at Ajit Printers, Delhi.
Typeset at In-house, Delhi.

NEW AGE INTERNATIONAL (P) LIMITED, PUBLISHERS
7/30 A, Daryaganj, New Delhi-110002
Visit us at **www.newagepublishers.com**
(CIN: U74899DL1966PTC004618)

Dedicated to

"My Master Parthasarathi Rajagopalachary (chariji garu)"
Shri Ram Chandra Mission, Chennai

Preface to the Second Edition

It is a great pleasure for me to present the revised edition of **Manufacturing Science and Technology.** This is a core subject for mechanical industrial and production engineering students at both degree and diploma levels. Some more chapters are added to fulfil the need of students.

This book covers the syllabus for 1st Semesters Part A, covers various manufacturing processes which is sufficient for 3rd B.Tech 1st Semester students. Part B covers various machine tools. This is more than sufficient for 3rd B.Tech 2nd Semester students. This book covers the syllabus prescribed by JNTU, Hyderabad and also satisfies the syllabus of all other universities. Objective type questions and answers are given in Appendices I & II at the end of Part A and Part B these are very very useful for students to score more marks in online examinations.

I wish to thank Sri M.V. Koteswara Rao (Chairman of Narasaraopet Engineering College, Mittapalli Engineering College, Narasaraopet Institute of Technology, MCA, MBA, Chairman B. Pharmacy College) for his encouragement. I wish to thank Secretary M. Satyanarayana Rao, Joint Secretary M. Ramesh Babu for their support.

I am extremely thankful to Dr. B.V. Rama Mohan Rao, Principal, Narasaraopet Engineering College for his encouragement.

My thanks to Prof. Raghu Ramulu, Sanjeev Reddy, Jayarami Reddy, Dr. B.D. Sarma, Dr. Veeranna of GPREC Kurnol, Smt. K. Lakshmi Chaitanya Asst. Professor of Narasaraopet Engineering College for their timely help.

My thanks to wife Jaya, daughter Vani and son Vijay for their tolerance, understanding and moral support in completing this book.

Last but not the least I am thankful to Mr. Saumya Gupta, Managing Director, M/s New Age International (P) Limited, Publishers and the editorial department for their untiring effort to publish the book in a short span of time with nice get up.

Prof. K. Vara Prasada Rao

Preface to the First Edition

Manufacturing Science and Technology is a core subject for mechanical, industrial and production engineering students at both degree and diploma levels. Keeping the requirements of these students in mind, this book has been written in simple language accompanied by the relevant specifications, description and with pictorial views for easy understanding of the conventional methods of production.

The highlights of the book are:

In Part A, various manufacturing processes like foundry, plastic deformation processes, welding and powder metallurgy are discussed in detail with examples and figures.

In Part B, various machine tools used in manufacturing like lathe, capstan and turret lathe, as well as milling, drilling, shaping and grinding machines are discussed with their constructional features, mechanics, operation details, various tools and attachments used.

This book covers the syllabus requirements satisfactorily for all universities having engineering courses.

I wish to thank Sri Sada Siva Reddy, Correspondent of Siddipeta Engineering College, Siddipeta and K. Krishna Reddy, Correspondent of Newton's Institute of Engineering College, Macherla for their encouragement to write this book.

I wish to record my indebtedness to Dr. P. Jayarami Reddy, Principal, GPREC for his encouragement and support in the preparation of the manuscript. I thank Dr. B.D. Sarma, Y.V. Mohan Reddy, V. Satish Kumar, V. Veeranna of I.P.E. department for editing the manuscript.

I am grateful to the publishers for their help in bringing out this book in a short time.

I thank my wife Jaya, my daughter Vani and my son Vijay for their understanding and moral support in completing this book.

Suggestions to improve this book are most welcome.

Prof. K. Vara Prasada Rao

Contents

PART B
MACHINE TOOLS

PART A

MANUFACTURING PROCESSES

1

Foundry

1.1 INTRODUCTION

Foundry is the most ancient industry deals with the manufacturing of metal castings. Metal casting is the process of pouring a material in a liquid form into a mould and allowing it to solidify to produce the desired product. Sand casting is a very old technique, but improved methods such as shell molding, investment casting, die casting, centrifugal casting, etc. are finding wider applications.

1.2 THE SEQUENCE OF STEPS INVOLVED IN CASTING

1. Pattern making
2. Mould and core making
3. Melting and pouring
4. Fettling
5. Inspection and testing

1.2.1 Pattern Making

Pattern is the replica or full size model of the casting to be made. It gives shape to the mould cavity where the molten metal solidifies to the desired its form and size. The process of making a pattern is known as 'pattern making'.

1.2.1.1 Difference between Pattern and Casting

(a) Pattern is slightly larger in size than casting because pattern is given shrinkage allowance and machining allowance.

(b) Pattern is slightly tapered because it is given draft allowance of 1° for external and 3° for internal surfaces.

(c) Pattern is provided with core prints to support the core to make holes in the casting.

(d) Pattern may be made in two or three pieces where as casting is a single piece.

(e) Pattern may not have all slots and holes of casting because they are machined afterwards.

1.2.1.2 Pattern Materials

The following materials are generally used for making patterns:

(a) wood
(b) metals
(c) plasters
(d) plastics
(e) wax

(a) Wood patterns

Wood is the most common material used for making patterns. As pattern material, it offers the following advantages:

(i) It is cheap
(ii) Easy of availability
(iii) Ease of shaping
(iv) Light in weight
(v) Its surface can be easily made smooth by sanding
(vi) Its surface can be preserved by shellac coating
(vii) Wood can be cut and fabricated into many forms.

However, it has the following disadvantages:

(i) It is affected by moisture, thereby it swells or shrinks.
(ii) It deforms on drying.
(iii) It wears out quickly as a result of sand abrasion.
(iv) If not stored properly, it may warp.
(v) Its strength is low and tends to break on rough usage.

The following kinds of wood are most commonly used for patterns.

(i) Pine wood
(ii) Teak wood
(iii) Mahogany
(iv) Deodar

(b) Metal patterns

Metal patterns are used for mass production of castings. Compare to wood patterns, metal patterns offer the following advantages:

(i) Strong and durable
(ii) They do not deform in storage
(iii) Wear resistance and maintains dimensional stability

On other hand metal patterns have the following disadvantages:

(i) Metal patterns are heavy
(ii) More difficult to repair and modify
(iii) They are liable to rust

The metals mainly used for making patterns are:

(*a*) *Steel*: It posses excellent wear resistance and strength, but it has poor resistance to corrosion. These can be easily repaired and are used for mass production.

(*b*) *Cast Iron*: Cast Iron is cheap and strong. It posses good cast ability and machinability and gives good smooth mould surface.

(*c*) *Brass*: Brass pattern is strong, easy working, corrosion resistant and posses high machinability and wear resistance. But brass patterns are very heavy and expensive. These are extensively used for making small patterns.

(*d*) *Aluminium*: Aluminium and its alloys are the most commonly used for pattern. It is probably the best metal because of low melting point, soft, and easy to shape, and corrosion resistant. However, due to low strength, it is subjected to damage by rough usage.

(*e*) *Plaster Patterns*: Gypsum cement know as plaster of paris is used for making patterns. It has a high compressive strength (upto 300 kg/cm^2) and controlled expansion. When the plaster is mixed with water, it forms a plastic mass capable of being cast into a mould made by a sweep pattern or a master pattern in order to obtain a plaster pattern. Plaster is used for making small and intricate patterns and core boxes.

(c) *Plastic patterns*

Plastics have replaced other materials and finding their place as a modern pattern materials. Thermo-setting resigns (phenolic resign, Epoxy resign) have the desired properties of a pattern material.

Advantages

(i) Durable
(ii) Provides a smooth surface
(iii) Moisture resistant
(iv) Light weight
(v) Wear and corrosion resistant
(vi) Plastic patterns are easy to make
(vii) They posses good compressive strength
(viii) Good resistance to chemicals
(ix) Better adhesive qualities.

Limitations

(i) Plastic patterns are fragile and thus light sections may need metal reinforcements.
(ii) Plastic patterns may not work well when subject to shocks as in machine molding (jolt machines).

(d) Wax patterns

Wax patterns provide high degree of surface finish and dimensional accuracy to castings. After being molded, the wax pattern is not taken out of the mould like other patterns, rather the mould is inverted and heated, the molten wax comes out of the mould. Thus, there is no chance of the mould cavity getting damaged while removing the pattern. Wax patterns are made in water cooled moulds or dies.

The most commonly used waxes are paraffin wax, carnauba wax, shellac wax, bees wax, ceresin wax.

Wax patterns are excellent for the investment casting process.

1.2.1.3 Factors Affecting the Selection of Pattern Materials

The selection of pattern materials depends on the following factors:

(a) The number of castings required

(b) Dimensional accuracy required (e.g. minimum thickness desired, intricacy of parts, finish required in the casting).

(c) Moulding process used (e.g. Hand moulding or Machine moulding).

(d) Size and shape of the casting.

1.2.1.4 Pattern Allowances

A pattern differ from the casting in certain dimensions. When the pattern is prepared, certain allowances are given on the sizes of casting. These are known as pattern allowances. Pattern allowances are as follows:

(a) Shrinkage or Contraction allowance

(b) Draft or taper allowance

(c) Finishing or machining allowance

(d) Shaking or rapping allowance

(e) Distortion or camber allowance.

(a) Shrinkage or Contraction allowance

Almost all metals used for casting shrink or contract volumetrically after solidification and cooling in the mould. Hence to compensate this shrinkage, the pattern must be made larger than the finished casting by an amount known as 'Shrinkage Allowance'. Although contraction is volumetrically, the correction for it usually expressed linearly as a ratio, a percentage or in mm per meter.

Pattern makers use special measuring rules (Shrinkage Rules) that take account of different contraction that occur when casting various metals. The rule have slightly larger divisions so that they measure over size. Shrinkage allowances for different cast metals are given in Table 1.1.

Table 1.1: Shrinkage Allowances for Various Metals

Metals/Alloys	*Shrinkage Allowance (mm/m)*
Gray Cast Iron	10.5
White Cast Iron	20.0
Plain Carbon Steel	21.0
Chromium Steel	20.0
Manganese Steel	25.0–38.0
Aluminium	17.0
Aluminium Alloys	12.5–15.0
Brass	15.5
Bronze	15.5–20.8
Copper	16.0
Magnesium	17.0
Zinc	24.0

Example 1.1: Calculate the dimensions of the pattern for the casting shown in Fig.1.1(a) for shrinkage allowance. The casting is made of white cast iron.

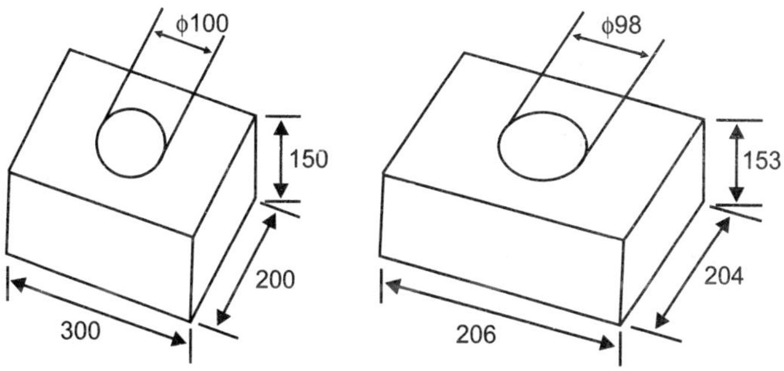

(a) Casting Dimensions (b) Pattern Dimensions

Fig. 1.1 Provision of Shrinkage Allowances

Shrinkage allowance from Table 1.1, for white cast iron = 20 mm/m
– 20/1000 = 0.02 mm

Allowance on Dimension 300	= 300 * 0.02	= 6 mm
Dimension	= 300 + 6	= 306 mm
Allowance on Dimension 200	= 200 * 0.02	= 4 mm
Dimension	= 200 + 4	= 204 mm
Allowance on Dimension 150	= 150 * 0.02	= 3 mm
Dimension	= 150 + 3	= 153 mm
Allowance of Q100	= 100 * 0.02	= 2 mm

As the dimension of wall thickness will be reduced, this allowance should be subtracted from the actual dimension.

Dimension = 100 – 2 = 98 mm

The actual dimensions of pattern after taking the shrinkage allowances are shown in Fig. 1.1 (b).

(b) Draft or Taper allowance

When a pattern is removed from the mould, there is always some possibility of damaging (tearing of edges) the edges of the mould around the pattern. This is greatly reduced if the vertical surfaces of the pattern are tapered slightly inward. This is known as 'draft'. The draft is expressed in mm per meter on a side or in degrees (see Fig. 1.2). The amount of taper depends on (i) Method of moulding, (ii) Shape and size of the pattern (iii) Moulding material.

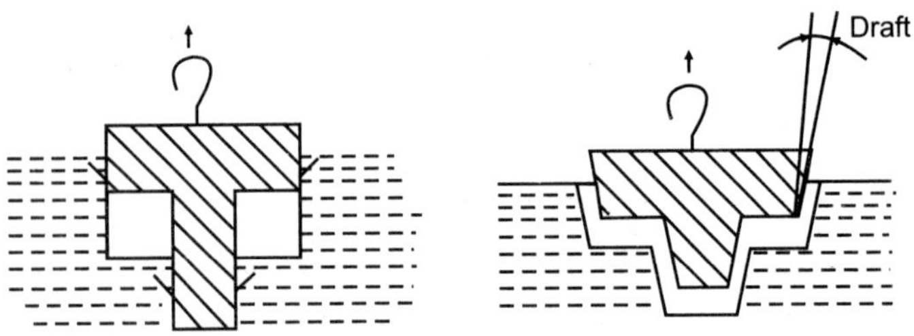

Fig. 1.2 Draft Allowance

Draft values of various pattern materials are shown in the Table 1.2.

Table 1.2: Draft Values for Patterns

Pattern Materials	*Draft (degrees)*	
	External Surface	*Internal Surface*
Wood	0.25 to 3.00	0.5 to 3.0
Metal	0.35 to 1.50	0.5 to 3.0
Plastic	0.25 to 1.0	0.35 to 2.25

Example 1.2: Provide draft allowances on wood pattern for Fig.1.3(a)

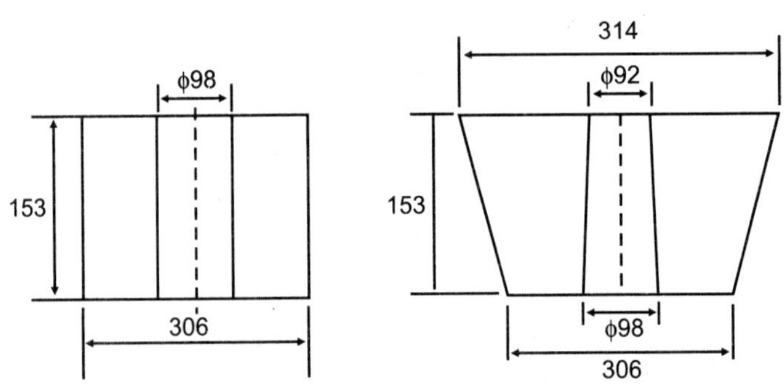

(a) Pattern (b) Pattern Dimensions

Fig. 1.3 Provision of Draft Allowances

Assume 1° taper for external and 3° for internal.

It is provided on the vertical face of the casting only.

External = 153*tan 1° = 4 mm

Internal = 153*tan 3° = 8 mm

Outer dimension = 306 + 4 + 4 = 314 mm

Inner dimension = 98 – 16 = 82 mm

After giving the draft allowance, the pattern is shown in Fig. 1.3(b).

(*c*) *Finishing or Matching allowance*

The finish of the casting obtained in sand casting is generally poor. To bring the casting to the desired level of quality, it has to be machined. For this some extra material has to be provided on the pattern. This is known as 'machining allowance'. Therefore the size of the pattern increases due to machining allowance. This allowance depends on casting metal, size and shape of the casting, method of machining and the degree of finish required. Table 1.3 gives approximate machining allowances on pattern for various metals.

Standard finishing allowance for ferrous metals is 3 mm and for non-ferrous metal it is 1.5 mm.

(*d*) *Shaking or Rapping allowance*

Before withdrawing the pattern, the pattern is shaked or rapped for easy withdrawal. By doing so, the cavity in the mould is slightly increased in size. This causes the size of casting also to increase. In order to compensate for this increase the pattern should be made initially smaller than the required size.

Table 1.3: Machining Allowances for Patterns

Metal/Alloy	*Machining Allowances (mm) (on surface)*
Cast Iron	
Medium castings	3
Large castings	10
Cast Steel	
Medium castings	4.5
Large castings	12
Non-Ferrous	
(brass, bronze, aluminium)	1.5
Medium castings	5
Large castings	

In small and medium sized castings, this allowance can be ignored, but for large sized castings or where high precision is desired, rapping or shaking allowance is provided by making the pattern slightly smaller.

(e) Distortion or Camber allowance

Some castings have a tendency to distort or warp during cooling. This is the result of uneven shrinkage due to uneven metal thickness or due to one surface being more exposed than the other causing it to cool more rapidly. To allow for this the shape of the pattern is modified in such a way that it bends in opposite direction of the distortion.

A 'U' shaped casting is an example of this feature. On cooling, the legs diverging instead of parallel. To compensate for this in the pattern, the legs are kept convergent so that on cooling the legs become parallel. The distortion allowance varies from 2 mm to 20 mm depending upon the size of the casting.

1.2.1.5 Types of Patterns

The type of pattern used depends upon the design of casting, complexity of shape, the number of castings required, moulding process, surface finish and accuracy.

The most common types of patterns are:

(a) Solid (single piece) pattern
(b) Split pattern
(c) Loose piece pattern
(d) Cope and drag pattern
(e) Match-plate pattern
(f) Gated pattern
(g) Sweep pattern
(h) Segmental pattern
(i) Skeleton pattern
(j) Built up pattern
(k) Follow board pattern
(l) Pattern colour code.

(a) Solid (single piece) pattern (Fig 1.4)

Solid pattern is made without joints. The mould cavity of this pattern is either in the drag or in the cope. Single piece pattern is generally used for large casting of simple shape.

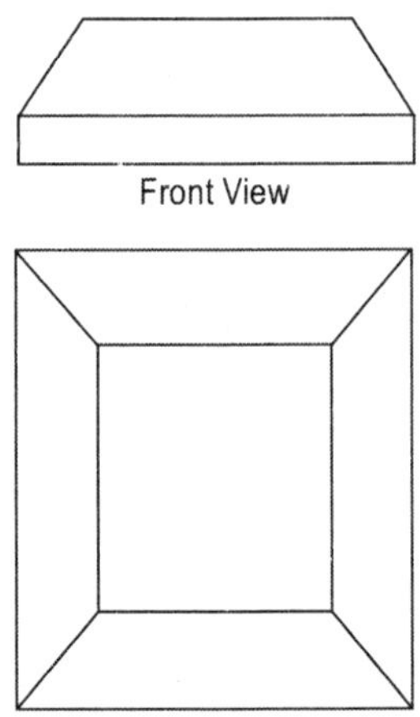

Fig. 1.4 Single Piece Pattern

(b) Split pattern (Fig 1.5)

Many patterns cannot be made in a single piece because of difficulties encountered in moulding operations (e.g. withdrawing the pattern from the mould etc.).

Suppose a spherical pattern is to be made, it cannot be done in single piece pattern. If the pattern is inserted in a cope or drag while withdrawing it, the cavity will become semi-sphere. Such objects can be cast with split pattern only.

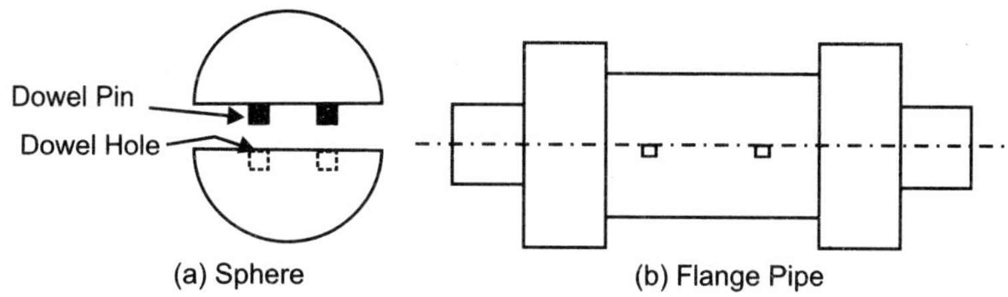

Fig. 1.5 Split Patterns

The split patterns are made in two parts so that one part may produce the lower half of the mould and the other, the upper half These are held in their proper relative positions by means of dowel pins. Sometimes it is necessary to construct three or more parts of pattern instead of two for complicated castings.

(*c*) *Loose piece pattern* (*Fig. 1.6*)

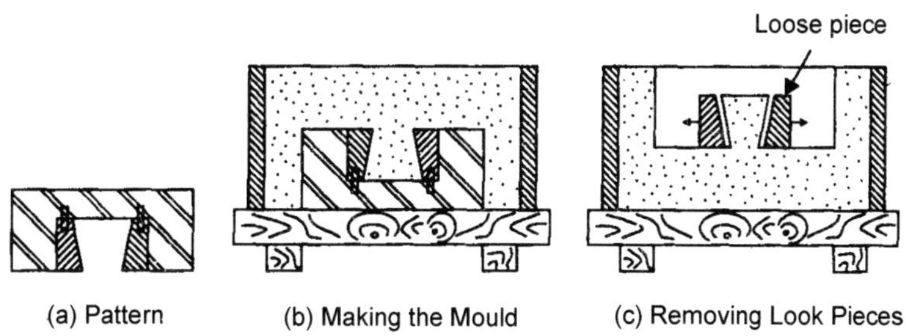

Fig. 1.6 Loose Piece Pattern

This type of pattern is required when it is not possible to withdraw the pattern as such from the moulding sand. In this case main pattern is removed first and then the loose pieces.

In these patterns, the projections or over hanging parts have to be fastened to the main pattern by means of wooden dowel pins. When the mould is made, such loose pieces remain in the mould until main pattern is withdrawn. Then the loose pieces are taken out separately through the cavity formed by the main pattern.

(*d*) *Cope and drag pattern*

A cope and drag pattern is another form of split pattern. This pattern is made up of two halves, which are mounted on different plates. In this case, cope and drag parts of the mould are made separately and then assembled. These are used for very large castings.

(*e*) *Match-plate pattern* (*Fig. 1.7*)

Match-plate patterns are mostly used in machine moulding as well as for producing large number of small casting by hand moulding. Cope and drag parts of pattern are mounted along with the gating system on opposite sides of wooden or metal plate.

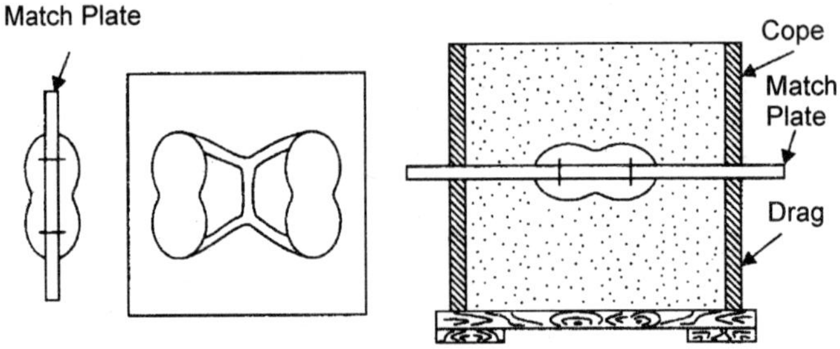

Fig. 1.7 Match Plate Pattern

(f) Gated pattern (Fig. 1.8)

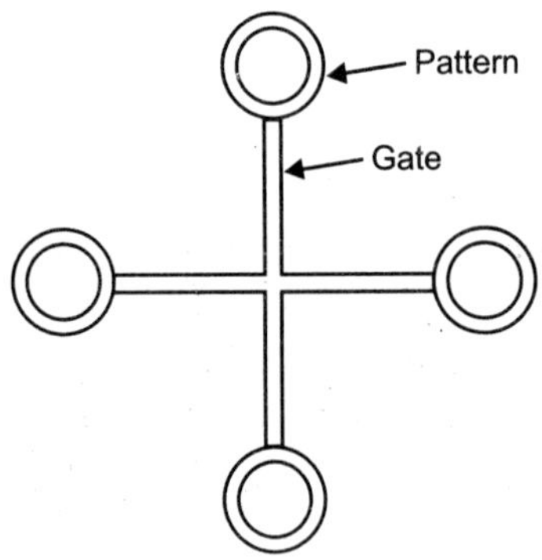

Fig. 1.8 Gated Pattern

In gated pattern, gates and riser for producing casting is included in the pattern itself. The use of gated pattern eliminate the time required to cut the gating system by hand. These are suitable for small quantity production.

(g) Sweep pattern (Fig. 1.9)

It is not a true pattern, but a template made of wood or metal revolving around a fixed axis in the mould, shapes the sand to the desired contour. It is suitable for producing large or medium size symmetrical castings.

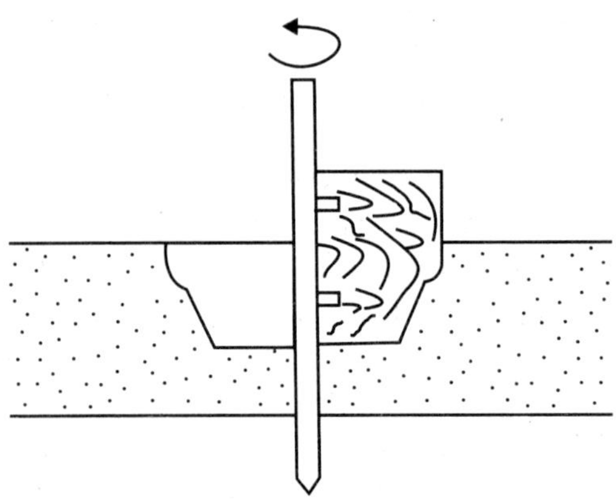

Fig. 1.9 Sweep Pattern

(h) Segmental pattern (Fig. 1.10)

These are also known as 'Part Patterns'. These are used for producing a large circular castings such as rings, Wheel rims and gears. This pattern revolves about centre and after ramming one section, it moves to another section to complete the mould.

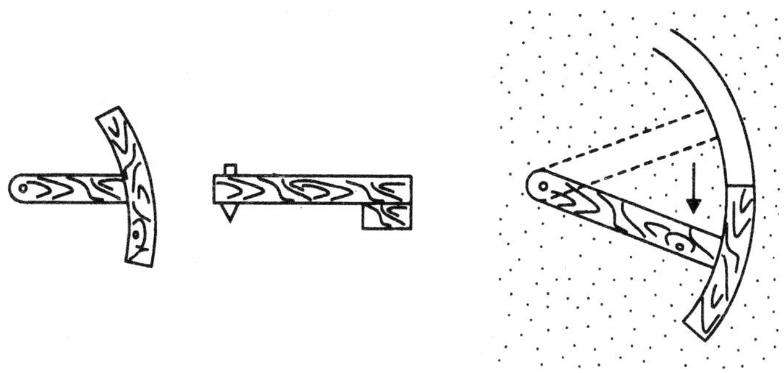

Fig. 1.10 Segmental Pattern

(i) Skeleton pattern (Fig. 1.11)

For very large castings, the patterns would require a large amount of timber for full solid pattern. If the number of castings required is small, it may not be economical to prepare solid pattern. In such cases, the pattern is made of wood frame and rib construction so that it will form a partially outline of the castings. This framework is called 'Skeleton'.

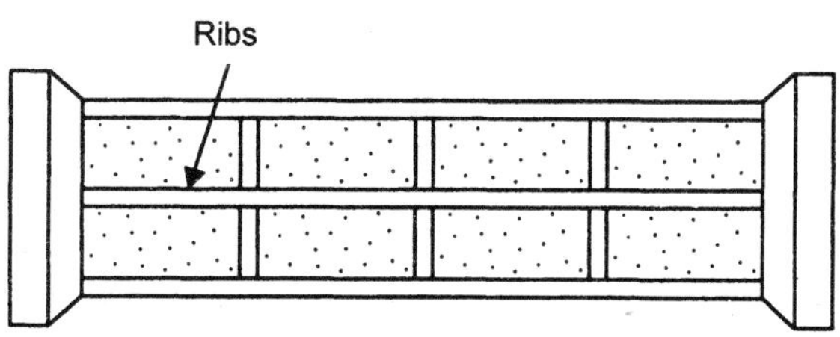

Fig. 1.11 Skeleton Pattern

(j) Built up pattern (Fig. 1.12)

Built up patterns are composed of two or more pieces. Patterns for special pulleys are built up of segments of wooden strips. These segments are made by cutting strips of wood to the curvature required.

Fig. 1.12 Built up Pattern

(k) *Follow board pattern* (*Fig. 1.13*)

A follow board pattern is a wooden board and is used for supporting a pattern which is very thin and fragile. With the follow board support under the pattern, the drag is rammed and the follow board is removed.

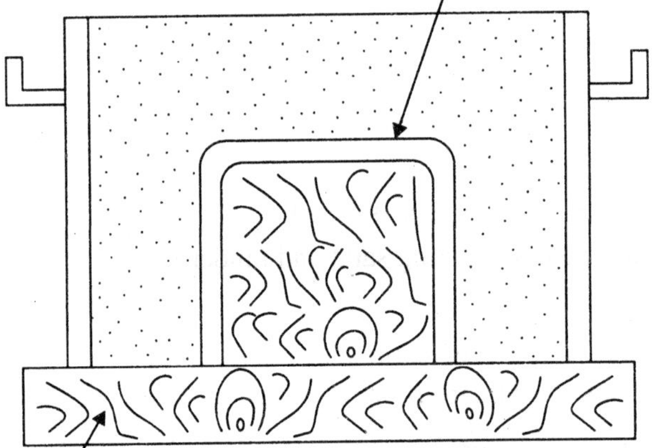

Fig. 1.13 Follow Board Pattern

(l) *Pattern colour code*

Patterns are painted with different colours to indicate different surfaces. There is no universal accepted standard for representation of different types of surfaces. The colour code adopted by most of the foundries are given in Table 1.4.

Table 1.4: Pattern Colour Code

S.No.	*Part of the Pattern*	*Colour*
1.	Surface to be left unmachined	Black
2.	Surface to be machined	Red
3.	Core prints and seats for loose core prints	Yellow
4.	Seats for loose pieces	Red stripes on yellow background.
5.	Stop-offs or supports	Black stripes on yellow background

1.2.2 Mould and Core Making

1.2.2.1 Moulding Sands

Moulding sand is one of the most important material in production of sand casting. Sand is formed by breaking up of rocks due to natural forces such as frost, wind, rain and action of water.

1.2.2.2 Classification of Moulding Sands

1.2.2.2.1 Classification according to the nature of its origin

(a) Natural sand (b) Synthetic sand

(*a*) *Natural Sand*: Natural sand is collected from the river beds or it is dug from pits. Natural sand contains sufficient amount of binding material (clay) in it so that it can be used directly. It has the following advantages:

Advantages

(i) Natural sand maintains moisture content for a pretty long time

(ii) They are cheap

(iii) The time for mixing the binder is saved

(iv) No extra equipment for mixing sand and binder.

Disadvantages

These are less refractory than synthetic sands because of impurities present. Natural sands are used for casting cast iron and nonferrous metals.

(*b*) *Synthetic Sand*: Synthetic sands are basically clay free high silica sands. They are mixed with desired amount of clay (3–5% bentonite) and water to, develop required moulding properties. It is used for steel castings. The advantages of synthetic sands over natural sands are:

(i) High permeability and refractoriness

(ii) Modulability with less moisture

(iii) Easier control of properties

However synthetic sands have the following disadvantages:

(i) It is more costly.

(ii) It needs extra time, equipment and men to prepare the sand.

1.2.2.2.2 Classification according to their initial conditions and use

(a) Green Sand, (b) Dry Sand, (c) Loam Sand, (d) Facing Sand, (e) Parting Sand, (f) Backing Sand.

(*a*) *Green Sand*: Foundry sand containing moisture is known as green sand. Green refers to the moisture content, it is a mixture of silica sand with 20 to 30% clay and water from 6 to 8%. This is suitable for moulding purposes without any further conditioning. Green sand is generally used for casting small or medium sized moulds.

(*b*) *Dry Sand*: Sand free from moisture is called dry sand. It passes greater strength than green sand and can be used for making larger castings.

(*c*) *Loam Sand*: Loam sand is a mixture of sand and clay (50%). It is used far making large castings such as large cylinders, paper rolls.

(*d*) *Facing Sand*: It forms the face of the mould and is in contact with the molten metal 25% of coal dust or graphite is used to prevent the metal from burning into the sand.

It may have the thickness of 20 to 30 mm. It should have sufficient strength and refractoriness.

(*e*) *Parting Sand*: Parting sand is sprinkled over the rammed drag to avoid the sticking of drag with cape. Similarly it is sprinkled over the pattern to avoid its sticking to the green sand, powder free from clay is used for this purpose. Very fine brick powder can be used as parting sand.

(*f*) *Backing Sand*: It is the sand which backs up the facing sand and to fill the rest of the flask it is the floor sand which is already used.

1.2.2.3 Properties of Moulding Sand

A moulding sand should possess the following properties:

(*a*) *Permeability or Porosity*: Molten metal always contains a certain amount of dissolved gases which are evolved when the metal solidifies, also when the molten metal comes in contact with the moisture sand, generates steam and water vapour. If these gases and water vapour do not find passage to escape completely through the mould they will form gas holes and pores in the casting. The ability of the sand to allow the gas to pass through it is called 'permeability'. It depends on size and shape of grains, moisture content and degree of ramming.

(*b*) *Plasticity or Flow ability*: This refers to the ability of the moulding sand to acquire a predetermined shape under pressure and retain the same when the pressure is removed. This will increase with increase in clay moisture content.

(*c*) *Adhesiveness*: The property to adhere with other materials is adhesiveness. Moulding sand particles should stick to the surface of the moulding boxes. This enables the mould to retain in a box during handling.

(*d*) *Cohesiveness*: Cohesiveness is the ability of sand particles to stick to each other. Lack of this property would result in breaking of the mould when molten metal is poured. This depends on grain size (decreases with grain size) and clay content (increases with clay) of sand.

(*e*) *Green Strength*: It is the strength of the sand in green or moist state. A mould having adequate green strength will not disturb or collapse even after removing the pattern from the mould box in the absence of green strength, dimensional stability and accuracy cannot be obtained.

(*f*) *Dry Strength*: It is the strength of the moulding sand in dry condition. A mould should possess adequate dry strength to withstand erosive force and pressure of molten metal.

(*g*) *Hot Strength*: It is the strength of sand of mould cavity above 100°C. If hot strength is inadequate, the mould is likely to enlarge, break or get cracked.

(*h*) *Refractoriness*: The capability of the moulding sand to withstand the high temperatures of the molten metal without fusing is known as refractoriness.

(*i*) *Chemical Resistivity*: The moulding sand should not react chemically with the molten metal, otherwise the shape of casting will be distorted and smooth surface will not be obtained.

(*j*) *Collapsibility*: It is the property of the moulding sand that permits it to collapse (break) easily during its knockout from the casting.

(*k*) *Fineness*: Finer mould sand resists metal penetration and produces smooth casting surface fineness and permeability are opposite to each other. Hence these should be balanced for optimum results.

(*l*) *Coefficient of Expansion*: Moulding sands should possess low coefficient of expansion.

(*m*) *Bench Lift*: It is the ability of mould sand to retain its properties during storing, handling or while standing (i.e. in case of any delay).

1.2.2.4 Principal Ingredients of Moulding Sand

(a) Silica Sand Grains (c) Water

(b) Clay (d) Additives

(*a*) *Silica Sand Grains*: They impart refractoriness, chemical resistivity and permeability to the sand. The sand grains may vary in size from a few micrometers to a few millimeters. The shape of the grains may be round, sub-angular, angular and compound. The size and shape of the sand grains effect the properties of moulding sand.

(*b*) *Clay*: Clay can be defined as natural earthy material that becomes plastic when water is mixed with. It's purpose is to impart necessary bonding strength to the moulding sand so that the mould does not loose its shape after ramming. Clay consists of flaked shaped particles about 20 microns in diameter. The most popular clays are kaolite and bentonite. Kaolite has a melting point of 175 to 1787°C and Bentonite has melting point of 1250 to 1300°C of the two, bentonite can absorb more water which increases its bonding power.

(*c*) *Water*: Clay acquires its bonding action only in the presence of the required amount of water (1.5 to 8%). When water is added to clay it penetrates into the mixture and forms a micro film which coats the surface of each flake. Too little water will not develop proper strength and plasticity. Too much water will result in excessive plasticity.

(*d*) *Additives*: Materials other than basic ingredients (sand binder and water) are also added to mould sand for improving existing properties. Additives include (i) Facing materials, (ii) Cushion materials.

(*i*) *Facing Materials*: Facing materials are used to get smoother and cleaner surfaces of castings and help easy peeling of sand from the casting surface during shake out, e.g. coal dust, graphite, silica flour.

(*ii*) *Cushion Materials*: Cushion materials burn when the molten metal is poured and thus give rise to space for accommodating the expansion of silica sand at the surfaces of mould cavity, e.g. wood flour cellulose, perlite.

1.2.2.5 Sand Conditioning

In general the natural and long used sands are not suitable for moulding directly. There is a need for conditioning the sand mixture for better results.

The basic steps in sand conditioning are as follows:

(i) The first step is remove all foreign and undesirable matters such as nails, fins, hard sand lumps from the moulding sand.

(ii) The second step is mixing of its ingredients, proper amounts of pure sand, clay and other additives are mixed and water is spread over the entire volume. Muller is used for mixing all the ingredients of sand.

(iii) In the third step, the sand is passed through a mechanical aerator to separate sand grains into individual particles. It is performed to increase the flow ability of sand.

1.2.2.6 Sand Testing

Periodic tests are necessary to determine the essential qualities of foundry sand. The sand can be tested either by chemical or mechanical methods. In most cases, the mechanical tests are employed. The most important tests to be conducted are grain fineness, permeability and strength test. In addition to them, moisture content, clay content and hardness tests are conducted.

(*a*) *Grain Fineness Test:* The grain size is determined by grain fineness number. It can be tested with the help of an equipment called sieve shaker (Fig. 1.14). It consists of set of standard sieves having varying number of meshes 6, 12, 20, 30, 40, 50, 70, 100, 140, 200 and 270. The sieve with minimum mesh number has largest aperture (opening) and so on.

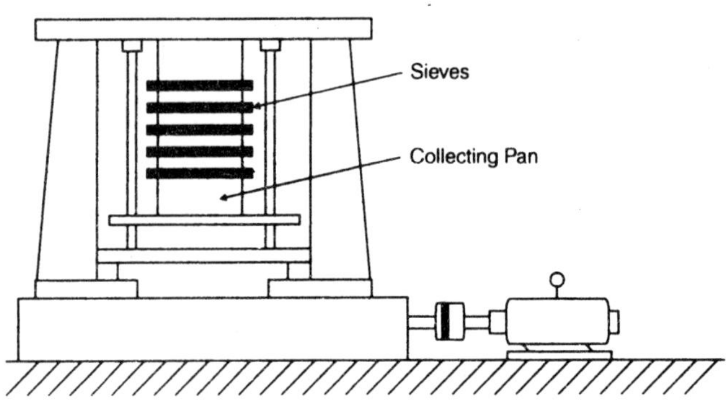

Fig. 1.14 Sieve Shaking Machine

The sample of sand is first washed to remove clay from it, then it is dried. A weighed quantity of this sand is now placed on the top sieve and the whole unit is shaken for a definite period with the help of electric motor. The sand falls through the apertures, and the sand of smallest size comes to the bottom pan. The sand in each sieve is collected and weighed separately and expressed as a percentage of the original sample weight. The percentage retained in each sieve is multiplied by it's own multiplier and all the products are added to obtain the total product. The grain fineness number is calculated by using the following equation.

$$\text{Grain Fineness number} = \frac{\text{Total product}}{\text{Total percentage of sand retained sieves}}$$

A typical example is shown in Table 1.5.

$$\text{Grain Fineness Number} = \frac{4800}{96} = 50$$

Table 1.5: Calculation of G.F.N.

Mesh	*Sand Retained in the Sieves*		*Multiplier*	*Product*
	By Weight (Wt)	*%Wt Retained =2*Wt*		
6	0	0	3	0
12	0	0	5	0
20	0	0	10	0
30	3.5	7	20	140
40	2.0	4	30	120
50	15.2	31	40	1240
70	18	36	50	1800
100	5	10	70	700
140	4	8	100	800
200	0	0	140	0
270	0	0	200	0
Pan	0	0	300	0
Total	96			4800

(*b*) *Permeability Test:* Permeability is the property of moulding sand which permits the escape of steam and other gases generated in the mould during metal pouring. It is measured by means of an apparatus called 'Permeability meter' (refer Fig 1.15).

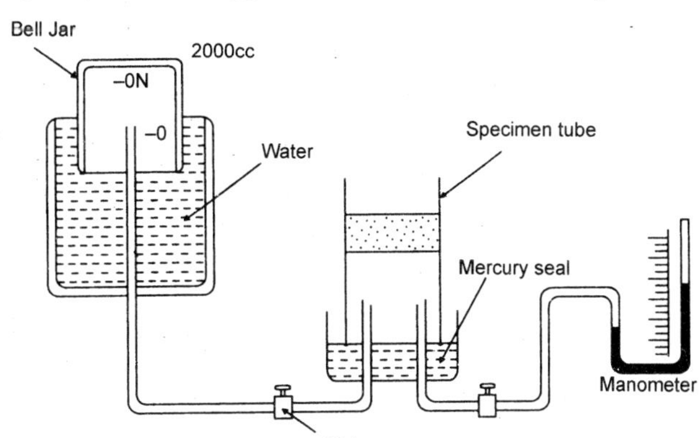

Fig. 1.15 Permeability Tester

Fix the proper orifice (1.5 mm dia or 0.5 mm dia) in position. Now keep the specimen in the tube over the orifice. Keep it tank on '0' mark and allow the air to pass through the sand specimen. When the tank comes to 0N position, note the manometer reading. From this manometer reading, find the corresponding permeability number from the calibrated chart attached to the equipment.

(*c*) *Strength Test*: To find out the holding power of various bonding materials in green and dry sand moulds, strength tests are performed. It is done on universal sand testing machine (Fig. 1.16). Moulding sand is tested for compressive, tensile strength, shear strength and traverse strength.

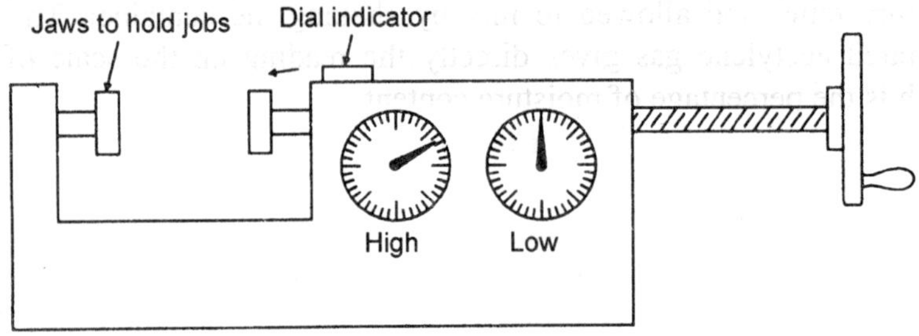

Fig. 1.16 Universal Sand Testing Machine

The specimen is held between the grips. Hand wheel when rotated, actuates mechanism to build pressure on the specimen. Dial indicator fitted on the tester measures the deformation occurring in the specimen. There are two manometer one for low strength sand and other for high strength sands. The shape of the specimen for measuring compressive, shear, tensile and traverse strengths are shown in Fig. 1.17.

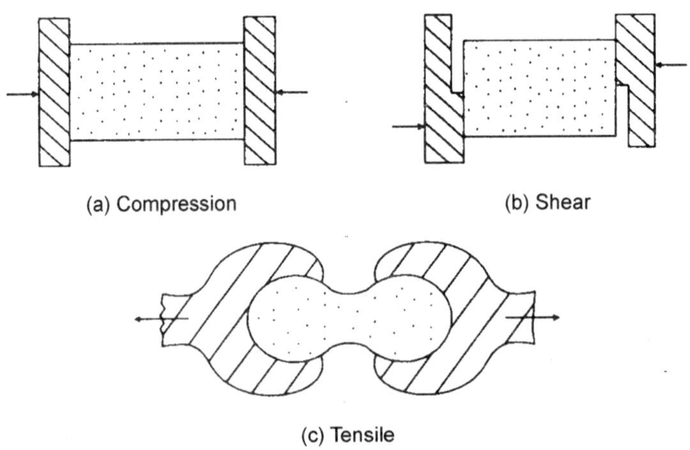

Fig. 1.17 Specimens for Sand Strength Test

(*d*) *Moisture Content Test*: Moisture content is determined by the loss of weight after evaporation. A sample of sand (50 m) is dried at 106°C–110°C. The drying is carried in an oven. The sand is cooled to room temperature and weighed again. The loss of weight gives the amount of moisture which can be expressed in percentage.

$$\% \text{ moisture content} = \frac{\text{loss of weight in the sample}}{\text{weight of sample before healing}} * 100$$

As the conventional method is time consuming, direct reading instruments are available. One such instrument is moisture teller. It works on the pressure of acetylene gas generated by the reaction of calcium carbide with the moisture present in the sand. Weighed amount of sand and calcium carbide are placed in the compartment of moisture teller and allowed to mix by shaking the container. The pressure of the generated acetylene gas gives directly the reading on the scale of pressure gauge which is the percentage of moisture content.

(*e*) *Clay Content Test*: (Fig. 1.18) Clay content in the sand is determined by the loss of mass of sand sample after washing 50 grams of dry moulding sand is taken into clay jar. Add 475cc of distilled water and 25cc of NaOH to it and shake it. Next the bottle is filled with water to a height of 150 mm and agitated vigorously by the stirrer and allowed to settle again for 10 minutes. The water in which clay is dissolved is removed by a syphon. The operation is repeated until water is clear after settling period (5 minutes). The remaining sand in the clay jar is dried and weighed. The percentage of clay is determined by the difference in the initial and final weight of the sample.

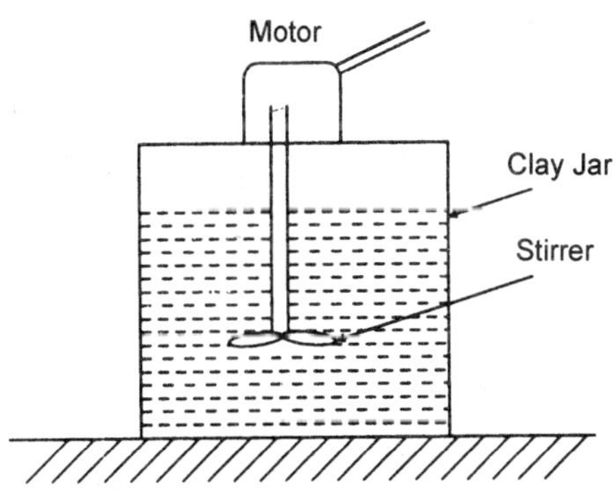

Fig. 1.18 Clay Content Tester

1.2.2.7 Moulding Processes

The moulding processes are classified as follows:

1.2.2.7.1 According to the method used

(a) Bench moulding
(b) Floor moulding
(c) Machine moulding

(a) *Bench moulding*: It is carried on a bench of convenient height. It is used for preparing small moulds. Green sand and dry sand moulds may be made by bench moulding.

(b) *Floor moulding*: It is used for preparing medium and large size castings. The mould is made in the foundry floor.

(c) *Machine moulding*: Moulding by hand is slow and laborious process and also does not yield good results as it does not impart uniform hardness to the mould. In machine moulding production becomes faster, labour is minimised. It is used in batch and mass production. Moulding machines are classified according to:

1. The method of compacting the moulding sand.
2. The method of removing the pattern.

1(a) Squeezer Machine (Fig. 1.19)

The pattern is placed on the mould board which is clamped on the table. The flask is placed on the mould and the sand frame on the flask. The flask and frame are filled with moulding sand and levelled off. Next the table is raised by the table lift mechanism against the platen of the stationary squeeze head. The platen enters the sand frame up to the dotted line and compact the moulding sand. After the squeeze, the table returns to its initial position.

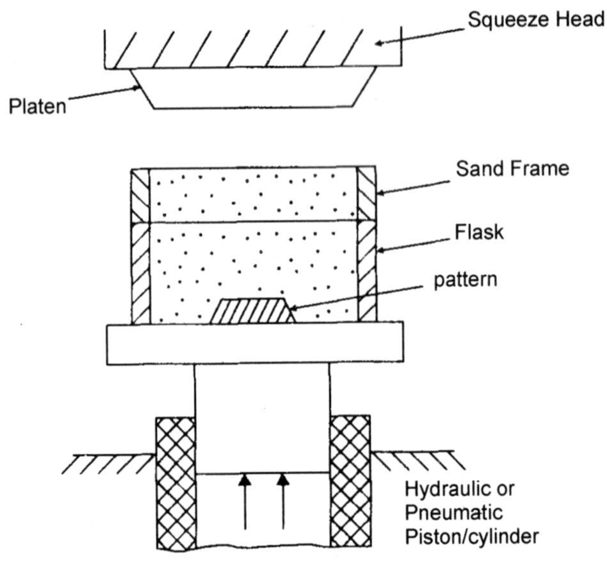

Fig. 1.19 Squeeze Machine

1(b) Jolt Machine (Fig.1.20)

The table with the flask having pattern and moulding sand is raised to 30 to 80 mm by the plunger when compressed air is admitted through the hole and channel. The air is next released through the hole (opening) and the table drops suddenly and strikes the guided cylinder at its bottom. This sudden action causes the sand to pack evenly around the pattern.

Springs are used to cushion the table blows and thus reduce the noise and prevents destruction of the mechanism and the foundation. The sand is rammed hardest at the parting plane and around the pattern and remain less dense in top layer. So hand ramming is necessary after jolting action is complete. Jolt and squeeze machine overcome drawbacks of jolt machine and squeeze machine.

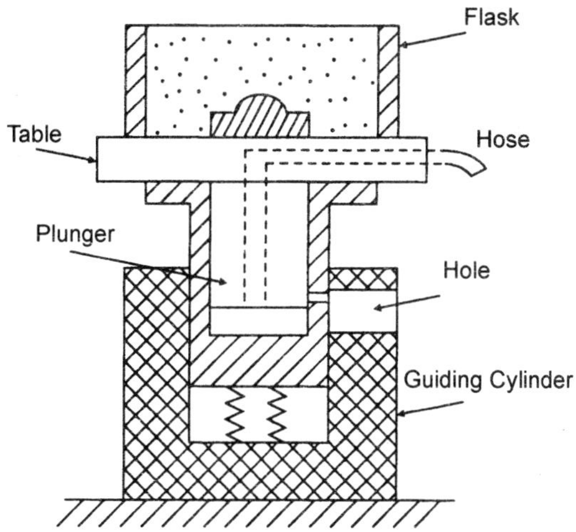

Fig. 1.20 Jolt Machine

1(c) Sand Slinger (Fig. 1.21)

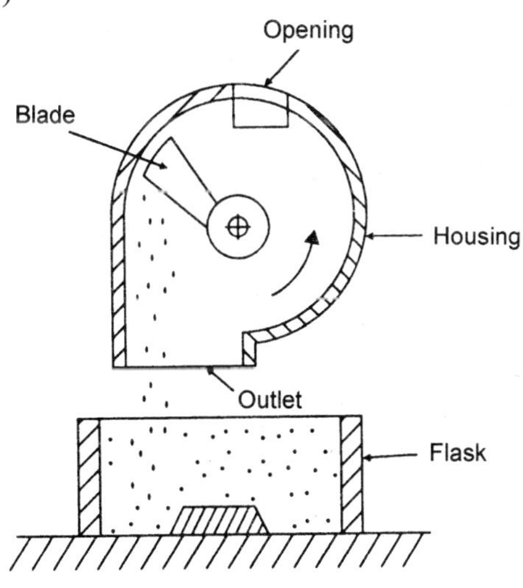

Fig. 1.21 Sand Slinger

The overhead impeller head consists of the housing in which the blade rotates at a very high speed. The sand is delivered to the impeller through the opening by means of a belt conveyor. The impeller head by the rotation of the blade throws the sand through the outlet down into the flask over the pattern. The density of sand can be controlled by the speed of the blade.

Classification of moulding machines according to the method of removing the pattern from the mould.

(*a*) *Straight-Draw Moulding Machine* (Fig. 1.22)

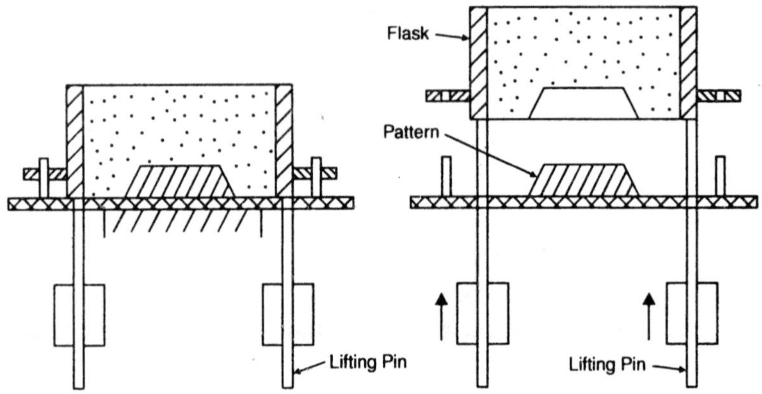

Fig. 1.22 Straight-draw Moulding Machine

In this machine the pattern is fixed on the table and mould box is placed over it and filled with sand. The squeeze head is next swung over in position and it squeezes the mould. The flask is then lifted from the pattern by stripping pins.

(*b*) *Turn-over Moulding Machine* (Fig. 1.23)

This is used for large sized moulds. The flask rests on the pattern plate during the moulding operation. Then the flask together with the worktable is rotated 180° and pins lift the worktable together with pattern out of the mould.

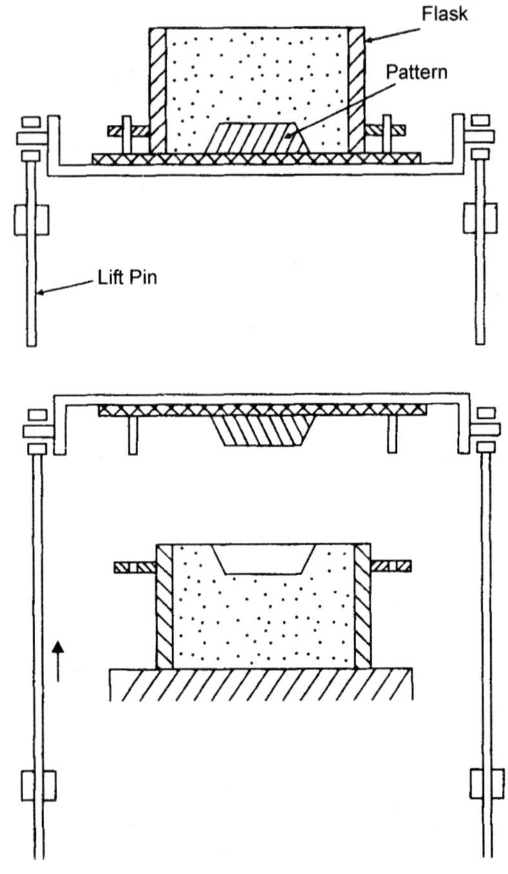

Fig. 1.23 Turn-over Moulding Machine

1.2.2.7.2 According to the mould materials used

(a) Green sand moulding
(b) Dry sand moulding
(c) Loam sand moulding
(d) Metallic moulding or Permanent moulding
(e) Carbon dioxide moulding

(*a*) *Green Sand Moulding*: The green sand is the sand with moisture only. It is the simplest of all the moulding sand as no other binding material is mixed.

If the mould is filled with green sand, the method is known as "Green sand moulding". These are used for small and medium castings.

Advantages

1. It is least expensive method
2. It requires less time to prepare

Disadvantages

1. Moisture in the sand may cause defects like blow holes.
2. Surface finish of the casting is poor.
3. It is not very strong hence, liable to be damaged during handling.

(*b*) *Dry Sand Moulding*: The method of making dry sand moulds is similar to that of making green sand mould except that the mould is dried before pouring molten metal. Drying (or baking) is carried in oven. The time of baking depends on the binders used in the sand mixture.

Advantages

1. Stronger than green sand moulds
2. The castings have better surface finish.

Disadvantages

1. The molding material is expensive
2. Extra equipment is required to mix the sand and binders.

(*c*) *Loam Sand Moulding*: These moulds are made of bricks and other materials to the approximate contour of the casting and a thick coating of loam sand. The correct shape of the mould is obtained by rotating sweep pattern as in (Fig. 1.24).

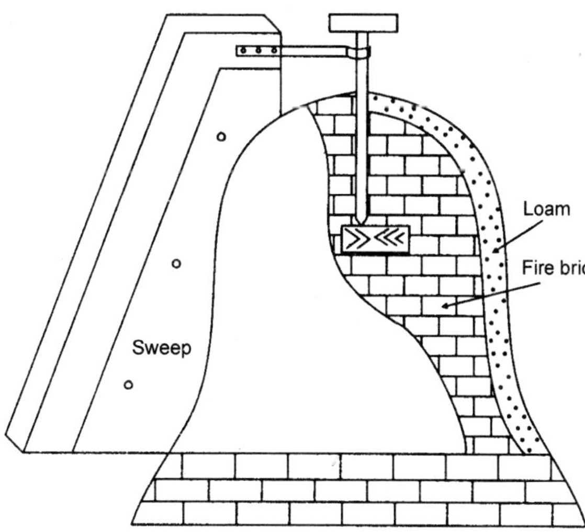

Fig. 1.24 Loam Moulding

The surface of the mould is dried by forced hot air or torches. The advantages of loam moulding is that large castings (large cylinders, paper making rolls and bells) can be made cheaply.

Disadvantages

1. Loam moulding is slow and laborious process.
2. Skilled moulders are required.
   (*d*) *Metallic Moulding or Permanent Moulding*: When the mould is made of metal, it is known as 'Metal mould'. These are used for the production of large number of identical castings.

Advantages

1. It has a very long life
2. It does not get eroded while pouring molten metal
3. Production rate is high

Disadvantages

1. These moulds are costly
2. The shape of the mould cannot be altered.
   (*e*) *Carbon Dioxide Moulding* (*CO, Process*): The process consists of thoroughly mixing clean, dry silica sand with 3 to 5% sodium silicate liquid base binder in a muller.

This mixture is packed around the pattern in the mould box by hand or molding machine. When the packing is complete, CO_2 is forced into the mould or core at a pressure of 1.5 kg/cm^2 for 10 to 30 seconds.

The sodium silica present in the mould reacts with CO_2 to form a substance called 'Silica gel'. This silica gel, hardens and acts as a cement to bond the sand grains together.

$$NO_2SiO_3 + CO_2 \longrightarrow NO_2CO_3 + SiO_2 \times H_2O$$

.... Silica gel

The major advantages of CO_2 process are:

1. Removing the pattern is easier since the mould may be hardened before the withdrawal of the pattern.
2. The process can also be used to make cores.
3. It is the fast process of hardening the moulds.
4. Accurate castings with sharp corners can be mould.
5. Surface finish of the casting is good.
6. Semi-finished labourers can be employed.

Limitations

1. CO_2 gas is expensive
2. The collapsibility of the mould is reduced.

Applications

CO_2 process can be used for both ferrous and nonferrous castings.

1.2.2.8 Special Moulding Processes

(*a*) *Shell Moulding*: It is a modification of the sand mould process. In this process, the mould is made up of mixture of dried silica sand and phenolic resin, formed into a thin half-mould shells which are clamped together for pouring metal.

Procedure (Fig. 1.25)

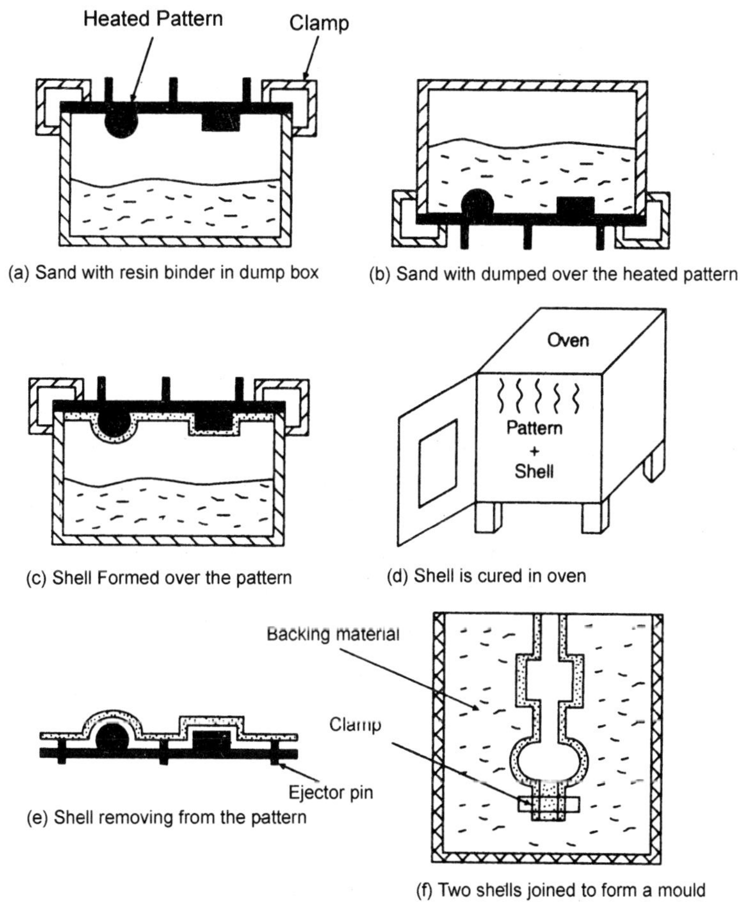

Fig. 1.25 Shell Moulding Process

The sand is first mixed with either urea or phenol formaldehyde resin in a muller. Metal pattern is heated to 205 to 230°C in an oven and sprayed with silicon grease and kept on the top of the dump box. The dump box contains sand mixed with thermo plastic resin. The box is inverted, causing the sand mix to fall on the hot pattern. The resin melts and flow in between the grains of sand, acting as a bond. After 30 seconds, a hard layer of sand is formed over the pattern. Then the dump box is inverted back to its original position. The pattern with a thin shell is cured for two minutes at 315°C. The shell is finally removed from the pattern by ejector pins. The two shells are clamped together to form the mould and placed in the flask with backing sand.

Advantages

1. High dimensional accuracy and good surface finish.
2. The chances of blow holes or pockets are reduced since the shells are highly permeable.
3. Thin wall sections can be produced.
4. Shells can be stored for long time.

Disadvantages

1. The metal patterns are costly than wood pattern.
2. Resin is an expensive binder.
3. Specialized equipments are to be used.

Applications

Cylinders of IC engines, automobile transmission parts, chain seat brackets, small crank shafts.

(*b*) *Investment Casting Process* (*Lost Wax Process*): In this process the wax pattern is melted from the mould, leaving the cavity.

Procedure: A master pattern is prepared from steel or brass. Using this pattern, bismuth alloy or lead alloy split mould is made. This mould is used for making wax pattern. Heated wax is injected into the mould (water cooled). Upon solidification, wax pattern is removed. Several patterns are assembled with necessary gates and risers. This assembly (tree) is dipped in a refractory slurry and then refractory fine sand is sprinkled over it to ensure smooth surface of the casting. The assembly is placed in a container and filled with coarser plaster mixture. After the investment material is set, the mould is placed upside down and heated in a oven to melt out the wax. The casting can be produced by pouring the molten metal. When the casting solidifies, the mould is broken and the casting is taken out.

Advantages

1. High dimensional accuracy.
2. Intricate forms having undercut can be cast easily.
3. A very smooth surface of casting can be obtained.

Disadvantages

1. The process is expensive
2. It is limited to small castings from few grams to 5 kg.
3. Skilled operators are required
4. Production rate is low.

Applications

Blades for gas turbines, parts of sewing machines, type writers, calculating machines etc.

(c) *Centrifugal Casting Process*: In the centrifugal casting process, the mould is rotated and the molten metal is distributed to the mould cavity with centrifugal force. The centrifugal casting process is classified as follows:

1. True centrifugal casting process
2. Semi-centrifugal casting process
3. Centrifuging process

1. *True Centrifugal Casting Process*: In this process, the axis of rotation of the mould coincides with the axis of the casting and the molten metal is pushed outwards because of the centrifugal force, no core is required for making the concentric hole. The axis of rotation may be horizontal, vertical or inclined. The most commonly cast parts are cast iron pipes, liners and cylindrical barrels. The mould may be permanent type or sand lined mould. A normal centrifugal casting machine used for making cast iron pipes in sand moulds is shown in Fig. 1.26.

The mould flask is rammed with sand to confirm to the outer contour of the pipe to be made. The mould is arranged between rollers (two at the bottom and two at the top) to revolve freely. At the end of the mould is fitted with a gear which meshes with a gear on a motor driven shaft. The ends of the hollow mould are partially covered by covers which can be detached when the casting is to be pushed out of the mould. At both end covers, a central hole is provided. From one side, the molten metal is poured, from a ladle and from other, the hot gases escape out.

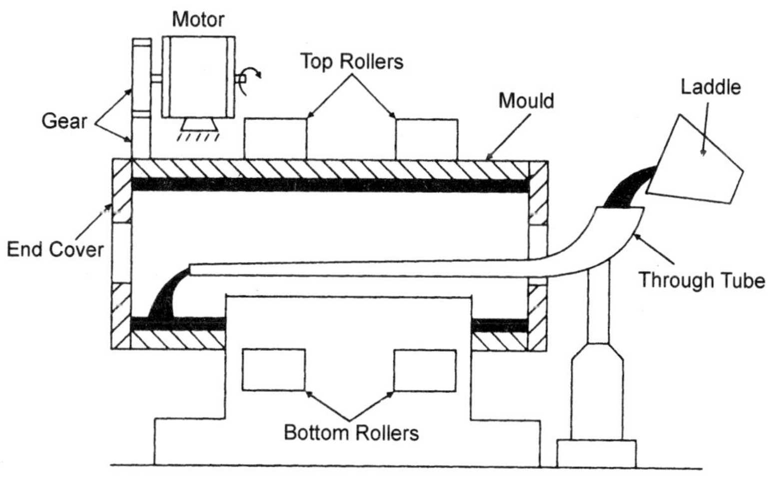

Fig. 1.26 True Centrifugal Casting Process

Advantages

1. The inclusions get segregated towards the centre and can be easily removed by machining.
2. The castings have better mechanical properties.
3. No central core is required.
4. No gates and risers are required.

Limitations

1. This method is limited to certain shapes with axis symmetric.
2. Equipment is costly.

Applications

Cast iron pipes, alloy steel pipes, liners of L.I.C. engine.

2. *Semi-centrifugal Casting Process*: In this process, the mould is rotated about the vertical axis and the metal is poured into a central sprue from where it first enters the hub and then is forced outwards to the rim by centrifugal force. If a central hole is required, core is used. The rotating speed is not as high as in the case of true centrifugal casting.

   The method is used for making large sized castings which are symmetrical about their own axis such as pulleys, spoked wheels, gears and propellers. For high production rates, the moulds can be stacked one over the other and fed simultaneously through a common central sprue.

3. *Centrifuging Process*: In this process, also the mould is rotated while the metal is poured. The difference between true centrifugal or semi-centrifugal and centrifuging is that in case of true centrifugal or semi-centrifugal casting process the axis of mould coincides with the axis of rotation where as in case of centrifuging the axis of rotation and the axis of the mould are not same. A number of mould cavities are arranged on the circumference of a circle and are connected to a central down sprue through radial gates. The process is similar to semi-centrifugal casting. This is suitable only for small jobs of any shape.

   (*d*) *Die Casting Process*: The term 'die' is used for permanent mould. In this process the molten metal is forced into the permanent mould under high pressure.

   Types of die casting machines:

1. Hot chamber die casting machine
2. Cold chamber die casting machine

1. *Hot Chamber Die Casting Machine*: (Fig 1.27(*a*) and (*b*))

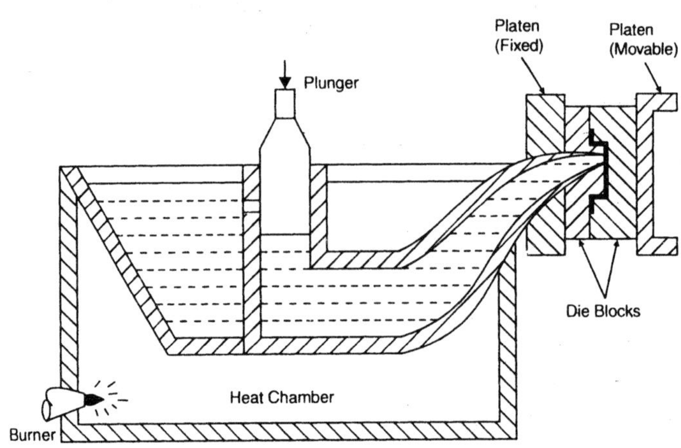

Fig. 1.27 (a) Plunger Type Hot Chamber Die Casting Machine

Fig. 1.27(a) illustrates the working of a hot chamber die-casting machine. The melting unit of metal forms an integral part of the machine. When the plunger is raised, it uncovers an opening in the cylinder wall, through which the molten metal enters, filling the cylinder. The molten metal is forced into the die either by hydraulic pressure or by air pressure applied to plunger. As soon as the metal solidifies, the pressure on the metal is relieved and the plunger travels upwards to its original position. The casting is ejected from the die by means of ejector pins.

In another type (see Fig. 1.27(b)) direct air is applied to force the molten metal into the die. The bottle is tilted to immerse the nozzle in the pot of hot metal and is filled by gravity. Then the bottle is raised so that the air nozzle comes in contact with top opening of the bottle. Compressed air is then applied directly on metal so that metal is forced into the die cavity. When solidification is complete, the air pressure is stopped. The die is opened and casting is ejected.

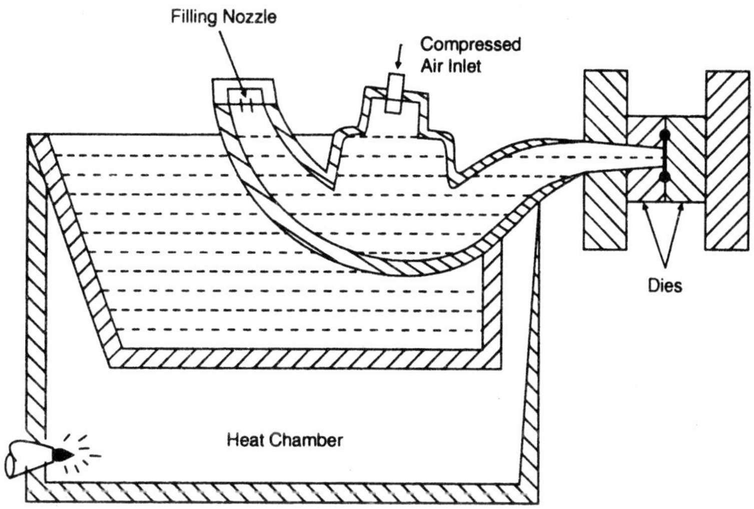

Fig. 1.27 (b) Air Blown Hot Chamber Die Casting Machine

2. *Cold Chamber Die Casting Machine*: Fig. 1.28 shows the cold chamber machine. The metal is melted in a furnace and transferred to the cylinder from where it is forced into the mould by means of plunger.

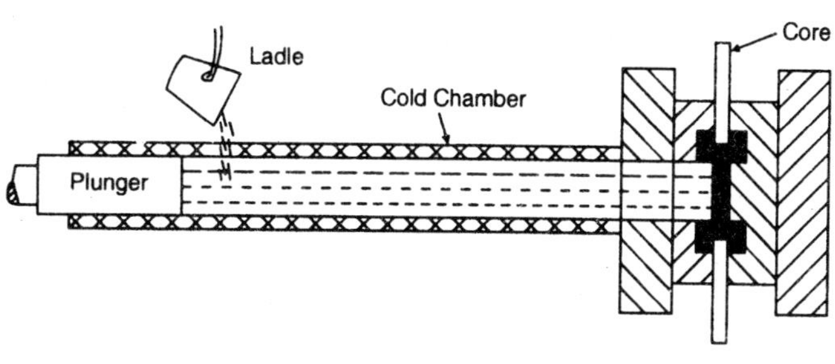

Fig. 1.28 Cold Chamber Die Casting Machine

Advantages

1. Very high production rates are possible.
2. Thin sections can be cast.
3. Close dimensional control can be maintained.

Disadvantages

1. Equipment is costly.
2. Due to high temperature of molten metal, dies life will decrease.
3. Die casting is limited to low melting nonferrous alloys.

(*e*) *Slush Casting*: Hollow castings are produced without the use of core in this method. The mould is filled with molten metal and held stationary until a thin skin of solid metal freezes against the mould walls. The mould is then inverted and the unfrozen metal runs out from the casting. Thus a thin walled casting is obtained.

1.2.2.9 Cores

Core is a mass of sand that is put into the mould to form holes and cavities in the casting.

Characteristics of cores and core sands:

1. The core should have sufficient strength to withstand the force of the molten metal.
2. It should be highly permeable to allow gas to escape.
3. The core should withstand high temperatures of the molten metal
4. It should have good collapsibility so that the core should be disintegrated easily after solidification.

Types of Cores

The various types of cores are as follows:

(*a*) *Horizontal Core*

This is the most common type. Horizontal cores are laid down horizontally at the parting line of the mould (see Fig. 1.29). The ends of the core rest in the seats provided by the core print of the pattern.

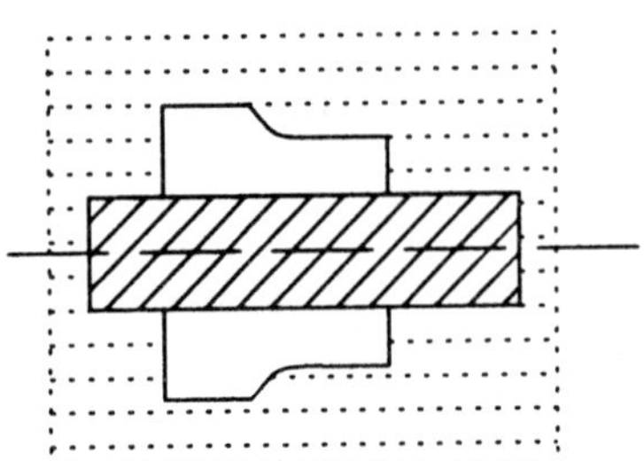

Fig. 1.29 Horizontal Core

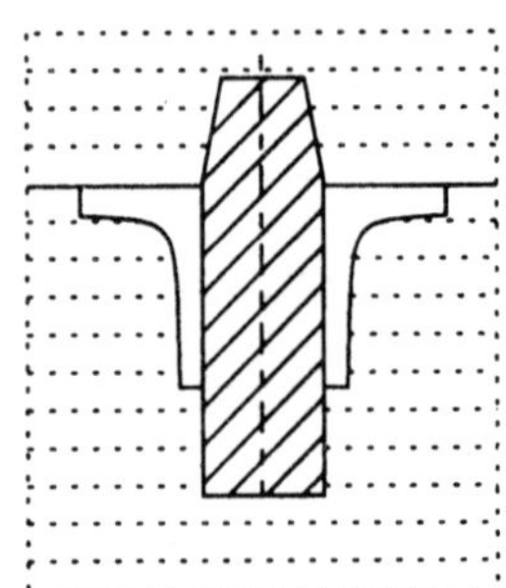

Fig. 1.30 Vertical Core

(b) Vertical Core

Vertical core is similar to the horizontal core except that it is standing vertically in the mould (see Fig. 1.30). It is a usual practice to have greater part of the core in the drag position of the mould.

(c) Balanced Core

A balanced core is one which is supported and balanced from its one end only. In such cases the core print should be large to support the weight of the core. This is used when a casting does not want a through cavity. For supporting the core in the mould, chaplets are used.

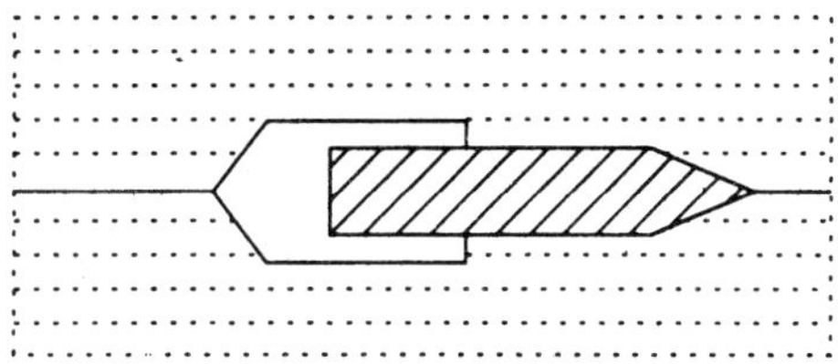

Fig. 1.31 Balanced Core

(d) Hanging Core

If the core hangs from the cope and does not have any support at the bottom in the drag (see Fig. 1.32), it is called as hanging core. In such cases, it is necessary to fasten the core with wire or rod extending through the cope.

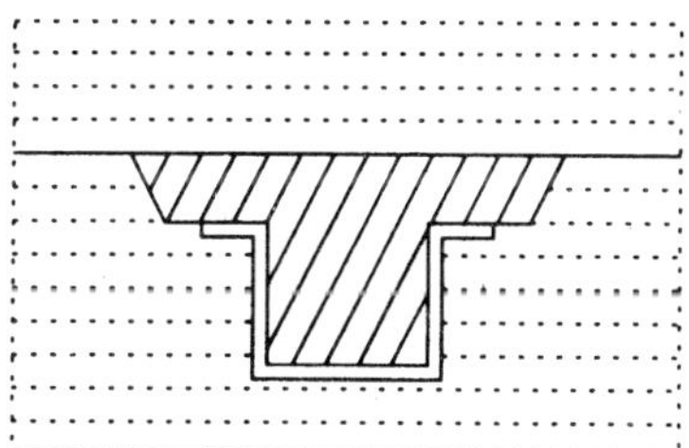

Fig. 1.32 Hanging Core

(e) Wing Core

Wing core is used when a hole is desired in the casting either above or below the parting line (see Fig, 1.33).

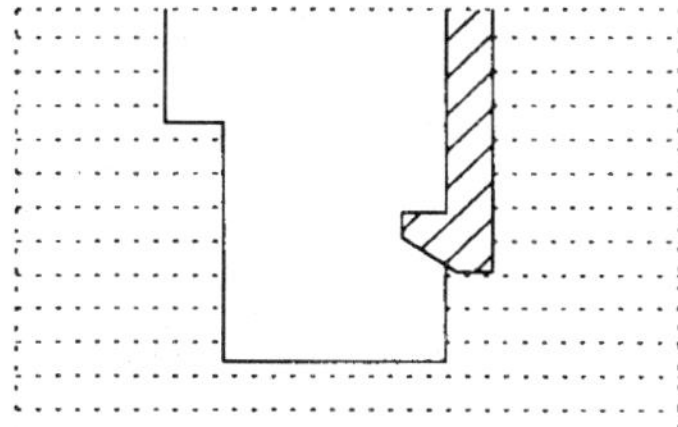

Fig. 1.33 Wing Core

1.2.2.10 Core Making Machines

A number of machines have been developed for the rapid production of cores. A particular type depends on factors such as the number of cores required, size of the core and the intricacy and design of cores.

(a) Core blowing machine

The core sand is forced into the core box from a sand reservoir with a stream of high velocity air, at a pressure of 6-8 kg/cm^2. The core box has a number of vent holes, as the sand is introduced, the air is escaped through these holes. Due to the high velocity air, the sand is packed in the core box.

(b) Core shooter

Another version of core blower in which core sand is ejected from the shooter head and is made impinge into the core box cavity under impact.

1.2.3 Melting and Pouring

After moulding, the molten metal is poured into the mould to get the casting. Various types of melting furnaces are available for the melting purpose. The choice of the furnace depends on the amount and the type of alloy being melted.

Furnaces most commonly used are:

(a) Cupola furnace (for melting cast iron)

(b) Crucible furnace (for melting nonferrous metals)

(c) Electric furnace (for melting steel and special alloy steels).

(a).1 Cupola Furnace (Fig 1.34)

Description: The cupola is a shaft type furnace for producing molten cast-iron. It is a vertical cylindrical shell made of 6 to 12 mm thick boiler plate rivetted and lined inside with acid refractory bricks. Diameter varies from 1 to 2 metres. Height 3 to 5 meters diameter. The whole shell is mounted on brickwork foundation or on steel columns. The bottom of the cupola is provided with drop bottom door, through which debris consists of slag, coke etc. can be removed at the end of melt. Towards the top of the furnace, there is an opening (charging door) through which the charge is fed. Air for combustion is blown through the tuyeres located at a height of 0.6 to 1.2 metres above the bottom of the furnace.

(a).2 Cupola can be divided into the following zones

1. *Crucible Zone*: It is between the top of the sand and the bottom of the tuyeres. The molten metal comes here.
2. *Tuyeres Zone*: It is between the bottom of tuyeres to the top of tuyeres.
3. *Combustion Zone*: This zone is located above the tuyeres where the combustion of the fuel occur by oxygen of the air blast and produces lot of heat in the cupola.

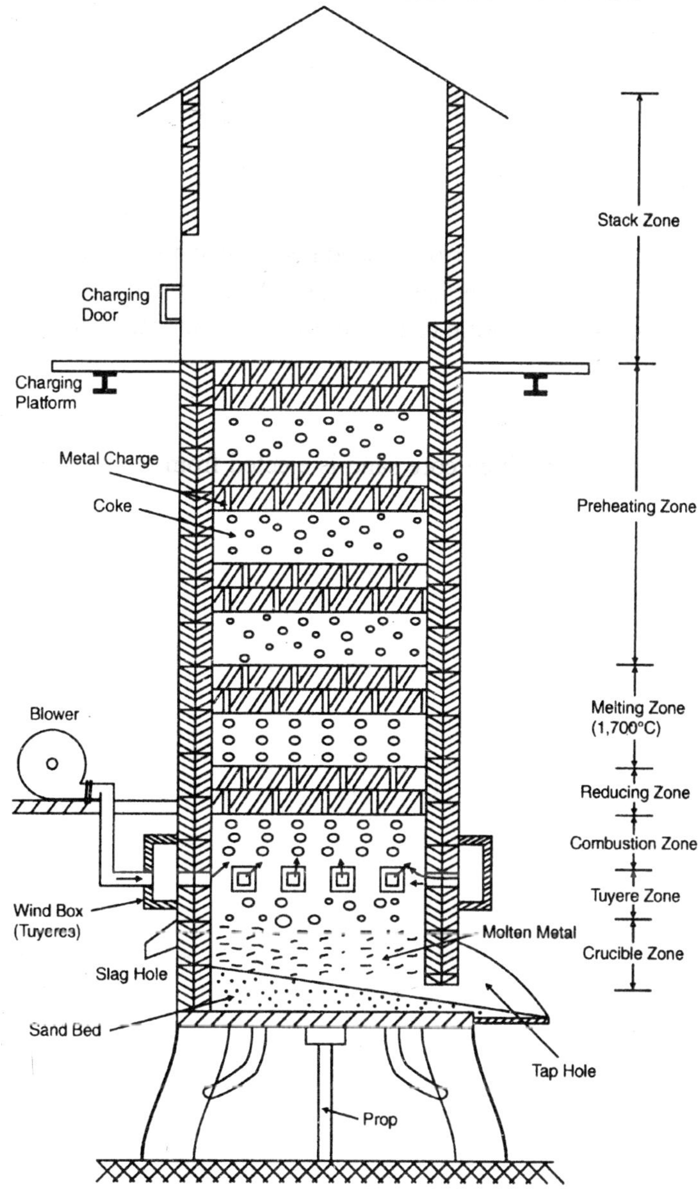

Fig. 1.34 Cross-Section of Cupola Furnace

4. *Reduction Zone*: This zone extends from the top of the combustion zone to the top of the coke bed. CO_2 produced in combustion zone comes in contact with hot coke and is reduced to CO. In this zone iron and other elements are protected from oxidizing influence.
5. *Melting Zone*: It is the first layer of iron above the coke bed. The temperature in this zone is as high as 1700°C. Iron is melted in this zone.

6. *Preheating Zone*: It is located above melting zone to the charging door. Iron and coke are preheated to this in zone.
7. *Stack*: Carries gases from preheating zone to atmosphere.

(*a*).3 Cupola Operation

Preparation of Cupola: After each heat, the slag and refuse are cleaned as soon as the patching of the lining is completed, the bottom doors are raised and held in position by metal props. The sand bottom is made such that it slopes towards the tap hole.

Firing the Cupola: Small pieces of wood are ignited on the sand bottom when the wood burns well, coke is added. Air necessary for coke combustion from tuyeres. Coke is added until the desired height is reached. Instead of placing wooden pieces, the initial coke may be ignited by gas burners or electric spark igniters.

Charging: After coke bed is properly ignited, coke and pig iron are charged in alternative layers until the cupola is full from charging door.

In addition of iron and coke, a certain amount of limestone is added to the first metal charge. Besides limestone fluorspar (CaF_2) and soda ash (Na_2CO_3) also used as fluxing material. A flux removes the impurities in the iron and protects the iron from oxidation. Limestone reduces the melting point of the slag and increases fluidity.

Soaking Iron: After it is charged, it is kept about 45 minutes. The charge gets preheated. This causes the iron to get soaked.

Opening the Air Blast: At the end of the soaking period, the blast is opened. As the melting proceeds, the molten metal appears at the tap hole.

Pouring the Molten Metal: When sufficient metal is collected, the slag hole is opened and the slag is run off. Then the tap hole is opened. Molten metal is collected in ladles and carried to moulds for pouring.

(*b*) Crucible Furnace

Crucible furnaces are used to melt nonferrous metals like bronze, brass, aluminium and zinc alloys. Crucibles are made of either refractory material or alloy steels. Refractory crucibles can be of clay graphite, either ceramic bonded or silicon carbide bonded types:

The crucible furnaces are the following types:

(a) Pit crucible furnace
(b) Tilting furnace

(*a*) *Pit Crucible Furnace* (See Fig. 1.35)

As the name implies, it is constructed in a pit dug in the ground. It may be coke, oil or gas fired furnace, but usually it is fired with coke. The coke bed is formed, ignited and allowed to bum. Now coke from the centre of the coke bed is removed and crucible with lid containing the metal charge is placed. Coke is again added surrounding the crucible on all sides. When the metal reaches the desired temperature, the crucible is lifted out with tongs and the metal is transferred to the mould.

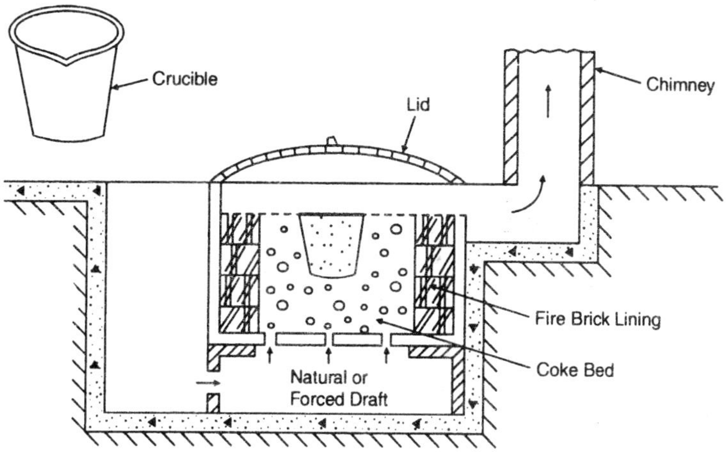

Fig. 1.35 Pit Crucible Furnace

(*b*) *Tilting Furnace* (see Fig. 1.36)

As compared to stationary furnaces, tilting type furnaces are preferred where large amounts of metal are melted. In tilting furnace, crucible is permanently cemented in place. The furnace made of sheet metal is lined with refractory (fire brick). It is mounted on two pedestals and is tilted with a geared hand wheel or power. Oil or gas is used as fuel in the furnace. Combustion air is supplied by a blower.

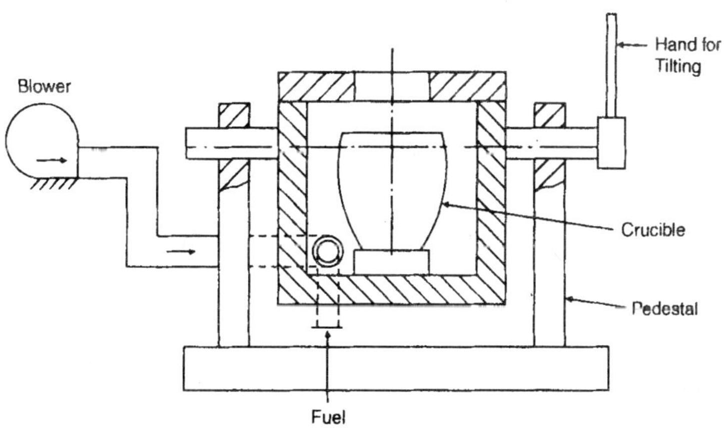

Fig. 1.36 Tilting Furnace

(c) *Electric Furnace*

These are used for the production of high quality castings:

1. Direct arc furnace
2. Indirect arc furnace

1. Direct arc furnace (see Fig. 1.37)

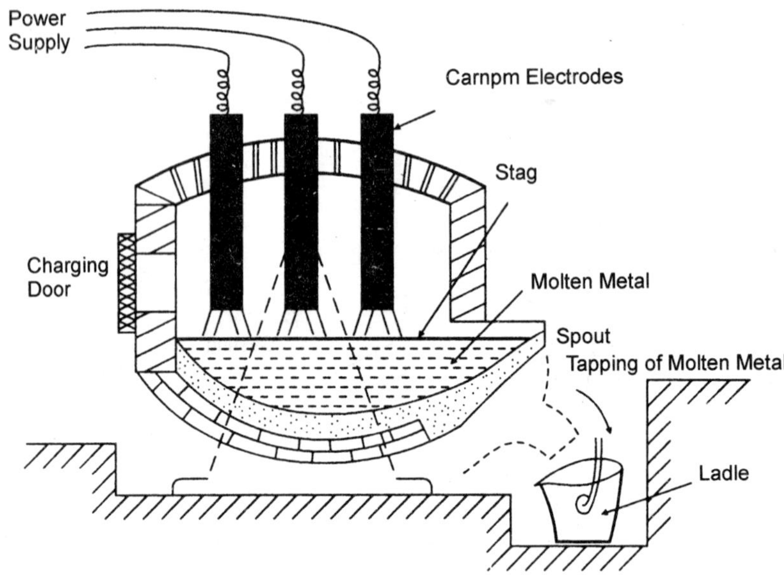

Fig. 1.37 Direct Arc Furnace

Three-phase direct arc furnace is the most popular one for melting steel in the foundry. In operation, scrap steel is placed on the hearth of the furnace. An arc is drawn between the electrodes and the surface of the metal charge by lowering the electrodes down till the current jumps. Slag is maintained on the molten metal to reduce oxidation. Before pouring the liquid metal into the ladle, the furnace is tilted back and the slag is removed from the charging doors. Now the furnace is tilted forward to pour the molten metal into ladle.

2. Indirect arc furnace (see Fig. 1.38)

This is a single-phase electric furnace. This differs from the direct arc furnace that the electrodes do not come in contact with the molten metal, but form an arc above the molten metal. The furnace is mounted on rollers which is driven by rocking unit to rock the furnace back and forth during melting. While the furnace rocks, liquid metal washes over the heated refractory linings and absorb heat from them. Thus the charge is heated by radiation from the arc and conduction from the lining.

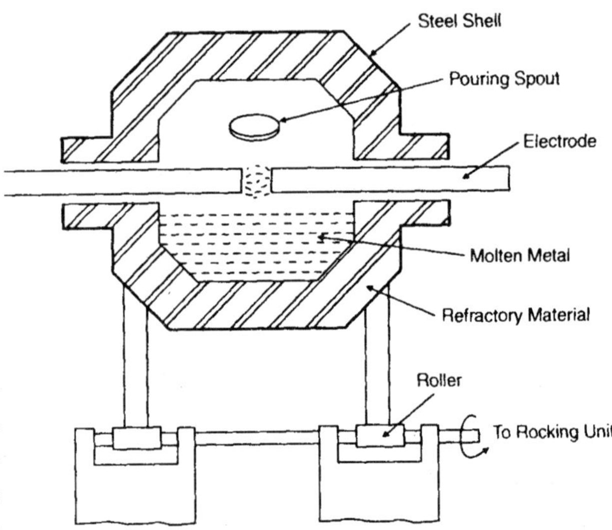

Fig. 1.38 An Indirect Arc Furnace

1.2.3.1 Pouring Molten Metal

The molten metal is poured into the mould through the gating system.

1.2.3.2 Gating System

It refers to all passages through which the molten metal enters into the mould cavity. A good gating system avoids the turbulence of flow and erosion of mould walls and it prevents entering of loose particles of sand into the moulds. The various parts of gating system are shown in Fig. 1.39.

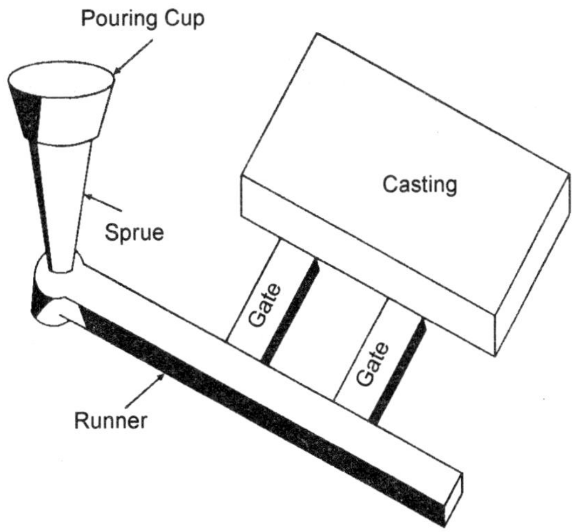

Fig. 1.39 Components of Gating System

(a) Pouring basin

It is a funnel shaped opening made at the top of the sprue in the cope. It is used to (i) make it easier for ladle operator to maintain the required flow rate, (ii) minimize turbulence, thereby the molten flow is smooth, (iii) aids in separating dross and slag from the molten metal before it enter the runner system. Fig 1.40 shows various designs of pouring basin.

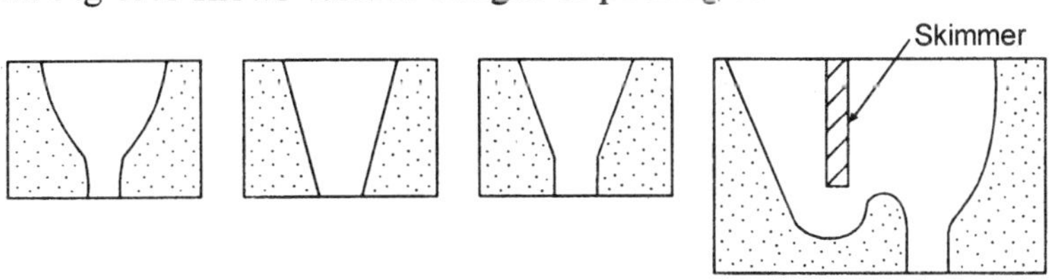

Fig. 1.40 Typical Designs of Pouring Basins

(b) Sprue

It is a vertical passage through the cope and connects the pouring basin to the runner or gate. The sprue cross-section may be circular or square or rectangular. Figure 1.41 shows the effect of sprue design on metal turbulence. If the sprue is straight and has sharp corners, there is severe aspiration (see Fig. 1.41a), thereby causing turbulence in the metal. If the tapered corners are round, dam type of pouring basin (see Fig. 1.41(b)) is used, aspiration is negligible and there is no turbulence.

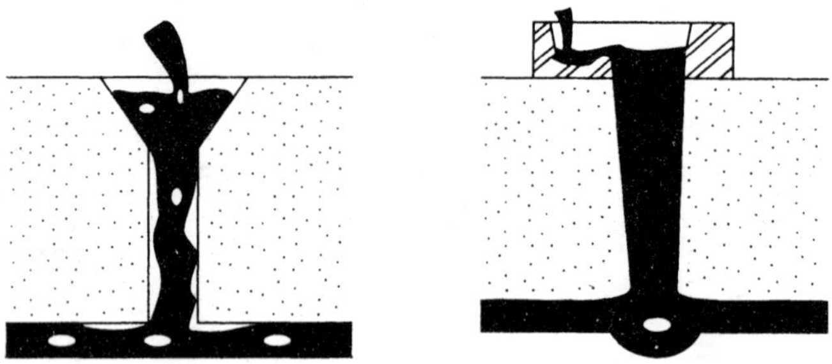

(a) Straight sprue sharp corners serve aspirations (air entrapment)

(b) Tapered sprue rounded corners negligible aspirations

Fig. 1.41 Effect of Sprue Design on Metal Turbulence

1.2.3.3 Gates

Gate is a channel which converts runner with the mould cavity. *Types of Gates:* The gates are classified as:

(a) Parting Line Gate, (b) Bottom Gate, (c) Top Gate

(*a*) *Parting Line Gate*: In parting line gate, the metal enters the mould cavity at the parting line (see Fig. 1.42).

The gate may contain such as Skim bob, Skimming gate, Shrink bob, Whirl gate.

A hollow recess in the cope is known as 'Skim bob' is used to trap the slag and foreign matter in the metal due to its curvature Fig. 1.42(a).

Skimming gate is a vertical passage through the cope. The purpose is similar to that of a skim bob. In this case, the foreign matter being lighter in weight than the metal can rise through the vertical passage Fig. 1.42(b).

Shrink bob may be provided, if there is a tendency for shrinkage defect to occur near the ingate Fig. 1.42(c).

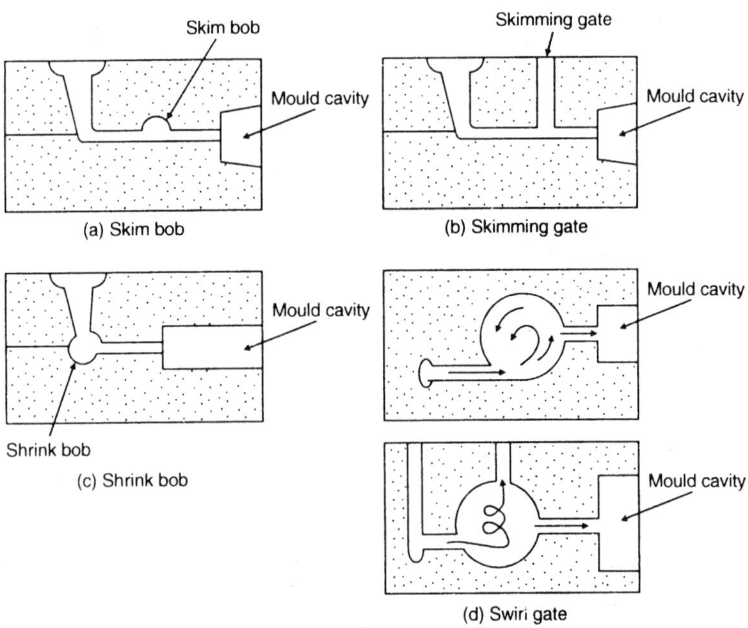

Fig. 1.42 Parting Line Gate

Whirl gate employs centrifugal force to aid in the slag come to the centre from where it rises up in the skimming gate Fig. 1.42(d).

(*b*) *Bottom Gate*: (Fig. 1.43)

A bottom gate is made in the drag portion of the mould. In this metal fills the bottom first and rises steadily up the mould. The main advantage is the turbulence is minimum hence mould erosion is prevented.

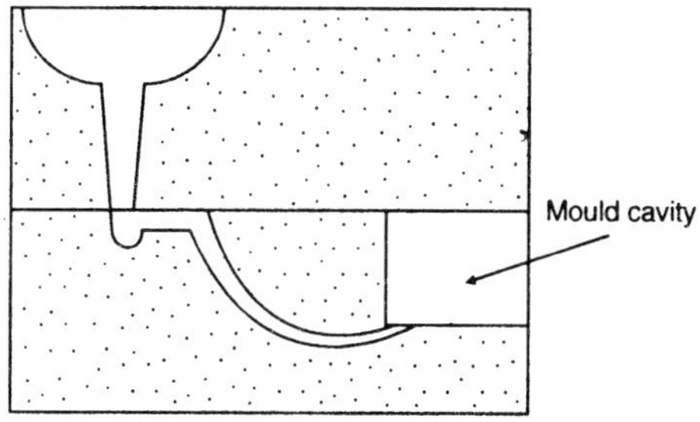

Fig. 1.43 Bottom Gate

(*c*) *Top Gate*: (Fig. 1.44)

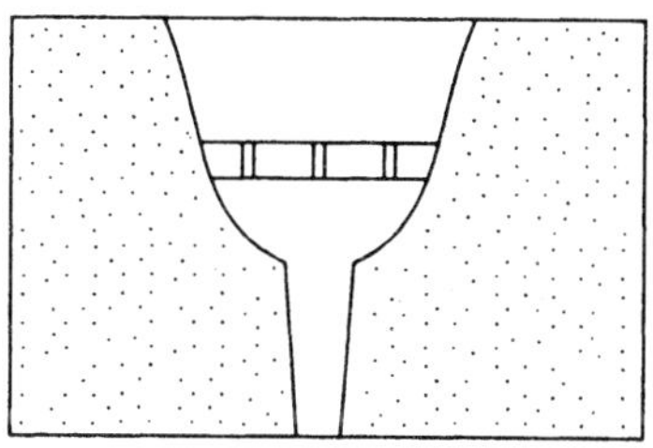

Fig. 1.44 Top Gate

In top gate, the metal is poured down directly into the mould cavity. The advantage is that the molten metal at the top of the casting is always hot.

1.2.2.10.3 Gating Ratio

It describes the relative cross-sectional area of sprue: total runner area: total gate area. For example a gating system having 4 sq cm total gate area, 8 sq cm runner area, and 8 sq cm total gate area, the gating ratio is 1:2:2. The gating ratios are classified as pressurised system and unpressurised system.

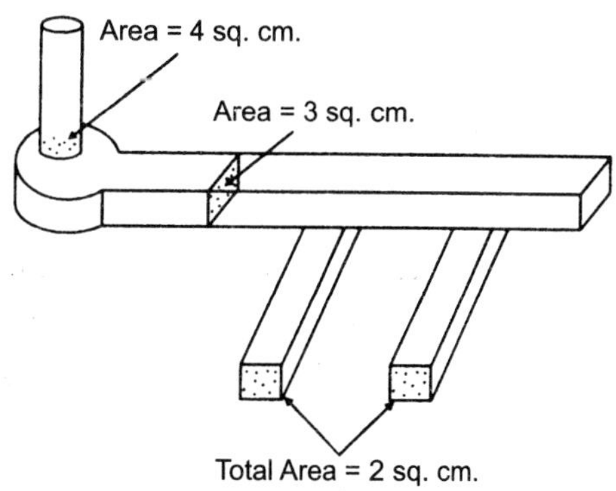

Fig. 1.45 Pressurised System with Gating Ratio 2:1.5:1

(*a*) *Pressurised Gating System*: If the total gate area is smaller than the area of sprue, back pressure is maintained on the gating system due to restriction of metal flow at the gates and the system is called pressurised gating system. Gating ratio 2: 1.5: 1 indicates a pressurised gating system (see Fig. 1.45).

(*b*) *Unpressurised Gating System*: The unpressurised gating system on the other hand has metal flow restriction at the sprue. A system having gating ratio of I :2:3 indicates an unpressurised gating system (see Fig. 1.46)

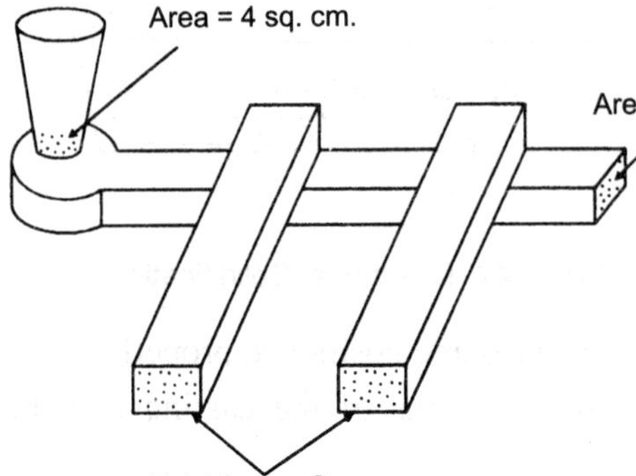

Fig. 1.46 Unpressurised System with Gating Ratio 1:2:3

1.2.3.4 Runner

Runner is a common passage for molten metal to flow into the mould cavity from sprue. The branches from runner to the mould cavity are called ingates.

1.2.3.5 Riser

Riser is a hole cut in the cope to permit the molten metal to rise above the highest point in the casting.

Functions of Riser

(i) It enables the pourer to see the metal in the mould cavity. If the metal is not seen in the riser, it indicates that either the metal is not sufficient to fill the mould cavity or there is some obstruction to the metal flow between the sprue and riser.

(ii) The riser gives passage to the steam, gas and air from the mould cavity while filling the mould with the molten metal.

(iii) It serves as feeder to feed the molten metal into mould cavity to compensate for it's shrinkage.

1.2.3.6 Directional Solidification (see Fig. 1.47)

Since the casting has various sections, all the parts do not cool at the same rate, some parts tend to solidify more quickly than others. This contraction causes voids and cavities in certain regions of the casting. These voids must be filled by liquid metal from riser. Hence the solidification should continue progressively towards the risers which should be the last to solidify. This is known as directional solidification. In general, the following ways are adopted for controlling the directional solidification.

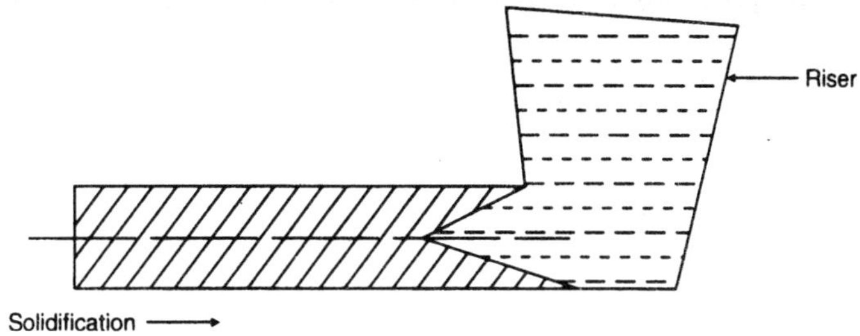

Fig. 1.47 Directional Solidification

(i) The risers should be designed and positioned properly.

(ii) The thickness of certain sections of the casting may be increased by the use of padding.

(iii) Exothermic materials in the risers.

(iv) Chills may be used in the moulds.

(*i*) *Design and Positioning of Risers*: The riser should keep the metal in a molten state as long as possible. This can be achieved when the riser is spherical in shape so that its surface area is a minimum. For the same volume, the next best shape is cylindrical shape and a square shape. Since it is difficult to mould a spherical shape, a cylindrical shape is the best shape to adopt.

(*ii*) *Use of Padding*: (see Fig. 1.48)

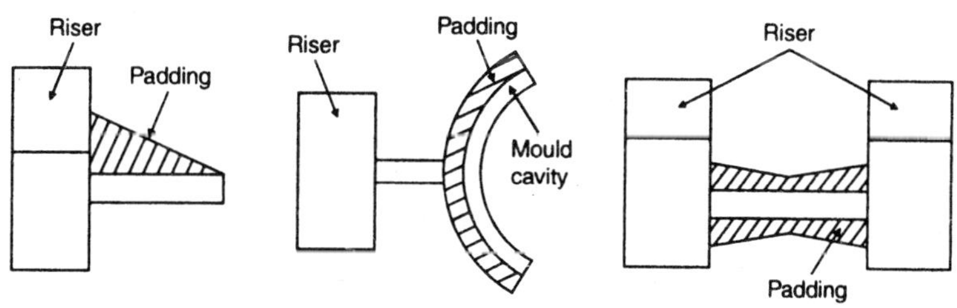

Fig. 1.48 Use Padding

Padding is another method of promoting directional solidification. The padding is simply extra metal added to the original uniform section of the casting. This extra metal, if not desired, can be removed later by machining.

(*iii*) *Use of Exothermic Materials*: These are used to provide directional solidification. These are provided in the riser on the top of poured metal or to the sand in the riser walls. When the molten metal contact with these exothermic materials, chemical reaction takes place, producing substantial heat. Thus the molten metal gets superheated and remains molten for a longer period.

(*iv*) *Use of Chills*: The thin sections of the casting solidify faster than the thicker sections. Due to this, there will be uneven contraction, thereby giving rise to internal strains in the casting. Even develop cracks in the casting. Hence for rapid solidification of heavy sections and to achieve directional solidification, chills are commonly used.

Chills are classified as:

1. External Chills.
2. Internal Chills.

1. *External Chills*: These may be direct or indirect type. The direct type forms mould face and contact with the molten metal. In indirect type, the chill is embedded below the mould and no contact with the molten metal (see Fig.1.49).
2. *Internal Chills*: These are within the mould cavity and go into the casting when the metal is poured.

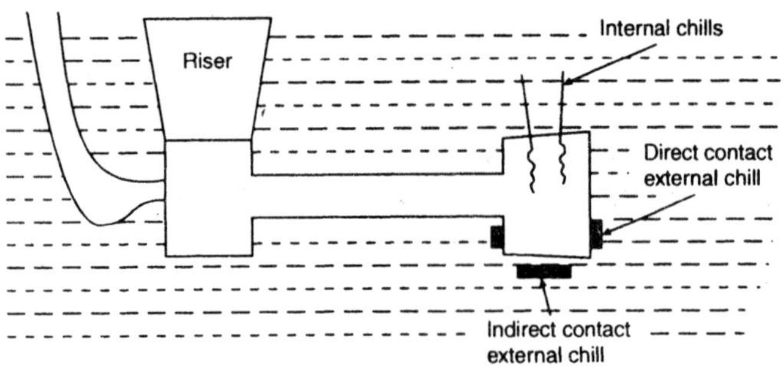

Fig. 1.49 Use of Chills

1.2.3.7 Pouring Time

High pouring rate means turbulent flow in the mould, leads to mould erosion. Low pouring rate may not fill the mould cavity completely, leads to defect like cold shut. Hence, optimum pouring time is required.

Pouring times for some of the materials are given below:

(a) Gray Cast Iron, mass less than 450 kg.

$$t = K(1.41 + T/14.59)\sqrt{w}$$

where t = Pouring time in seconds

K = Fluidity of iron in inches/40

T = Average section thickness in mm

w = Mass of the castings in kg

(b) Gray Cast Iron mass greater than 450 kg.

$$t = K(1.236 + T/16.65)3\sqrt{w}$$

(c) Steel Castings

$$t = (2.4335 - 0.3953 \log w)\sqrt{w}$$

Example 1.3: Calculate the optimum pouring time for casting cast iron whose mass is 50 kg and a thickness 50 mm. Fluidity is 2.2 inches.

Solution:

$$\text{Pouring time } t = 22/40(1.41 + 50/14.59)\sqrt{50}$$
$$= 18.81 \text{ seconds.}$$

Choke Area: It is the area at the sprue exit. It can be calculated based on Bernoulli's theorem.

$$CA = W/cdt\sqrt{2gh}$$

Where CA = Choke Area, mm^2

W = Casting mass, kgs

C = Efficiency factor – a function of gating system used

D = Mass density of liquid metal, kg/mm^3

T = Pouring time, seconds

G = Acceleration due to gravity mm/sec^2

H = Effective liquid metal head.

The effective sprue height H, of a mould depends on the type of gating used as shown below:

Top gate $H = h$

Bottom gate $H = h - hm/2$

Parting gate $H = h - hc^2/2hm$

Where h = height of sprue

hc = Height of mould cavity design in cope

hm = Total height of mould cavity

The values of h, hc and hm are shown in Fig. 1.50 for the various types of gating.

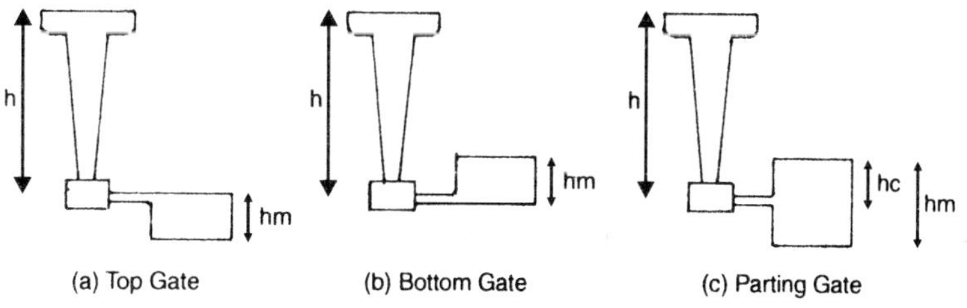

Fig. 1.50 Different Gating Systems

Example 1.4: Determine the diameter of taper sprue at the bottom to fill G.I. casting neglecting the directional flow and flow losses. Casting weight 30 kg, pouring time 20 seconds, density of melt $7gm/mm^3$, height of sprue is 60 mm and top gate.

Solution:

Choke Area $CA = W/cdt\sqrt{2gh}$

$W = 30$ kg

$H = h = 160$

$T = 20$ secs

$D = 7$ gm/mm^3 $=7 \times 10^{-3}$kg/1000 m^3

$= 7 * 10^{-6}$kg/mm^3

C = for taper sprue is 0.85

$Ca = 30/0.85 \times 7 \times 10^{-6} \times 20\sqrt{2} \times 980 \times 160$

$= 142.36$mm

$\Pi D^2/4 = 142.36$

$\therefore$ $D = 13.46$mm

Example 1.5: A casting of 200 × 100 × 70 mm^3 size solidifies in 10 minutes. Estimate the solidification time for 200 × 100 × 10 mm^3 casting under similar conditions (Gate Problem).

Solution:

Solidification time $t = K(V/A)^2$

Where K = mould constant; V = volume of casting; A = surface are casting

$V = lbh$; $A = 2(lb + bh + lh)$

$V1 = 200 \times 100 \times 70 = 14 \times 10^5$ mm^3

$V2 = 200 \times 100 \times 10 = 2 \times 10^5$ mm^3

$Al = 2(200 \times 100 + 100 \times 70 + 200 \times 70) = 82000$ mm^2

$A2 = 2(200 \times 100 + 100 \times 10 + 200 \times 10) = 46000$ mm^2

$tl/t2 = (Vl/Al)^2/(V2/A2)^2 = (14 \times 10^5/82 \times 10^3)/(2 \times 10^5/46 \times 10^3)$

291.48/18.896%= 15.42

10/12 = 15.42

$\therefore$ $t^2 = 0.648$ minutes

1.2.4 Fettling

Fettling includes:

(a) Removal of cores

(b) Removal of gates, risers and runners

(c) Cleaning the surfaces

Sand cores are generally removed by shaking. Gates and risers are removed by grinding operation. Flame cutting using oxyacetylene gas is used for cutting gates and risers of steel castings.

Next is to remove the fins and sand adhering the casting. This is carried out by tumbling. In this method, the castings to be cleaned are placed in the large shell or barrel which contains small, hard, star shaped pieces of cast iron. The barrel is rotated at about 30 rpm. As the barrel rotates, the castings tumble over each other, rub off the adhering sand.

Sand blasting is other process of cleaning the castings. In this method, sharp silica sand is blown against the castings at high velocity.

Defects in Castings

Casting defects are usually not accidental, but due to improper control of manufacturing. The major defects generally found in the sand castings are as follows:

(i) Gas defects
(ii) Shrinkage cavities
(iii) Moulding material defects
(iv) Pouring metal defects
(v) Metallurgical defects
(vi) Moulding and core box defects

(*i*) *Gas Defects*: These are blowholes and open blows, air inclusions and pin hole porosity. These are due to lower permeability of the mould.

(*a*) *Blow holes and Open blows*: These are in the form of spherical, flattened or elongated cavities present inside the cavity or on the surface as shown below Fig. 1.51.

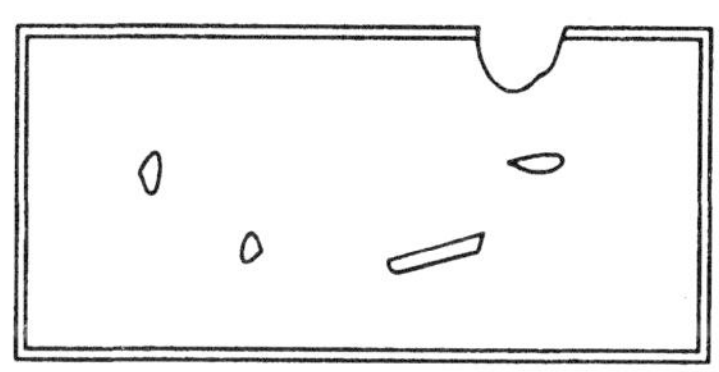

Fig. 1.51 Blow holes

On the surface, they are called open blows holes. These are due to the moisture left in the mould and the core. Due to heat of the molten metal, the moisture is converted into steam, a part of which may entrapped in the casting. Apart from the moisture, they occur due to the lower venting and lower permeability of the mould.

(*b*) *Air Inclusions*: The atmospheric and other gases absorbed by the molten metal in the furnace, in the laddle and during the flow in the mould when not allowed to escape, would be trapped inside the casting. The main reason for this defect is the higher pouring temperatures which increase the amount of gas absorption.

(*c*) *Pinhole Porosity*: This is due to the hydrogen in the molten metal. The hydrogen while leaving the solidifying metal would cause very small diameter and long pinholes.

(*ii*) *Shrinkage Cavities*: These are caused by the liquid shrinkage occurring during the solidification. To compensate this, proper feeding of the liquid metal is required. Also proper casting design.

(*iii*) *Moulding Material Defects*: Scabs, swell, run out and drop. These defects occur because of the moulding materials are not having required properties or due to improper ramming.

(*a*) *Scabs*: These are projections on the casting which occur when a portion of the mould face lifts and metal flows underneath in a thin layer. In other words, liquid metal penetrates behind the surface layer of sand (see Fig. 1.52).

These scabs are of two types: 1. Expansion scabs, 2. Erosion scabs. Expansion scab is caused by the expansion of the surface layers of the sand. It may occur where metal has been agitated or has partly eroded the sand.

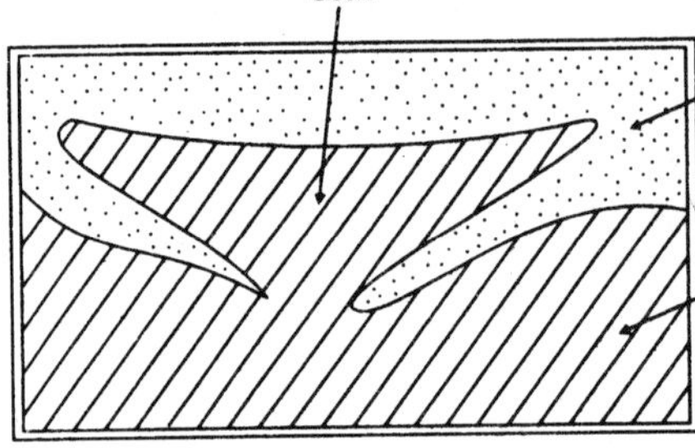

Fig. 1.52 Scab

Scabs occur due to uneven ramming, excess moisture in the sand.

(*b*) *Swell*: Under the influence of the metallostatic forces, the mould wall may move back causing a swell in the casting.

It may occur due to in sufficient ramming of the sand, pouring the molten metal too rapidly.

(*c*) *Run out*: This occurs when the molten metal leaks out the mould cavity. This may be caused either due to faulty mould making or because of the faulty moulding flask.

(*d*) *Drop*: A drop occurs when cope surface cracks and breaks, thus the pieces of sand fall into the molten metal. This is due to either low green strength or improper ramming of the cope flask.

(*iv*) *Pouring Metal Defects*: The defects in this category are misrun, cold shut, poured short.

(*a*) *Misrun*: When the mould is not completely filled with molten metal due to some obstruction in the form of metal, the defect is known as misrun.

(*b*) *Cold shut*: Sometimes metal is poured from opposite directions in the mould. If the two streams of metal approach each other, make physical contact, but do not fuse together thus leaving a gap, the resulting defect is called 'cold shut'. The reasons for cold shut or misrun may be too thin sections, improper gating system, slow and intermittent pouring, poor fluidity of metal.

(*c*) *Poured Short*: When the mould cavity is not completely filled because of insufficient metal, the defect is called 'poured short'.

(*v*) *Metallurgical Defects*: Hot tears and Hot spots.

(*a*) *Hot Tears or Hot Cracks*: These are internal or external ragged discontinuities or cracks on the casting surface, due to hindered contraction occurring immediately after the metal has solidified. These will occur when the casting is poorly designed.

(*b*) *Hot Spots*: These are caused by the chilling of the casting. For example, with gray cast iron having small amounts of silicon, very hard white cast iron may result at the chilled surface. This hot spot will interfere with the subsequent machining of this region. Proper metallurgical control is essential for elimination of the hot spots.

(*vi*) *Moulding and Core Box Defects*

(*a*) *Mould Shift* (*Mismatch*): A shift in a mismatch of cope and drag flasks at the parting line. Fig. 1.53 shows the mismatch of the sections of a casting at the parting line. This is due to worn or loose dowels in the pattern made in halves. This defect can be prevented by ensuring proper alignment of the patterns or die parts. Core shift may also occur due to misalignment of cores or core halves during assembly.

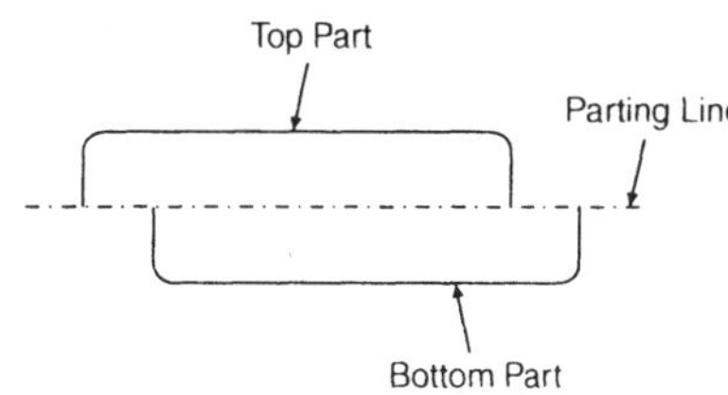

Fig. 1.53 Shift

(*b*) *Fins*: Thin projections of metal on the surface of the casting usually at the parting of mould or core sections. Moulds and cores incorrectly assembled will cause fins. Insufficient weight on the mould or improper clamping of the flasks may produce fin defect.

1.2.5 Inspection and Testing

The aim of inspection is to reject those castings which do not meet with the specifications and also determines the location and magnitude of various defects in the casting.

The inspection and testing of castings are broadly classified as:

(a) Destructive Testing and (b) Non Destructive Testing

The destructive testing include tensile, compressive and shear testing. In addition microscopic examination to determine physical and metallurgical qualities of castings. The disadvantage is that the component under test becomes unserviceable.

Non-Destructive testing consists of the following tests:

(a) Visual Inspection
(b) Penetrant Test
(c) Sound Test
(d) Magnetic Particle Test
(e) Ultrasonic Testing
(f) Radiography Testing

(*a*) *Visual Inspection*: Visual inspection is the simplest, fastest and most commonly used method. This is used to detect defects on the surfaces of the casting like cracks, blow holes, swells, swifts etc. This is carried out with naked eye or using a magnifying glass.

(*b*) *Penetrant Test*: The casting is sprayed with a liquid penetrating agent having low viscosity. The penetrant enters the cracks. The casting is then wiped and cleaned. A dry developer is sprayed on the casting. This draws some of the suspension from the cracks to the surface where it flourescences and is readily visible under ultraviolet light. This is used only for surface defects.

(*c*) *Sound Test*: In this method, the casting is given blows with hammer and listen the sound waves produced. The defect free casting emits a clear ringing sound where as the defective casting gives a dull sound.

(*d*) *Magnetic Particle Test*: This test consists of magnetising the casting and then sprinkling the fine powder of magnetic material. This powder tends to be held and bridge over defects, thus forming a visible indication and location and magnitude of the defect.

(*e*) *Ultrasonic Testing*: In Ultrasonic testing, high frequency sounds (frequency of sound beyond audible range are passed through one end of the surface of the casting, the waves travel through the casting to the opposite surface and are reflected back to the original point. Any defect in the part of the waves scatter the waves and are reflected back from the defect and return in a shorter period of time. The advantage of this method is not only detecting but also locating accurately.

(*f*) *Radiography Testing*: This method is used to detect internal defects of the casting. Radiant energy from the X-Ray tube is passed through the casting or section of the casting and recorded on a film held against the opposite surface. Defects in the form of cracks or voids are recorded as blackened areas on the film, since the radiant energy moves more easily through the less dense regions. Defects like cracks, internal and external, hot tears, shrinkage, gas or pin hole porosity are detected by this method.

1.3 SOLIDIFICATION OF CASTINGS

Sound Casting is one which is free from defects like porosity, shrinkage and cracks etc. In order to produce a casting free from such defects, it is essential to know cast structure developed during solidification of metals and alloys.

1.3.1 Recrystallization

(a) In this process, old distorted grains are replaced by new equiaxed stress free, strain free grains by a process a nucleation and growth. This is called recrystallization.

(b) Nucleation occurs at the points of high energy and subsequently these nuclei grow at the expense of old grains. The probability of forming a nucleus is the same at every place a state suitable for homogeneous nucleation is said to be in the system.

(c) Nucleation occurs at the surface near the mould wall contains equidistanced grains and the nucleation is based on the volume of the particles formed i.e., free energy and the energy need to join the surface layer of the particle. The microstructure at the end of recrystallization process is similar to the original structure. The grains become equiaxed and the dislocation density gets reduced to a value of strain free metal.;

(d) Due to the change in the microstructure of the casting metal, mechanical properties increases, internal stresses are reduced almost to the original level with increase in corrosion resistance.

(e) The grain size of the material obtained at the end of recrystallization depends on the temperature of heating, time of heating, heating rate and type end level of impurities.

(f) Insoluble particles lock the grain boundaries and prevent their migration. They also reduce the energy of grain boundaries. Due to this, a fine grain size rate sharply decreases.

1.3.2 Formation of Grains

All the metals are crystalline and crystals are made up of several atoms. The individual crystals or grains together form the visible mass of a solid metal.

A grain is a crystal with almost external shape, but with an internal atomic structure based on the space lattice with which it was formed. The mechanical properties of the metal varies with the arrangement of grains. The solidification process is shown in Fig. 1.54.

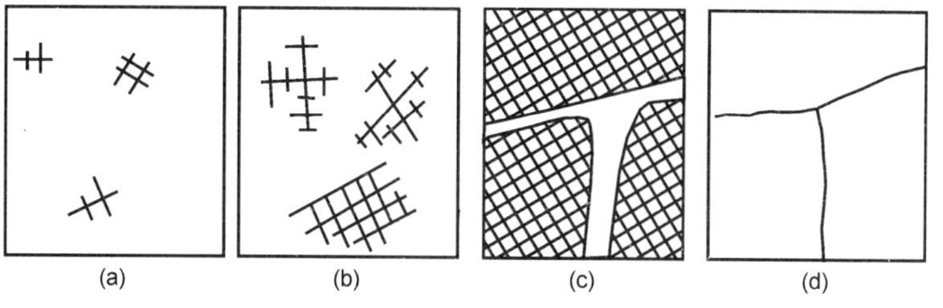

Fig. 1.54 Formation of Crystal Grains

The metal begins to solidify when the temperature of the liquid metal drops below the critical temperature. When two or more atoms associated to form a small crystal called "Nucleus". It happens in number of locations throughout liquid metal. They are simultaneously cooled. Slow cooling favours growth of crystals uniformly in all directions of growth and give equiaxed crystals i.e., the crystals with equal dimensions in all the direction. Rapid cooling always favours tree like crystals called "dendrites" which consisting of unit cells, with straight line branches.

Crystal grows until it come in contact with the adjacent crystal of proper geometrical form and having different orientations. They can be distorted by interference of each crystal with its neighbours.

The boundary formed between two adjacent crystalline growth because of different orientations of the grain is known as grain boundary.

- As shown in Fig (1.54) The formation of nucleus in straight line branches is shown in (a)
- Crystals having the same geometrical form but different in orientations can be seen in (b) & (c)
- Grain boundaries formed between adjacent crystals can be seen in (d)

1.3.3 Grain Size

The rate of whole crystallization process is determined by the rate of nucleation (N) and rate of crystal growth (G). The size of the grain is determined by the rate of crystal growth (G) and rate of nucleation (N). At high value of G and low value of N, coarse grains are formed. Besides rate of cooling, the grain size also depend on factors.

- Temperature of liquid metal
- Impurities in metal
- Chemical composition.

1.3.4 Solidification of Pure Metals

Solidification process differs from pure metal to the alloys as it is the transformation of the molten metal back to the solid state.

As the pure metal have sharply defined freezing temperature which is the same melting point composed of the small group of atoms oriented into common crystal pattern (Fig. 1.55). The process occurs in the overtime called a cooling curve. During the process of solidification nuclei spring up, each nucleus grow and able to form the crystal which is visible to the eye.

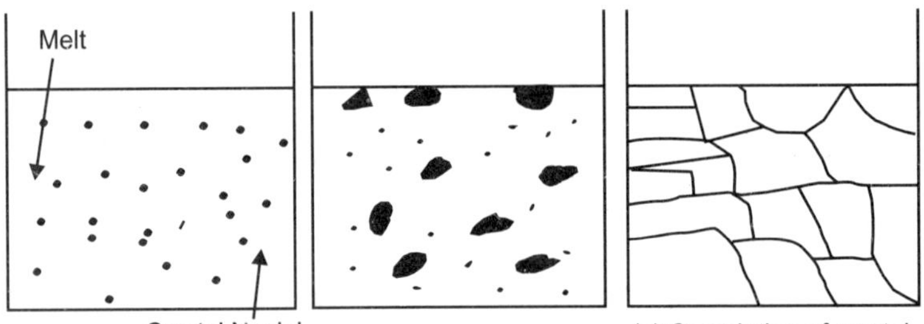

Fig. 1.55 Progressive Freezing of a Uniformly Cooled Metal

As each nucleus grows, the atoms within it are having the same orientation. When the nucleus have grown to the point, they absorb all the liquid atoms and come in contact with each other along their boundaries. The boundaries do not line up the plan of atoms, change directions from one crystal to other so solid state composed of a number of crystals of different orientation i.e., mixed crystals.

The actual freezing takes time called "Local solidification time" in casting during which the metals' latent heat of fusion is released into the surrounding mould. The total solidification time is the time taken between pouring and complete solidification. After the casting is completely solidified, cooling gradually increases by decreasing the slope of the cooling curve as shown in Fig. 1.56.

Because of the chilling action of the mould wall, a thin skin (a thin layer) is formed at the interface immediately after pouring. Thickness of the skin increases to form a shell around molten metal as solidification progress inward toward the centre of the cavity. The rate of freezing depend on the heat transfer in the mould.

As the conductivity of the mould is high, fine, equiaxed, random orientation atoms of small crystal grows near the mould face. As the cooling progress, the grain formation in the direction

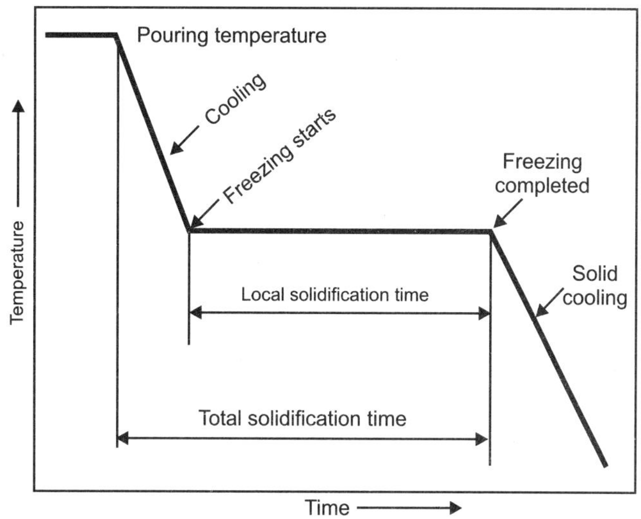

Fig. 1.56 Temperature as a Function of Time for Solidification of Pure Metal

away from the heat transfer gradually long columnar crystals, with the axis perpendicular to the mould face are formed.

The beginning of solidification and end of solidification takes place at constant temperature in pure metals. These two points are called congruent melting points. Perfect crystals of proper external shape can be obtained only if crystallization develops under the degree of super cooling is very low and the metal is having high purity. In most of the cases it leads to the formation of branches form at right angles to the first branch (Tree-like crystals) called dendrites as shown in Fig. 1.57.

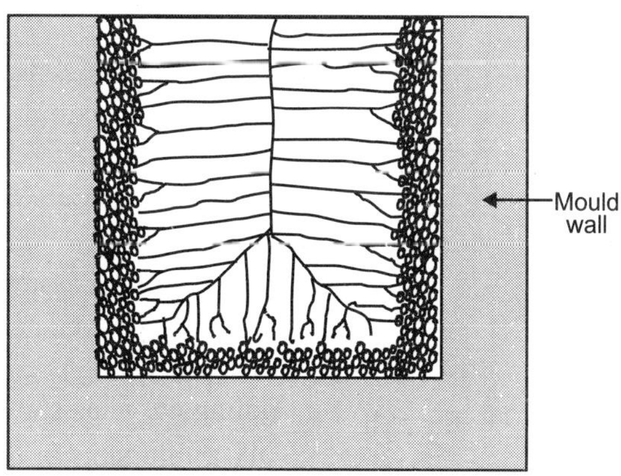

Fig. 1.57 Grain Structure of Casting of Pure Metal Showing Randomly Oriented fine, Equiaxed Grains near the Mould Wall and Large Columnar Grains Oriented towards Centre of Casting.

1.3.5 Solidification in Alloys

Most alloys freeze over a temperature range rather than at a single temperature. The exact range depends on the alloy stem and the particular composition. This can be explained with reference to the phase diagram as shown in Fig. 1.58.

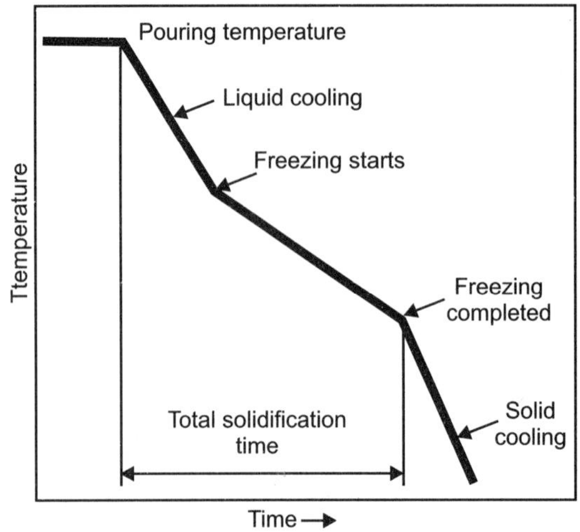

Fig. 1.58 Temperature as a Function of Time for Solidification of an Alloy

Just below the solidification starts the solid phase start separating out from the liquid. As the temperature decreases, the freezing begins from the liquidus line and is completed when the solidus is reached. As similar to the pure metal the freezing starts by forming a thin skin at the mould wall due to large temperature gradient at the surfaces and dendrites grow away from the surface of the mould wall where both liquid and solid metal together. This solid region has a sort of consistency leading to its name as "Mushy Zone". As the freezing progress the mushy zone is relatively narrow, and exists throughout casting. As the temperature difference increases the dendrite matrix solidify as casting drops to the solidus line for the given alloy composition.

Metals having the higher melting points favours the formation of the dendrites composition in the solidification of alloys.

Composition imbalance can be seen in dendrites growth depending upon the segregation of the elements (see Fig. 1.59).

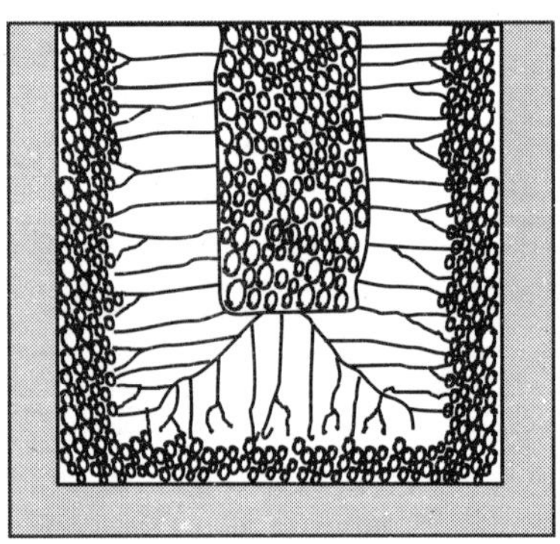

Fig. 1.59 Grain structure in an Alloying Casting Showing Segregation of Alloying Components in the Centre of Casting

The segregation is of two types

(a) Microscopic level

(b) Macroscopic level

Microscopic level: Composition varies throughout each individual grain. Each dendrite has a higher portion of one of the elements in the alloy.

Macroscopic level: Composition varies throughout the casting as the regions of the casting freeze first and richer in one component than the other.

QUESTIONS

1. Explain various types of pattern materials used. State its advantages and disadvantages.
2. Briefly explain the allowances provided on a pattern.
3. Sketch and explain different types of patterns used in a foundry.
4. What are the properties of a good moulding sand? Explain.
5. List out the various tests performed on a moulding sand and explain in detail.
6. Enumerate the various types of moulding machines used in a foundry.
7. Define a core. Explain various types of cores used in moulding practice.
8. Briefly describe the step by step procedure for CO_2 moulding process.
9. Explain the shell moulding process with neat sketches.
10. Describe centrifugal casting process. What are its applications?
11. With a neat sketch describe the working of a pit furnace.
12. Sketch and explain die casting process.
13. What are the functions of a Riser?
14. Describe the working of a cupola.
15. Explain various castings defects.
16. Explain the inspection and testing methods used in foundry.
17. Discuss the various elements that comprise the gating system.
18. What is the difference between the solidification of pure metals and metal alloys? Explain.

2

Plastic Deformation Processes

2.1 INTRODUCTION

Plastic deformation occurs when the metal is stretched or compressed beyond the elastic limit. During the deformation, the metal flows plastically and the shapes of grains are changed. If the deformation is carried out at high temperature, the growth of new grains is accelerated and continues till the metal comprises fully of only the new grains. This process of formation of new grains is known as recrystallisation. The temperature at which this process is complete is known as the recrystallisation temperature. Plastic deformation of a metal above the crystallisation temperature, but below the melting temperature is called *hot working* where as plastic deformation of a metal below its recrystallisation temperature is known as *cold working*.

2.2 DIFFERENCES BETWEEN HOT WORKING AND COLD WORKING

Hot Working	*Cold Working*
1. Hot working is done at a temperature above recrystallisation but below, melting point. It can therefore be regarded as a simultaneous process of deformation and recovery. 2. Hardening due to plastic deformation is completely, eliminated by recovery and recrystallisation. 3. Mechanical properties such as elongation, reduction of area and impact values are improved. Ultimate tensile strength, yield point, fatigue strength, hardness are not affected by hot working.	1. Cold working is done at temperature below recrystallisation temperature. So no appreciable recovery can take place during deformation. 2. Hardening is not eliminated since working is done at a temperature below recrystallisation. 3. Cold working decreases elongation, reduction of area. Increases ultimate tensile strength, yield point and hardness.

(Contd...)

Hot working	Cold working
4. Surface finish of hot worked metal is not nearly as good as with cold working because of oxidation and scaling.	4. Good surface finish is obtained.
5. Refinement of crystals occurs.	5. Crytallisation does not occur. Grains are only elongated.
6. Cracks and blowholes are welded up.	6. Possibility of crack formation and propagation is great.
7. Internal or residual stresses are not developed in the metal.	7. Internal and residual stresses are developed in the metal.
8. Oxide forms rapidly on metal surface.	8. Cold parts possess less ductility.
9. Less force is required.	9. Higher forces are required for deformation.
10. Equipment used in hot working is light.	10. More powerful and heavier equipment's are required for cold working.
11. Handling and maintenance of hot metal is difficult and troublesome.	11. Easier to handle cold parts.
12. Hot working processes: (a) Hot forging (b) Hot rolling (e) Hot spinning (d) Hot extrusion (e) Welded pipe and tube manufacturing (f) Roll piercing (g) Hot drawing	12. Cold working processes: (a) Cold rolling (b) Cold extrusion (c) Press work (i) Drawing (ii) Squeezing (iii) Bending (iv) Shearing

2.3 FORGING

2.3.1 Introduction

Forging is the operation where the metal is heated and then a force (impact type or squeeze type) is applied to manipulate the metal in such a way that the required final shape is obtained.

Forging enhances the mechanical properties of metals and improves the grain flow, which in turn increases the strength and toughness of the forged component.

2.3.2 Forgeability of Metal and Alloys

It is important to know the deformation behaviour of the metal to be forged with regard to the resistance to deformation and any anticipated adverse effects such as cracking. Hence, forgeability can be defined as the tolerance of a metal or alloy for deformation without failure. It can be evaluated on the basis of the following tests:

(a) Hot twist test
(b) Upset test
(c) Hot-impact tensile test

(*a*) *Hot Twist Test*: In this test, hot bar is twisted and count the number of twists until failure. A large number of twists before failure indicate better forgeability.

(*b*) *Upset Test*: This test is widely used in the forging industry. In this test, a number of cylindrical billets are upset-forged to various thickness. The limit for upset forging without failure or cracking is considered a measure of forgeability.

(*c*) *Hot-Impact Tensile Test*: A conventional impact-testing machine fitted with a tension test attachment is used. The impact tensile strength is taken as measure of forgeability.

2.3.2.1 Forgeable Materials

In general, the selection of a forging material is made on the basis of certain desirable mechanical properties inherent in the composition and for those which can be developed by forging such as strength, resistance to fatigue, good machining characteristics, durability etc.

Following is a list indicating the relative forgeability of some alloys in a descending order (i.e. alloys with better forgeability are mentioned first):

1. Aluminium alloys
2. Magnesium alloys
3. Copper alloys
4. Plain carbon steels
5. Low-alloy steels
6. Martensitic stainless steels
7. Austenitic stainless steels
8. Nickel alloys
9. Titanium alloys
10. Tantalum alloys
11. Molybdenum alloys
12. Tungsten alloys

2.3.3 Forging Temperatures

For forging, the metal work piece is heated to a proper temperature to attain plastic properties before deformation which is essential for satisfactory forging. Excessive temperature may result in burning of the metal. Insufficient temperatures will not induce sufficient plasticity in the metal so that it is difficult to shape by hammering. Like wise finishing temperatures is also important to possesses a fine grained structure. The temperature ranges for forging some common metals are given in Table 2.1.

Table 2.1: Forging Temperatures

Metal/Alloy	*Forging Temperature °C*	
	Starting	*Finishing*
Mild steel	1300	800
Wrought iron	1275	900
Medium carbon steel	1250	750
High carbon steel	1150	825
Copper, brass and bronze	950	600
Aluminium and magnesium alloys	500	350

2.3.4 Hand Forging Tools and Equipment used in Smithy

In order to carryout forging with hand, certain tools are required.

(*a*) *Anvil* (see Fig. 2.1)

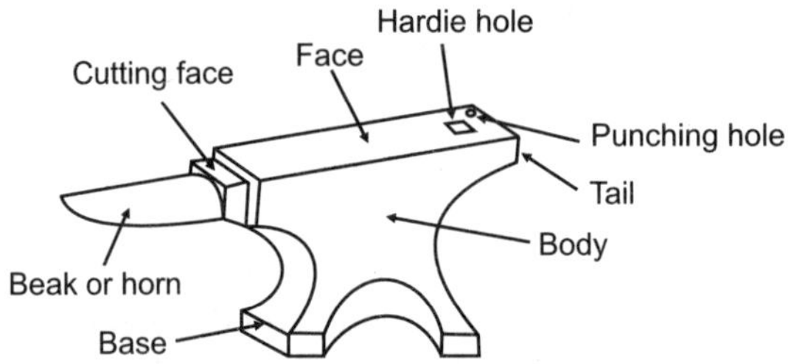

Fig. 2.1 Anvil

The anvil forms a support for black smith's work when hammering. It is made of wrought iron or cast steel. The top surface of the anvil has some square and other round shaped holes. These holes are used to bend the rods of small diameter and as a die for hot punching operations. The horn portion of the anvil is used for bending the work pieces.

(*b*) *Swage block* (see Fig. 2.2)

A swage block has a number of slots of different shapes and sizes along its four side faces and through holes of different shapes and sizes. This is used as a support while forming (swaging) different shapes and in punching holes. It is made of cast iron or cast steel.

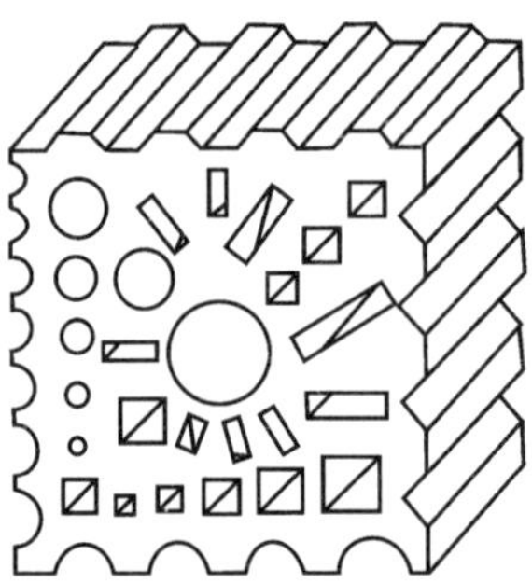

Fig. 2.2 Swage block

(*c*) *Hammers* (see Fig. 2.3)

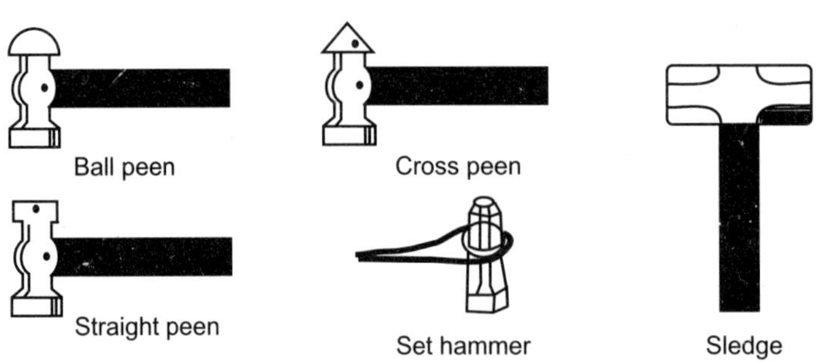

Fig. 2.3 Hammers

The hammers are used by a smith in order to give the required shape to the heated metal piece.
Types of hammers:

(i) Ball peen hammer
(ii) Cross peen hammer
(iii) Straight peen hammer
(iv) Sledge hammer

(*i*) *Ball Peen Hammer*: This is best suited to practically all-hard forging operations. It is made of cast steel or forged steel fitted to wooden handle. One end of the head is flat which is used for striking or hardening purpose. The end opposite to face is half ball shaped and is known as peen. This peen is used for riveting.

(*ii*) *Cross Peen Hammer*: The peen is cross. It is used for bending, stretching and hammering into the inside portions of a component.

(*iii*) *Straight Peen Hammer*: The peen is straight i.e. parallel to the axis of the handle of the hammer. The straight peen hammer is used for stretching the metal.

(*iv*) *Sledge Hammer*: A sledge hammer is heavier than ball peen hammer.

The weight of the hammer varies from 3 to 10 kg and are used where heavy blows are required on the job.

(*d*) *Tongs* (see Fig. 2.4)

Tongs are used by smith for gripping and turning hot metal work pieces during forging. Tongs have varieties of bit (mouth) shapes in order to accomodate different sized and shaped work pieces as shown in Fig. 2.4.

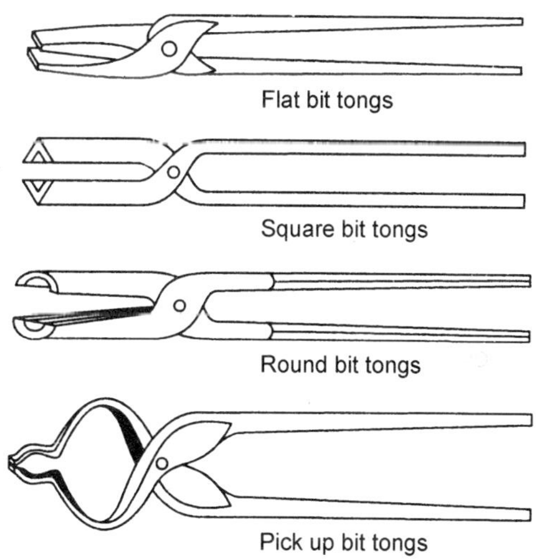

Fig. 2.4 Tongs

(*e*) *Swages* (see Fig. 2.5)

Swages are used in pairs (top part and bottom part). These two parts are either separate or connected by a strip steel handle. These are used to reduce and finish the job to the correct size and shape. The shape may be round or hexagonal. During swaging, the hot metal is rotated between swages, which are hammered to produce smooth round surface on the work-piece.

(*f*) *Fullers* (see Fig. 2.6)

Fullers are a set of tools (top and bottom). The top is provided with a handle and the bottom fits into the hole of the anvil. Fullers are used to form grooves. They spread the metal and reduce the thickness of the work-piece. These are made in various shapes and sizes according to the need.

(*g*) *Flatter* (see Fig. 2.7)

Flatters are used to give smoothness and accuracy to the work-pieces, which have been already shaped by fullers and swages.

(*h*) *Punch*: A punch is used for making holes in the heated work-piece.

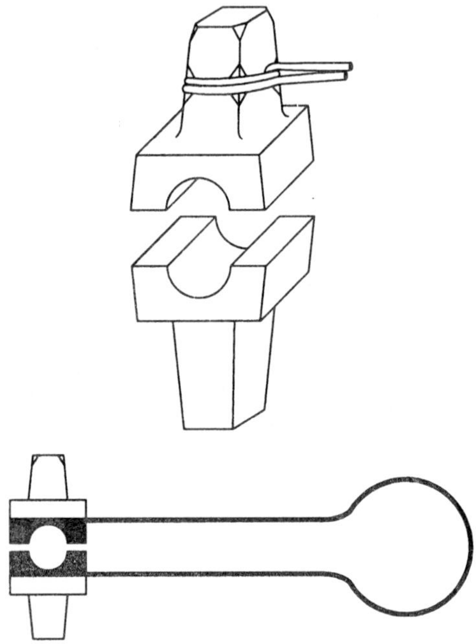

Fig. 2.5 Swages

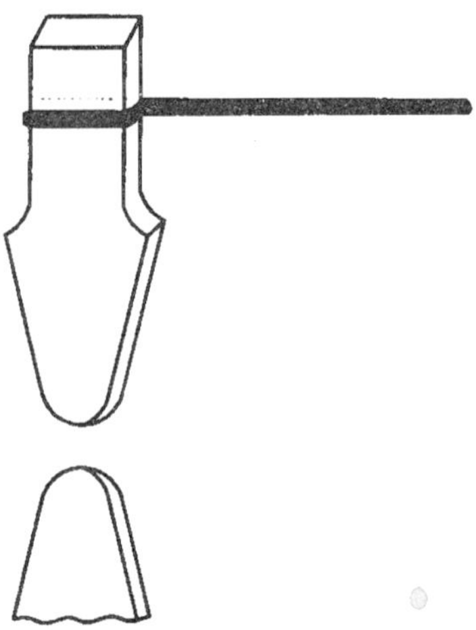

Fig. 2.6 Fullers

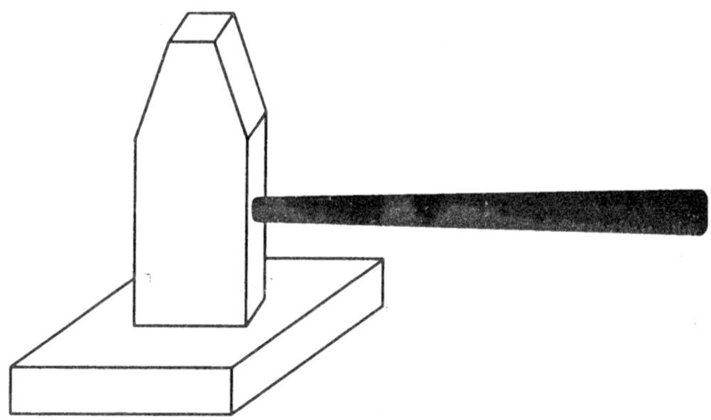

Fig. 2.7 Flatter

(*i*) *Drift*: The drift can be used to enlarge a hole to a particular shape and size, which is already made by a punch.

(*j*) *Forge or Hearth* (see Fig. 2.8)

A blacksmithy's forge is used as a fuel to heat the job. The required air for the fire is supplied under pressure by means of a blower through the tuyeres in the hearth. The blower may be hand operated or power driven. Fig. 2.8 illustrates the forge.

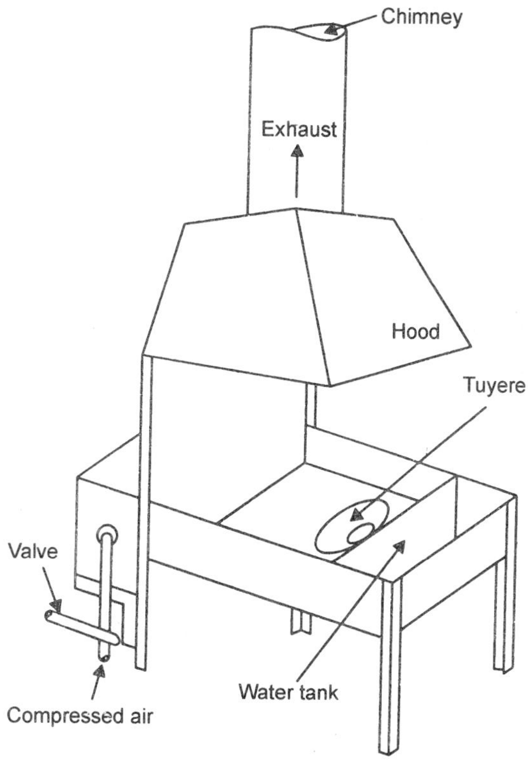

Fig. 2.8 Smith's Forge

2.3.5 Smith Forging Operations

For giving desired shape to the products the following operations are used in a smithy shop by hand on an anvil.

(a) Upsetting
(b) Drawing down
(c) Bending
(d) Cutting
(e) Punching
(f) Welding

(*a*) *Upsetting* (see Fig. 2.9)

Upsetting is the process of increasing the thickness of a bar with a corresponding reduction in length by end pressure. The pressure is applied at the end of the bar against the anvil or clamping in vice and then hammering. For this, force is applied in a direction parallel to the length axis.

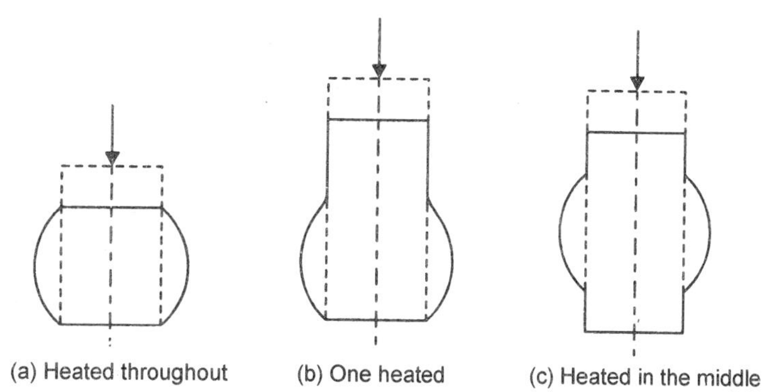

Fig. 2.9 Upsetting

(*b*) *Drawing Down* (see Fig. 2.10)

Drawing down is used to reduce the thickness of the bar to increase its length. For this purpose, the force is applied in a direction of the length axis.

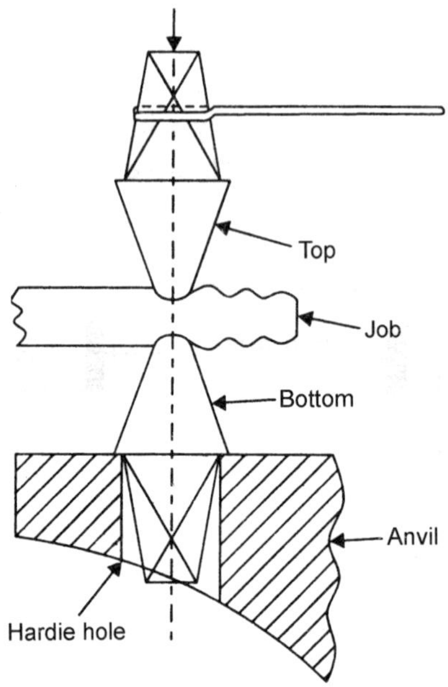

Fig. 2.10 Drawing Operation

(*c*) *Bending* (see Fig. 2.11)

Bending is very common forging operation. This may be angular or curvilinear. It is done on the edge of the anvil face, over the anvil horn or by inserting the end in the pritchel hole and bending the bar with tong. Fig. 2.11 shows the stages in bending bar over the horn of an anvil using a hammer.

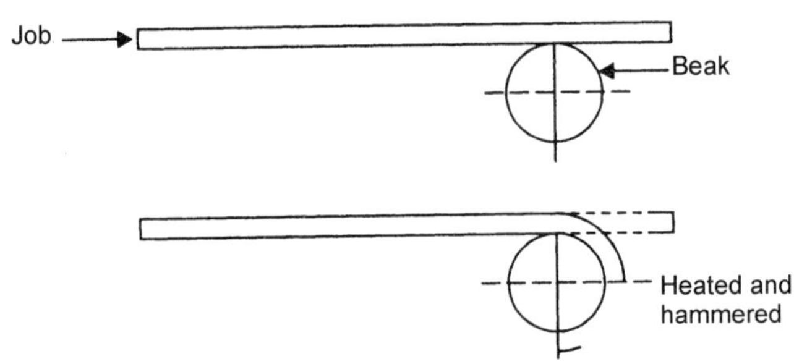

Fig. 2.11 Bending Operation

(*d*) *Cutting*

Cutting off is a form of chiselling to cut a long piece of stock into several pieces of specified lengths. For hot chiseling the work piece must be heated in a blacksmith's furnace. A notch is first made about one-half the thickness or diameter of the stock. Then the work must be tuned through an angle of 180° and the chisel is placed exactly opposite the notch and hit the chisel with hammer to cut the piece.

(*e*) *Punching* (see Fig. 2.12)

Punching operation is used for making holes in the work during forging. A punch is forced about half way through the work by striking it slightly with a hand or sledge hammer. The punch is removed, the work is turned over and the punch driven into the metal by a sledge hammer and thus hole is made.

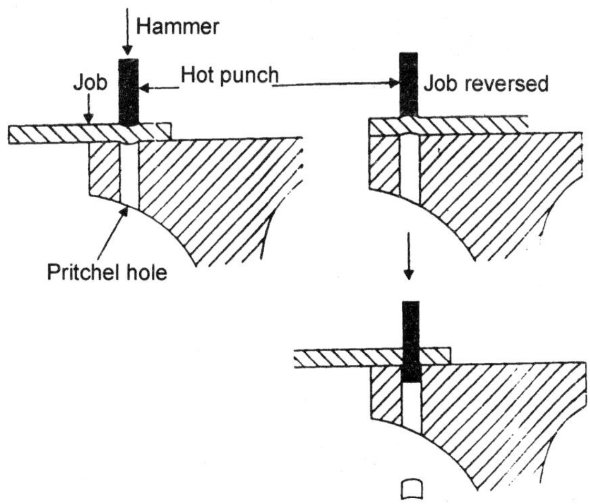

Fig. 2.12 Punching Operation

(*f*) *Welding* (see Fig. 2.13)

A forge weld is made by hammering together the ends of thc two bars. In thc lap weld, the ends of the pieces to be joined must be upset and shaped slightly convex, so that when put together the junction takes place first at the centre and extends to the edges (see Fig. 2.13). Wrought iron and mild steel can be satisfactorily forge welded.

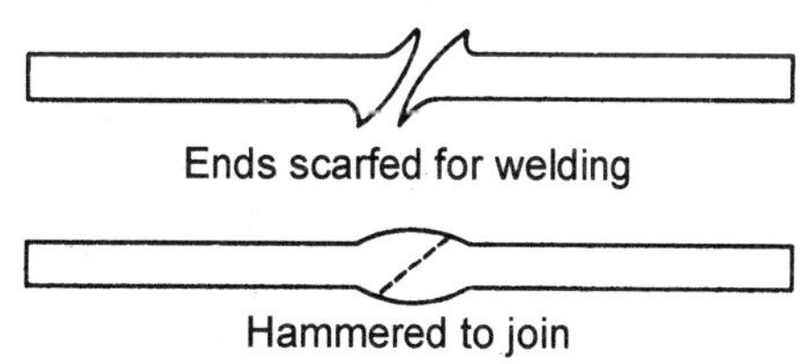

Fig. 2.13 Forge—Welding a Lap Joint

2.3.6 Types of Forging Processes

The process of reducing a metal billet between open dies or in a closed impression dies to obtain the required shape are called smith forging or impression-die forging respectively. Depending on the equipment used, they are further sub divided as hand forging, hammer forging, press forging, drop hammer forging, mechanical press forging, upset or machine forging.

In general the methods of forging, may therefore be classified as follows:

(1) Smith Forging (Open-Die Forging)

(a) Hand forging
(b) Power forging
(c) Hammer forging
(d) Press forging

(2) Impression-Die Forging

(a) Drop forging
(b) Press forging
(c) Machine forging or upset forging

2.3.6.1 Smith Forging

It is also called open-die forging. The accuracy of the component produced by this process is less.

(*a*) *Hand Forging*: Smith forging is known as hand forging. It is used to produce a small number of light forgings.

(*b*) *Power Forging*: Large components cannot be forged by hand.
Moreover hand forging is lengthy process and requires repeated heating of metal. Machines which work on forgings by blow are called hammers, while those working by pressure are called presses.

(*c*) *Hammer Forging*: In hammer forging the hammer is lifted upto a certain distance, and then it is allowed to fall by gravity. Depending upon the lifting mechanism the hammers may be classified as:

(i) Mechanical hammers
(ii) Pneumatic hammers
(iii) Steam or air hammers.

(*i*) *Mechanical hammers* (see Fig. 2.14)

The mechanical hammers are helve hammers, trip hammers and lever spring hammers. Lever spring hammer has constant lift. The arm is driven from a rocking lever acting on an elastic rod. The rocking lever consists of a leaf spring. Mechanism is illustrated in Fig. 214.

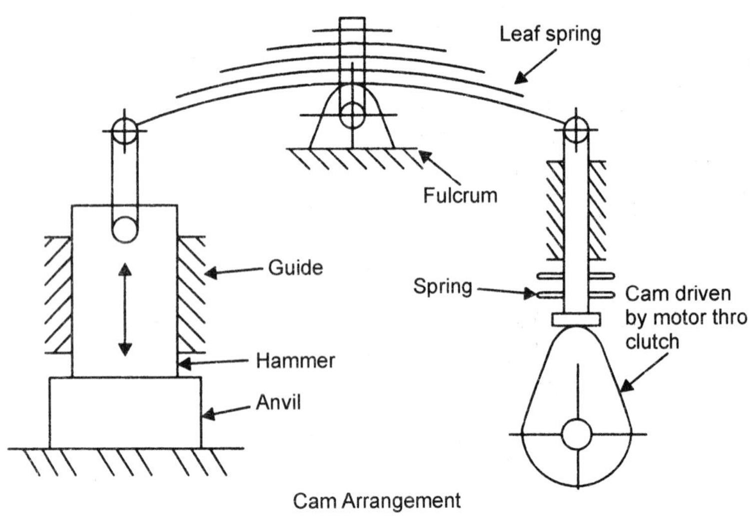

Fig. 2.14 Mechanical Hammer

(*ii*) *Pneumatic hammers* (see Fig. 2.15)

Pneumatic hammer is used for smith forging of small parts. A pneumatic hammer has a built in compressor to provide a compressed air to ram cylinder. The upper die is connected to the lower end of the ram. The lower die is supported on the anvil.

When compressed air enters the ram cylinder from the top of the ram (see Fig. 2.15) strikes the hot metal placed between the dies. To move the ram up, the compressed air enters from the bottom of the cylinder. Repeated blows are struck until the desired shape is obtained. These are operated at 70 to 190 blows per minute.

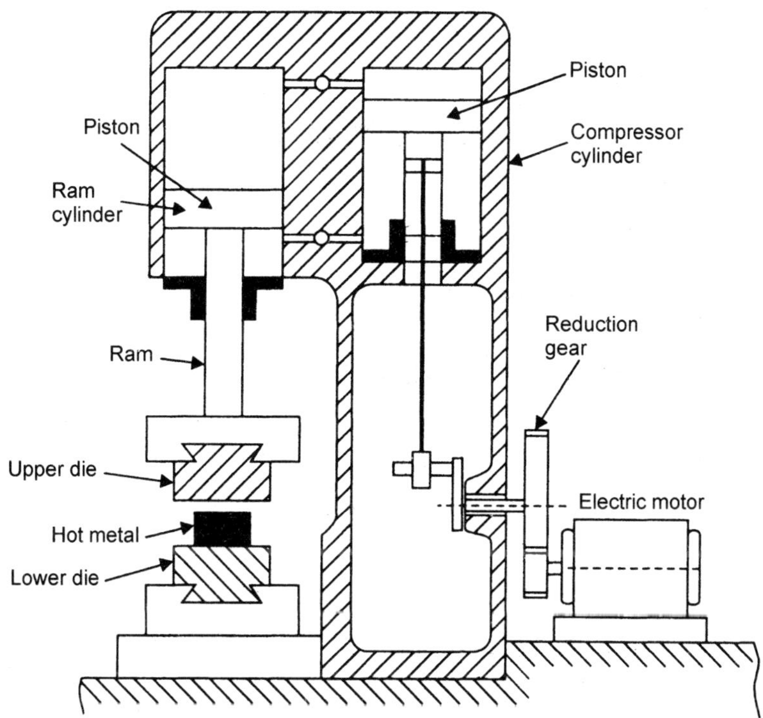

Fig. 2.15 Pneumatic Hammer

(*iii*) *Steam or air hammers* (see Fig. 2.16)

A steam hammer operates with the help of steam and air hammer requires compressed air for its operation. A steam or an air hammer may be (i) single acting (ii) double acting. Steam or air pressure is usually between 6 to 8 kgf/cm^2.

As shown in Fig. 2.16 steam entering from the top exerts pressure on the piston which moves downwards and upper die applies force on the hot metal to get deformed. The steam then enters from the bottom of the piston so that the piston moves upward. This cycle is repeated till the required shape is obtained.

2.3.6.2 Impression-Die Forging

It is used to make more complex shapes of products with greater accuracy. After forging operation, the product should be trimmed to remove flash.

(a) Drop Forging

The ram is raised to a definite height and then it is allowed to drop or fall freely under its own weight. The commonly used drop hammers are:

(i) Board hammer or gravity drop hammer
(ii) Air lift hammer
(iii) Power drop hammer

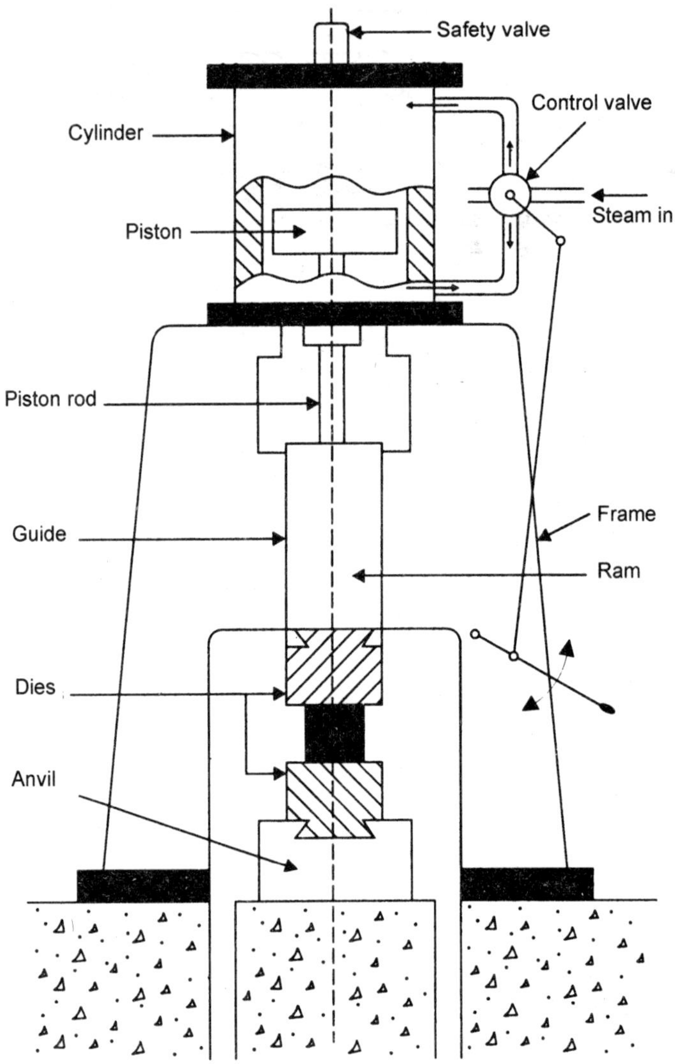

Fig. 2.16 Steam Forging Hammer

(i) Board hammer or gravity drop hammer (see Fig. 2.17)

In these hammers, ram is fastened to a hard board as shown in Fig. 2.17. The board is lifted up by two counter revolving rolls. When the rolls are released, the ram falls down producing a working stroke. The height to which the board is lifted determines the striking force of this gravity hammer.

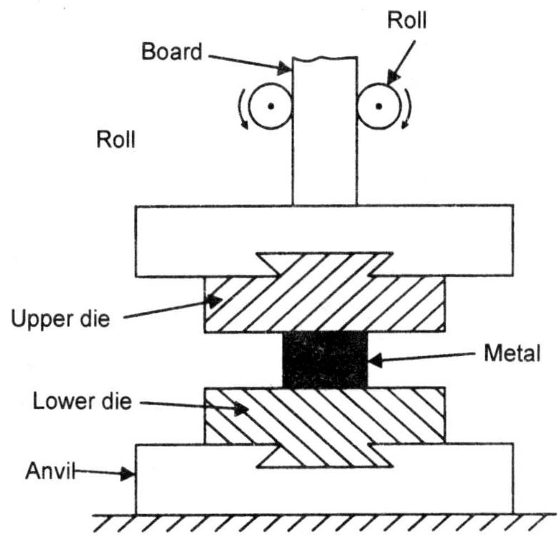

Fig. 2.17 Board Drop Hammer

(ii) Air lift hammer

These hammers use compressed air to lift the ram which is then allowed to fall by gravity similar to the board drop hammers.

(iii) Power drop hammer

They use air or steam. They are similar to board drop hammers except that steam or air piston and rod are substituted for board lifting mechanism. They are the largest of forging hammers and are made from 450 to 25,000 kgs falling weight hammers.

(b) Press Forging

The press forging is also done in impression dies. In press forging, the metal is shaped not by means of a series of blows as in drop forging, but by means of continous squeezing action. The manner in which the metal deformation takes place in press forging substantially differs from that of hammer forging. Blow of hammer works only in the surface layer of forging and deformation does not penetrate into the volume of the metal. Squeezing pressure of a press applied to the forging gradually increases and penetrates deep into the metal. Two types of presses are used:

(i) Hydraulic press (ii) Mechanical press

(i) Hydraulic Press (see Fig. 2.18)

Fig. 2.18 illustrates a hydraulic press. The press is operated by pump which increases the pressure in the oil or water. This pressure is transmitted to main cylinder to move the piston (ram) downward

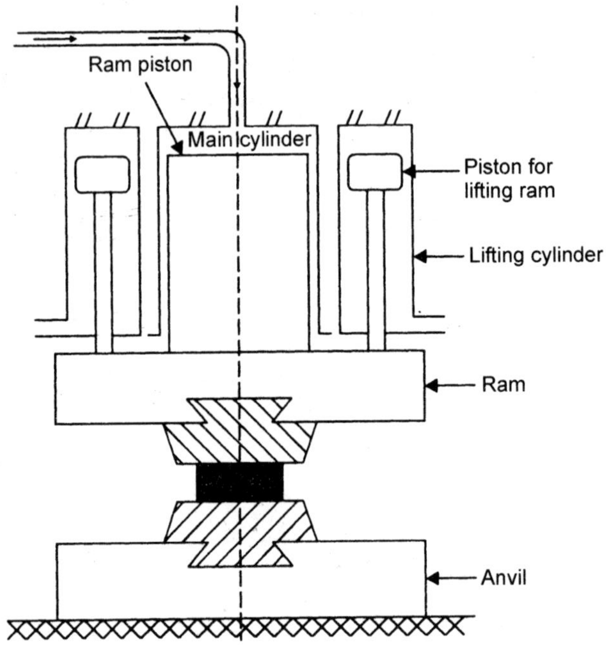

Fig. 2.18 Hydraulic Press

to squeeze the hot metal between the dies. The lifting cylinders raises the ram up. In a hydraulic press, pressure can be changed as desired at any point in the stroke by adjusting the pressure control valve. This will help in controlling the rate of deformation according to the metal being forged. But in hydraulic press, the contact time between the work piece and the dies is more, hence the die life is less.

(*ii*) *Mechanical Press* (see Fig. 2.19)

Crank type mechanical press is shown in Fig. 2.19 an electric motor drives the flywheel mounted on the counter shaft by means of a belt drive. Torque from the counter shaft is transmitted to the crankshaft by gearing. From the crankshaft, the reciprocating motion is given to the ram with the help of connecting rod. The bottom die is locked in position by means of wedge mechanism. Disk clutch is used to start and stop motion of ram, which is brought to a gradual stop by means of a brake. Mechanical press is faster than hydraulic press and operate at about 25 to 100 strokes per minute.

(*c*) *Machine Forging or Upset Forging* (see Fig. 2.20)

Unlike press forging, it operates in horizontal direction. As it involves the upsetting operation, it is simply called as upset forging. Upset forging was originally developed for heading operations. But today its scope has been widened to perform a large variety of operations such as punching, bending, cutting and squeezing etc.

The forging machine consists of a heavy cast steel body in which three main components, stationay die, moving die and heading punch are properly secured. The sequence of operation of

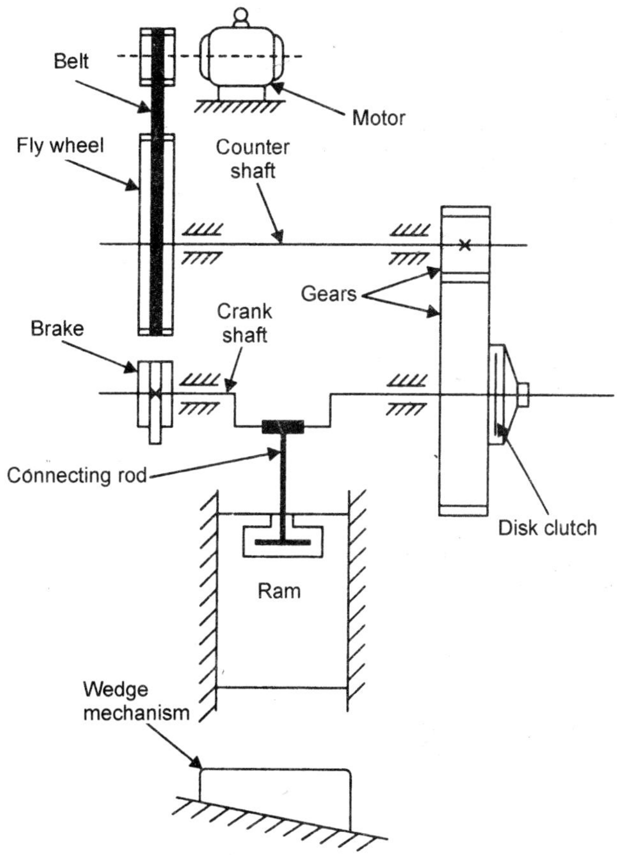

Fig. 2.19 Crank Type Mechanical Press

machine is explained in Fig 2.20. First the bar stock of one end heated is placed between the fixed and movable halved of the set of dies up to stop. Next the moving die grips the bar stock and at the same time, a recess is formed in the closed dies for shaping the projected stock. Stop is then brought to its initial position. Now the heading punch advances to upset the bar end and forms the finished forging.

2.3.7 Fibrous Structure of Forgings (see Fig. 2.21)

In forging, the fibrous structure and the grain structure or the flow lines of the metal are not interrupted, but are made to flow the contour of the forged part. The main objective of good forging design is to control the lines of metal grain flow, so that a part with greatest strength and resistance of fracture is produced. In addition, certain mechanical properties like elongation percentage, resistance to shock and vibration are improved. A typical example is shown in Fig. 2.21, which illustrates fibrous structure. The crankshaft produced by casting Fig. 2.21(a) has no grain flow and so has mechanical properties. In Fig. 2.21(b), the crankshaft has been made by machining from a bar stock and the fibre of the metal gets interrupted and for this reason the mechanical properties of the crankshaft will be poorer than the crankshaft made by forging Fig.

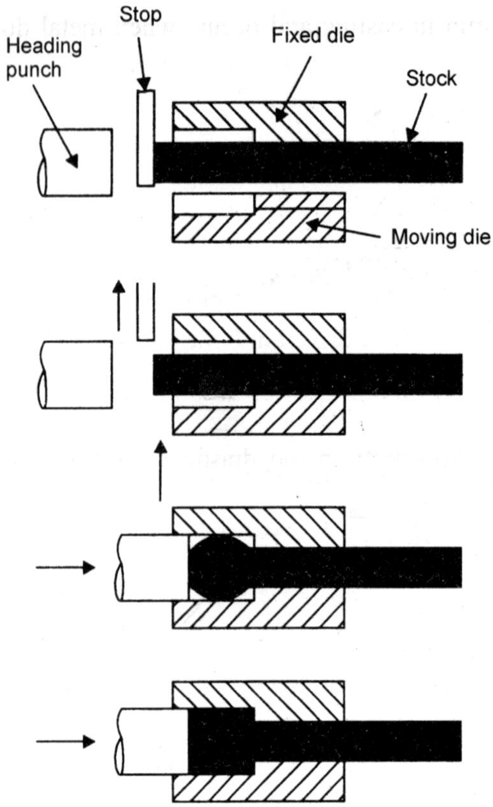

Fig. 2.20 Upset Forging, Steps

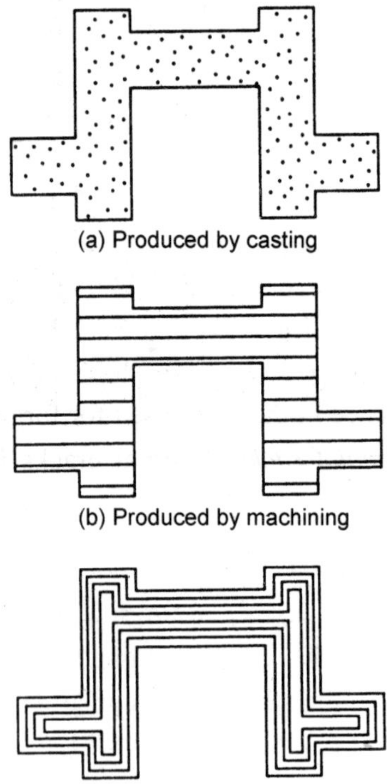

Fig. 2.21 Fibrous Structure

2.21 (c), where the fibre of the metal has not been interrupted and continues along the entire length of the shaft.

2.3.8 Defects in Forged Parts

Various surface and body defects may be observed in forging. The kind of defect depends upon a lot off actors such as forging process, poor quality of stock, improper heating, incorrect die design, uneven cooling of stock after forging etc. The most commonly found forging defects are as follows:

1. *Mismatch*: This is due to the misalingnment between the top and bottom forging dies. This may be caused due to loose wedges. This results in a lateral displacement between the portions of the forging.
2. *Scale Pits*: These are shallow surface depressions caused by not removing scale from the dies. The scale is worked into a surface of the forging. When this scale is cleaned from the forging, depression remains which is known as scale pits.
3. *Cold Shuts or Laps*: Cold shuts or laps are short cracks, which usually occur at corners and at right angles to the surface. They are caused by metal surface folding against itself during forging. Sharp corners in dies can result in hindered metal flow, which can produce laps.

4. *Unfilled Section*: This defect is similar to misrun in casting and occurs when metal does not completely fill the die cavity. It is caused by using insufficient metal or insufficient heating of the metal.
5. *Dents*: Dents are the result of careless work.
6. *Burnt and over Heated Metal*: This defect is due to improper heating conditions and soaking the metal too long time.
7. *Cracks*: Cracks occur on the forging surface may be longitudinal or transverse. These are due to bad quality of ingot, improper heating, and forging at low temperature.
8. *Fins and Rags*: These are small projections or loose metal driven into forging surface.
9. *Dirt, Slag and Sand*: These may be present on the surface of the forging due to their presence in the ingot used for forging.
10. *Internal Cracks*: Internal cracks in forging can result from too drastic a change in the shape of the raw stock at too fast a rate.

2.3.9 Advantages of Forging

The various advantages of forging are as follows:

1. Forgings have a high strength and offer resistance to impact and fatigue loads.
2. Forging improves the grain structure of metal and hence its mechanical properties.
3. Close tolerances.
4. Less machining or no machining in some cases.
5. Smooth surface.

2.3.10 Limitations of Forging

1. High tool cost
2. High tool maintenance
3. The rapid oxidation of metal surfaces at high temperature results in scaling which wears thc dics.

2.4 ROLLING

2.4.1 Principle

The process of rolling basically consists of passing the metal between two rolls (Fig. 2.22) rotating in opposite directions at a uniform peripheral speed. The space between the rolls is adjusted to conform to the dcsircd thicknCss of thc rollcd scction.

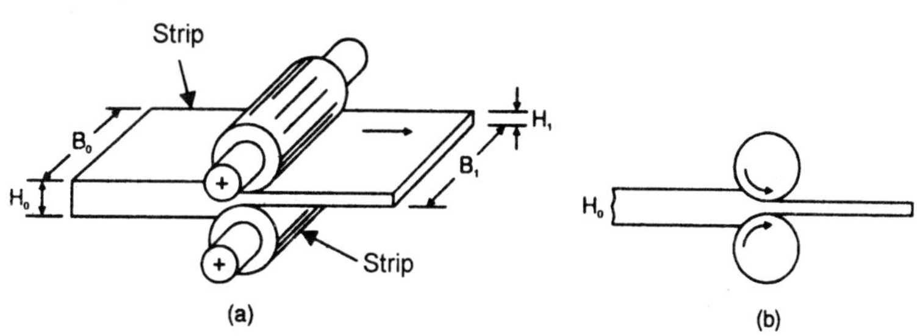

Fig. 2.22 Principle of Rolling

2.4.2 Hot and Cold Rolling

In hot rolling (Fig. 2.23(a)) the metal is fed between rolls after being heated above the re-crystallization temperature. This leads to grain refinement, thereby mechanical properties are improved. During hot rolling, the work hardening does not occur and coefficient of friction between the rolls and the metal is higher. Heavy reduction in area of the work piece can be obtained. Hot rolled parts does not have a good surface finish due to scaling.

In cold rolling (see Fig. 2.23(b)), the metal is fed to the rolls when it is below its recrystallization temperature. This results in elongation of grain structure. The metal shows work hardening effect after cold rolling. This increases hardness and decreases the ductility of the metal. Heavy reductions are not possible. The coefficient of friction between the rolls and the metal is lower. The cold rolled a surface is smooth and oxide free.

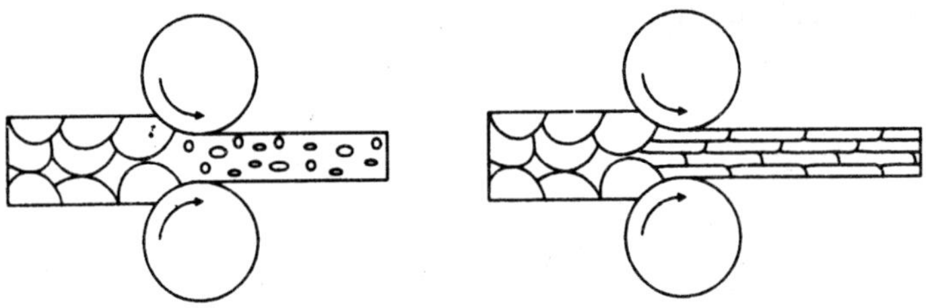

(a) Hot Rolling (b) Cold Rolling

Fig. 2.23 Hot and Cold Rolling

2.4.3 Rolling Mill (Fig. 2.24)

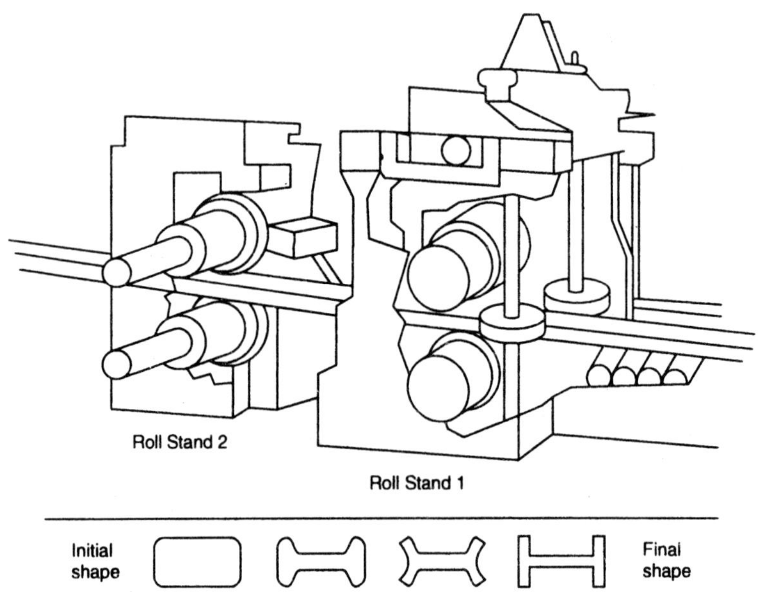

Fig. 2.24 Rolling Mill

A rolling mill consists of one or more roll stands, motor drive, reduction gears, flywheel and coupling gears between units. The roll stand is the main part of the mill, where the rolling process is performed. It basically consists of housings in which bearings are fitted, which are used for mounting the rolls. There is a screwdown mechanism to control the gap between the rolls to get the required thickness of product.

Depending upon the profile of the rolled product, the body of the roll may be either flat for rolling sheets (plates or strips) or grooved for making structural members (channel, I-beam, rail). Rolls are made from high quality steel or sometimes from high grade cast iron. This is to withstand the very severe service conditions which the rolls are subjected during the rolling process. Cast or forged steel are used in blooming, slabbing and section mills as well as cold rolling mills. Forged rolls are stronger and tougher than the cast one. Alloy steel rolls are made of chrome-nickel or chrome-molybdenum steels are used in sheet mills.

2.4.4 Classification of Rolling Mills

Rolling mills are classified according to the number and arrangement of rolls in a stand. They are as follows:

(a) Two-high rolling mill
(b) Three-high rolling mill
} Generally used for hot rolling of metals

(c) Four-high rolling mill
(d) Tandem rolling mill
(e) Cluster rolling mill
} Generally used for cold rolling of metals

(*a*) *Two-High Rolling Mill* (see Fig. 2.25)

It consist of two heavy horizontal rolls placed one over the other. The space between the rolls can be adjusted by raising or lowering the upper roll. The position of the power roll is fixed. The rolls rotate in opposite direction. The work can be rolled by feeding from one direction only. This is called non-reversing mill.

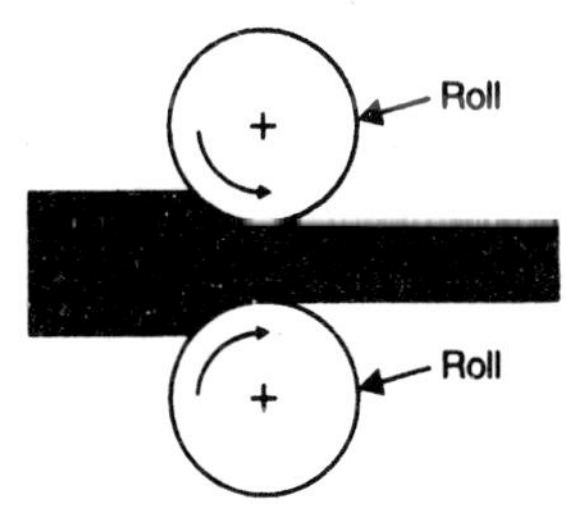

Fig. 2.25 Two-High Rolling Mill

There is another type of two-high mill, which incorporates a drive mechanism that can reverse the direction of rotation of the rolls. This is known as two-high reversing mill. In this, the rolled metal is passed backward and forth through several times. This type is used in blooming and slabbing mills and for rough work.

(*b*) *Three-High Rolling Mill* (see Fig. 2.26)

It consist of three horizontal rolls placed one over the other. The upper and lower rolls rotate in the same direction, whereas the intermediate roll rotates in direction opposite to the outer roll.

First of all the work piece passes through the bottom and the middle rolls and then returning between the middle and top rolls so that the thickness is reduced at each pass. Mechanically operated lifting tables are used which move vertically on either side of the roll stand. It may be used to make plates or sections.

(*c*) *Four-High Rolling Mill* (see Fig. 2.27)

It consist of four horizontal rolls, two of smaller diameter and two of large diameter. The bigger rolls are called back up rolls because they reinforce the smaller rolls to minimize roll deflection thereby minimizing the tendency of producing plates and sheets thicker at the centre than at the two outer edges. The two smaller rolls are called work rolls. It is used for both hot and cold rolling of plates and sheets.

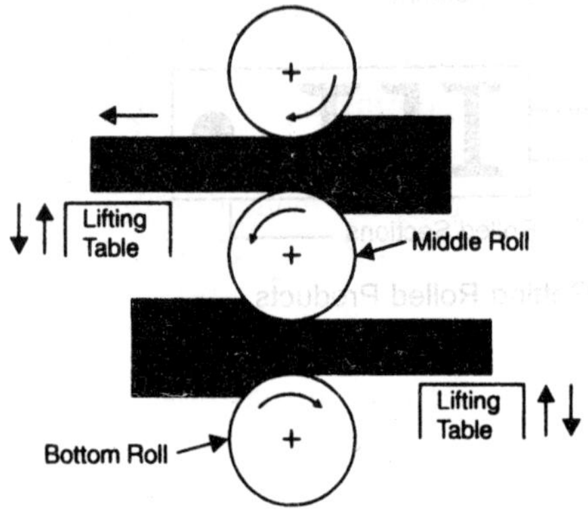

Fig. 2.26 Three-High Rolling Mill

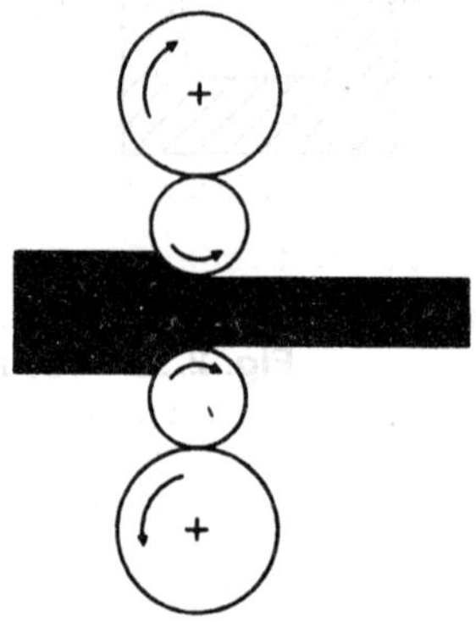

Fig. 2.27 Four-High Rolling Mill

(*d*) *Tandem Rolling Mill* (see Fig. 2.28)

It is a set of two or three stands of rolls set in parallel alignment so that a continous pass may be made through each one successively without change of direction of material.

(*e*) *Cluster Rolling Mill* (see Fig. 2.29)

It consist of two working rolls of smaller diameter and one or more backup rolls of larger diameter. The number of backup rolls may go up to 20 or more, depending on the amount of support needed for the working rolls during the operation. Cold rolling is employed for providing a smooth and bright surface finish.

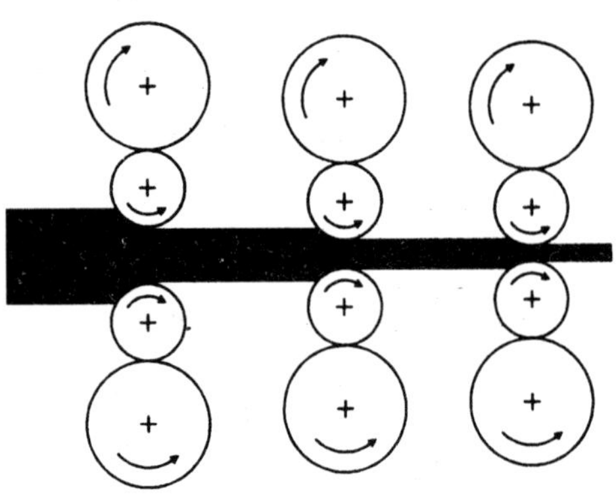

Fig. 2.28 Tandem Rolling Mill

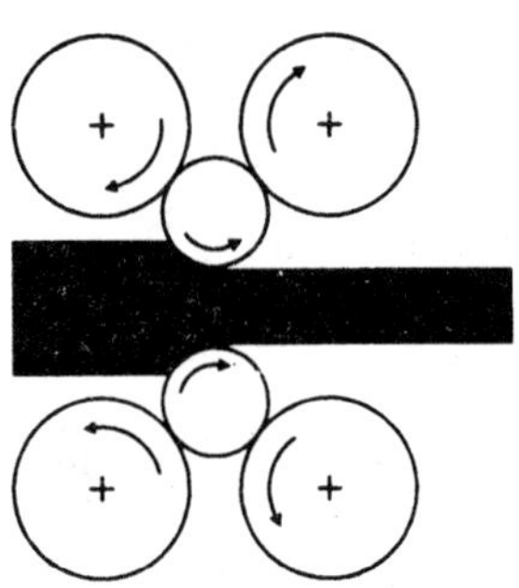

Fig. 2.29 Cluster Rolling Mill

2.4.5 Production Sequence in Getting Rolled Products (see Fig. 2.30)

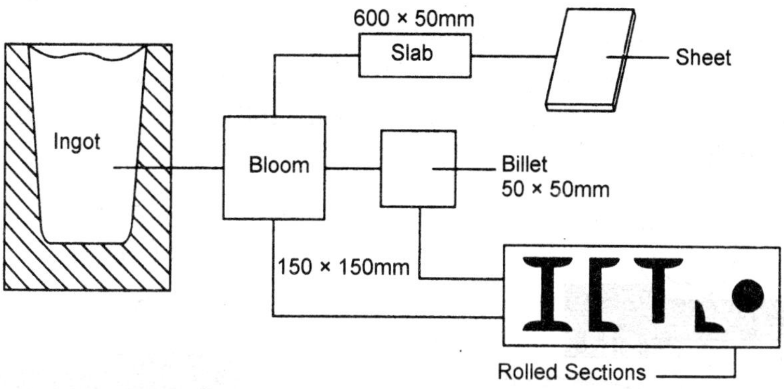

Fig. 2.30 Production Sequence in Getting Rolled Products

The ingot is rolled to intermediate shapes-blooms and slabs. These blooms, billets and slabs are further rolled into plate sheets, bar stock and structural shapes shown in Fig 2.30.

2.4.6 Roll Passes

An ingot or bloom is required to pass many times between the rolls before it is shaped into final shape. Grooved are made in the rolls according to the shape of the product. The shape formed when the grooves of mating rolls are matched together is called the pass.

Types of Passes

(i) Roughing passes
(ii) Leader passes
(iii) Finishing passes.

Roughing passes reduce the cross-section of the stock, leader passes also reduce the cross-section, but along with it, the shape of the rolled part comes nearer to the final shape. Finishing pass gives the required final shape of the rolled section.

2.4.7 Types of Roughing Passes (see Fig. 2.31)

(a) Box pass series
(b) Diamond square series
(c) Oval square series

(a) Box Pass Series (see Fig. 2.31(a))

Box passes are used for medium and large sections of blooming and billet mills. The coefficient of elongation in this series varies from 1.1 to 1.25.

(b) Diamond Square Series

The dotted square shows the previous shape of the shape of the stock and over lapping firm line diamond shows the reduced shape obtained in a particular pass. As shown in Fig. 2.31(b) in the first pass, the square shaped stock is changed into a diamond shape, then in the next smaller pass, the diamonds shape of the stock is made to change to a square and in this manner the rolling process continues. The coefficient of elongation ranges from 1.25 to 1.5.

(*c*) *Oval Square Series* (see Fig 2.31(c))

In the first pass of the square stock changed into an oval shape, then in the next smaller pass, the oval shape of the stock is made to change to a square one by turned over by 90 and vice-versa. The coefficient of elongation in this series varies from 1.5 to 2.5 and even higher.

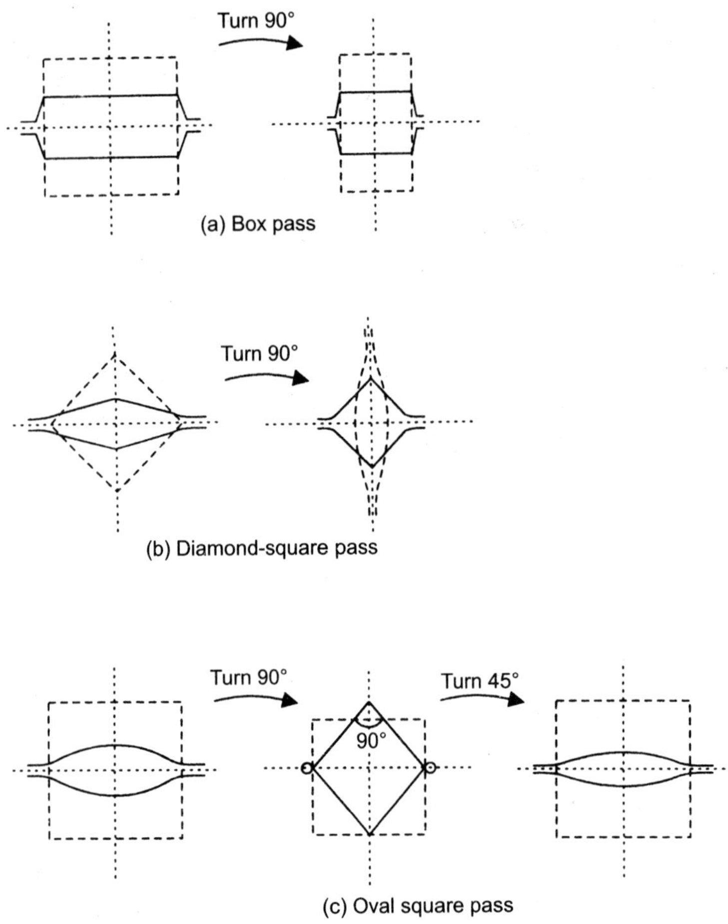

Fig. 2.31 Types of Roughing Passes

2.4.8 Rolling of Rounds (see Fig. 2.32)

Figure 2.32 shows a billet is reduced to a round bar in 10 passes. Oval square passes are used for shaping the bar.

2.4.9 Rolling of Angle Section (see Fig. 2.33)

Figure 2.33 shows the pass sequence for the rolling of angles.

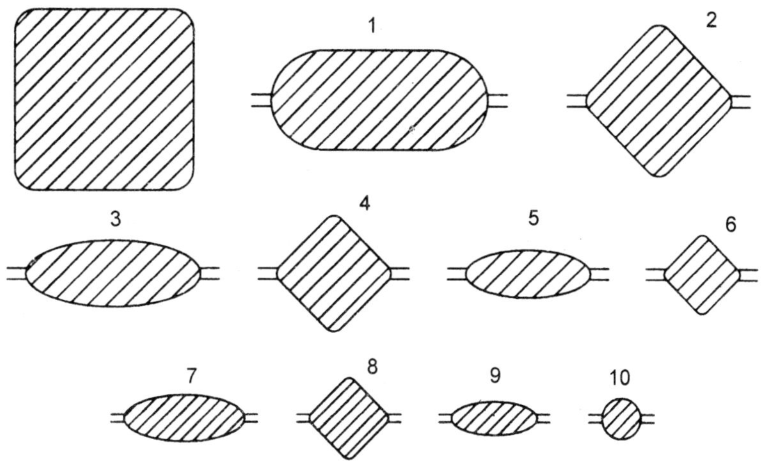

Fig. 2.32 Stages in Rolling a Billet into a Round Rod

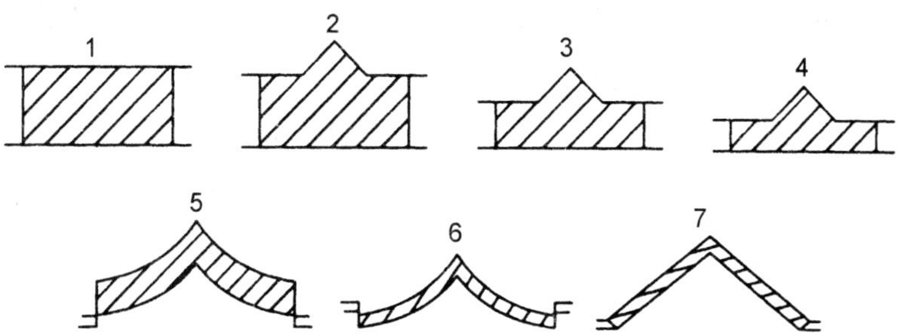

Fig. 2.33 Rolling of Angle Section

2.4.10 Range of Rolled Products

The whole range of rolled products can be divided into the following:

(*a*) *Structural Shapes or Sections*: This includes sections like round square, hexa-gonal bars, channels, H and I beams and special sections like rail sections. Fig. 2.34 shows some of the rolled structural shapes.

(*b*) *Plates and Sheers*: Plates and sheets are produced by rolling.

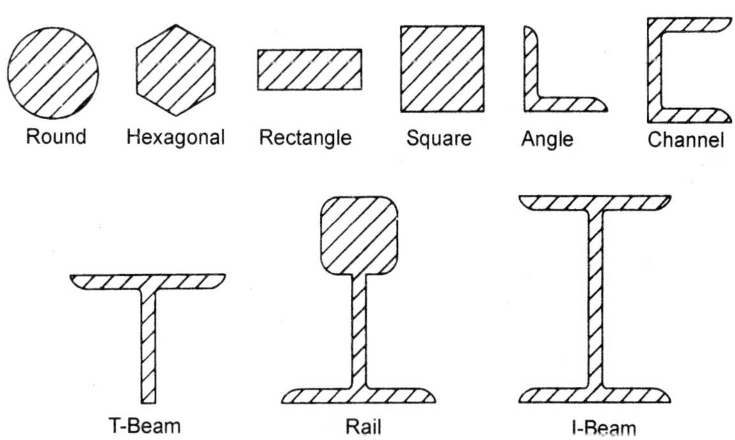

Fig. 2.34 Rolled Products

(c) *Special Purpose Rolled Products*: These include rings, balls, wheels and ribbed tubes.

2.4.11 Defects in Rolled Products

The various defects in rolled products are as follows:

1. *Edge cracking*: This defect occurs in plate or slabs because of either limited ductility or metal or uneven deformation, especially at the edges.
2. *Folds*: These defects occurs during plate rolling when reduction per pass is very small.
3. *Lamination*: These are small cracks, which develop when reduction in thickness is quite high.
4. *Alligatoring*: This defect takes place in rolling of slabs of aluminium alloys where the work piece splits along a horizontal plane on exit as shown in Fig 2.35.

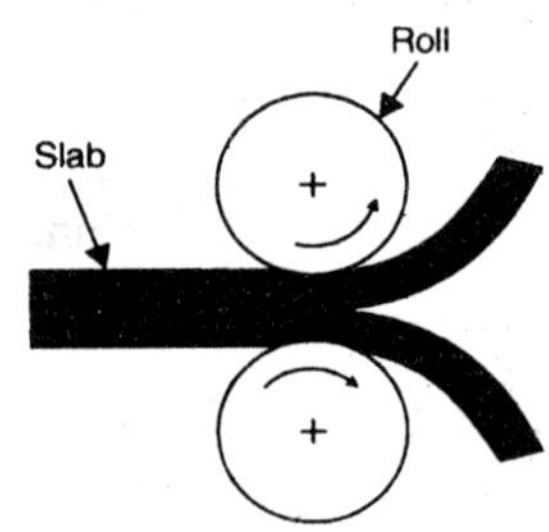

Fig. 2.35 Alligatoring

2.4.12 Lubrication in Rolling Process

Lubrication in rolling protects the rolls against wear, reduces friction and allows smooth flow of metal between rolls. It also protects the metal surface from scratching and peeling. The selection of lubricant depends on (i) Material (ii) Roll pressure (iii) Speed of rolling. The most commonly used lubricants are (i) High penetrating and wetting soluble oils (ii) Synthetic soluble (iii) Oils with excellent polarity.

2.5 EXTRUSION

The process of extrusion consists of forcing a heated billet inside a chamber through a small opening called die under high pressure. The high pressure is obtained by hydraulic press or mechanical press. In its cross-section, the extruded metal acquires the contour and dimensions of the die opening. Extrusion is more widely used in fabricating non-ferrous metals and their alloys.

The extrusion process can be classified as:

1. Hot extrusion process
2. Cold extrusion process.

2.5.1 Hot Extrusion Process

(*i*) *Direct Hot Extrusion* (Forward extrusion)

This is the most widely used method. A hot billet is placed in the container and the forced through the die with the help of pressure by a hydraulic driven ram. The extruded metal comes out of the die opening. In this process, the flow of metal through the die is in the same direction as the movement of the ram. The length of the extruded part will depend on the size of the billet and corss-section of die.

Direct extrusion is shown in Fig. 2.36.

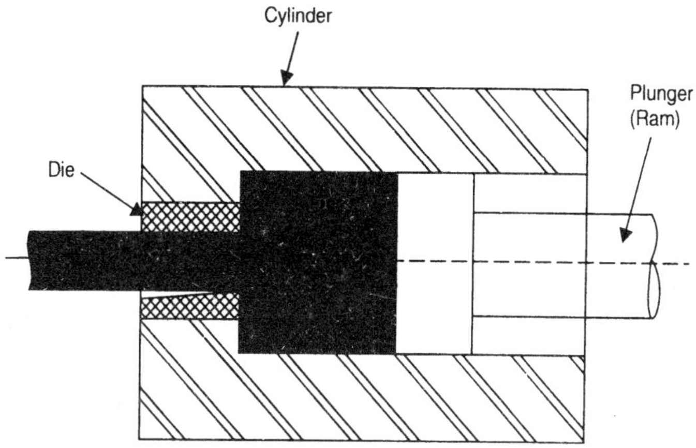

Fig. 2.36 Direct Hot Extrusion

(*ii*) *Indirect Extrusion*

For this type of extrusion, the ram used is hollow and the die is mounted over the bore of the ram. In this process, the billet remains stationary, while the die is pushed into the billet by hollow ram. The metal flows in the direction opposite to the movement of the ram (Fig. 2.37). Indirect extrusion does not require as much force as direct extrusion because no force is required to move the billet inside the chamber.

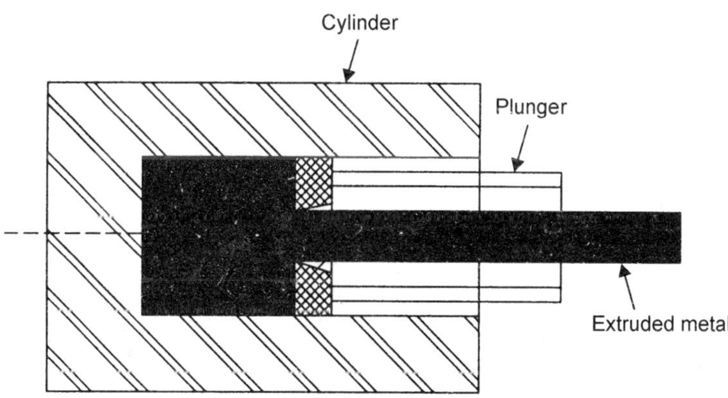

Fig. 2.37 Indirect Extrusion

(*iii*) *Backward Extrusion*

This is another indirect extrusion method used in manufacturing hollow sections as shown in Fig 2.38. in direct and indirect extrusion methods the ram is of the same diameter as the bore of the container, where as in backward extrusion the ram is smaller in diameter than the container. In this process, the metal is extruded through the gap between the ram as the container.

Advantages

(i) It is a very fast process.

(ii) Materials and shapes that are difficult by rolling can easily extruded.

(iii) Close tolerances can be achieved.

(iv) The mechanical properties ofthe products are superior than obtained by rolling process.

(v) Products of complex shapes can be easily extruded.

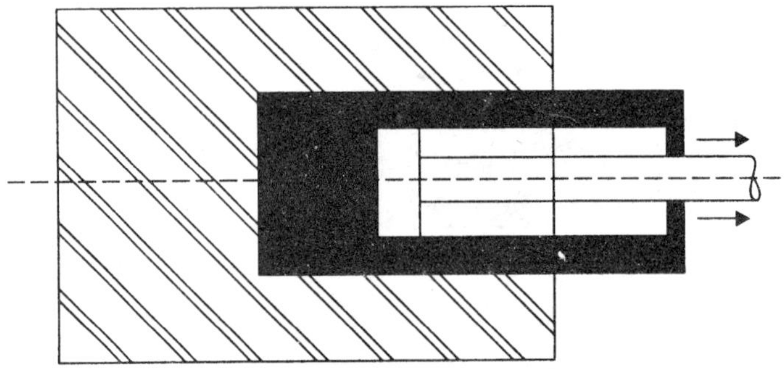

Fig. 2.38 Backward Extrusion

2.5.2 Cold Extrusion

(*i*) *Forward Cold Extrusion*: The forward cold extrusion is similar to that of forward hot extrusion (direct extrusion).

(*ii*) *Hydrostatic Extrusion*: In this extrusion process, the billet is surrounded by a working fluid, which is pressurized by ram to provide the extrusion force. When the plunger is pressed, it increases the pressure inside the container and the resulting high pressure forces the billet to flow through the die, Friction between billet and container is thus eliminated. This makes it possible to extrude very long billets.

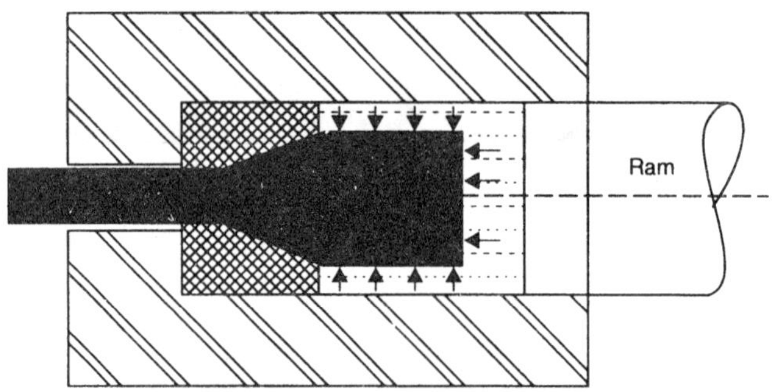

Fig. 2.39 Hydrostatic Extrusion

(*iii*) *Impact Extrusion*: The backward cold extrusion is called the impact extrusion. This process involves striking a cold slug of soft metal (like aluminium) which is held in a shallow die-cavity with a moving punch. The metal is then extruded through the gap between the punch and die opposite to the punch movement. The height of the sidewalls is controlled by the amount of metal in the slug. Various items of daily use such as tubes for shaving cream, toothpaste and paints are made by impact extrusion.

There are three types of impact extrusion processes:

(a) Reverse impact extrusion

(b) Forward impact extrusion

(c) Combination impact extrusion

(a) Reverse Impact Extrusion

Figure 2.40 indicated the process of reverse impact extrusion. In this process, the metal flows in reverse direction of the plunger. It is used for making hollow parts with forged bases and extruded walls. The flowing metal is guided only initially, thereafter if goes by its own inertia.

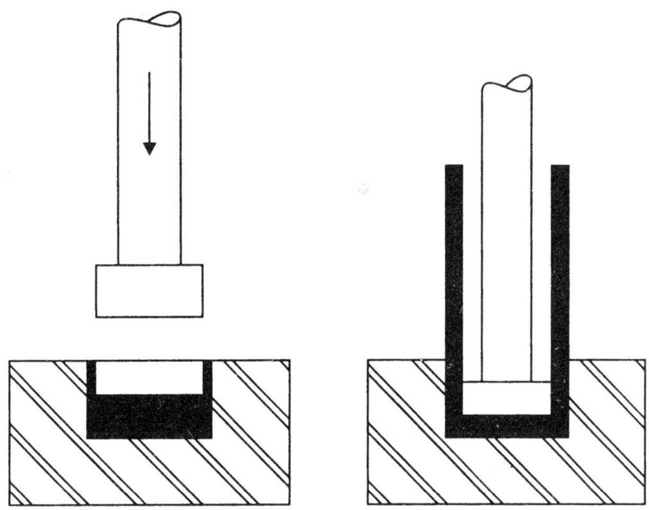

Fig. 2.40 Reverse Impact Extrusion

(b) Forward Impact Extrusion

The process of forward impact extrusion is shown in Fig. 2.41. It is mainly used for making hollow or semi hollow products with heavy flanges.

(c) Combination Impact Extrusion

Complex shapes can be produced by a combination of the two preceding procedures, which are performed simultaneously in the same single stroke as shown in (Fig. 2.42).

Advantages

(i) Complex parts can be produced.

(ii) Very little scrap is there in the process.

(iii) Mechanical properties of extruded parts are improved.

Disadvantages

(i) Tools are expensive.

(ii) Metal blank should be free from internal or external defects.

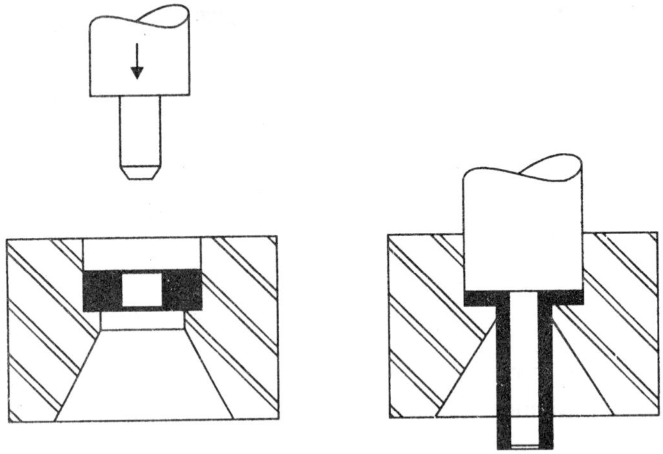

Fig. 2.41 Forward Impact Extrusion

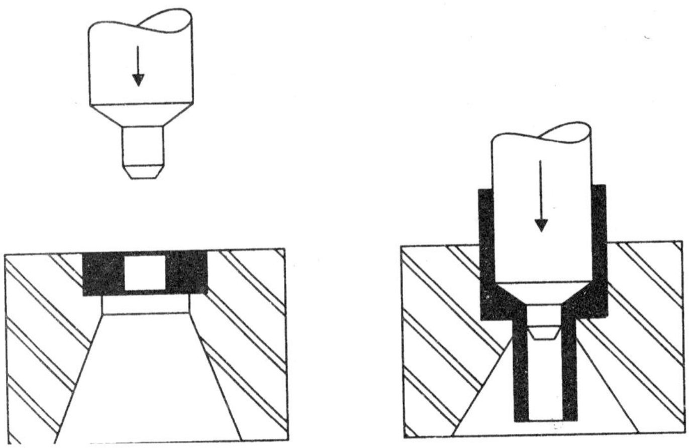

Fig. 2.42 Combination Impact Extrusion

2.5.3 Range of Extrusion Products

Extrusion process is used to manufacture:

(i) Rods

(ii) Tubes

(iii) A variety of circular, square, rectangular, hexagonal and other shapes both in solid or hollow form.

(iv) Channel. I T and other sections (Fig. 2.43).

2.6 METAL SPINNING

Metal spinning is the operation of shaping thin sheets by pressing against a revolving form. This process is generally applicable to symmetrical articles, which have circular corss-section.

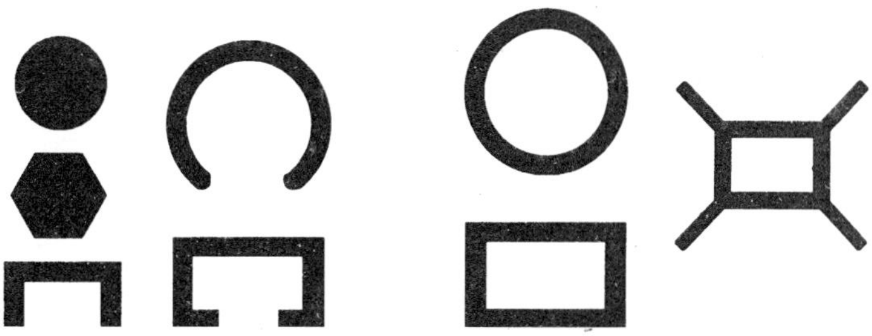

Fig. 2.43 Range of Extrusion Products

Deformation of the metal during spinning proceeds by a mixture of bending and stretching. The spinning may be (i) Hot spinning and (ii) Cold spinning.

Principle of Operation (see Fig. 2.44, p. 85)

Figure 2.44 shows spinning operation setup. The form block which has the shape of the desired object is fixed to the headstock of the spinning machine. The metal disc is held against the form block with the help of support (tail stock). After clamping, the metal disc is rotated at its operation speed. The metal disc is progressively formed against the form block by pressing by means of tool made of wood or metal or roller. Spinning is normally applied only to thin materials such as sheets of ductile metals/ alloys.

2.6.1 Spinning Lathe (see Fig. 2.45, p. 86)

The spinning lathe consists of :

1. Bed
2. Head stock
3. Form
4. Tail stock

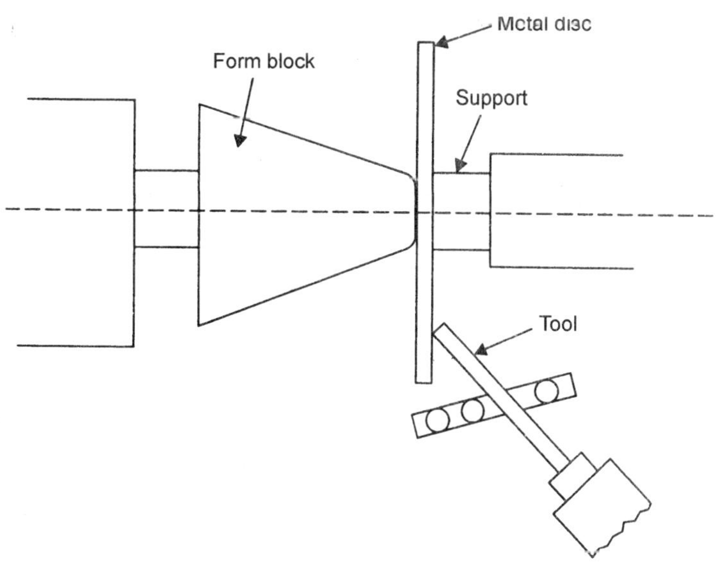

Fig. 2.44 Principle of Operation

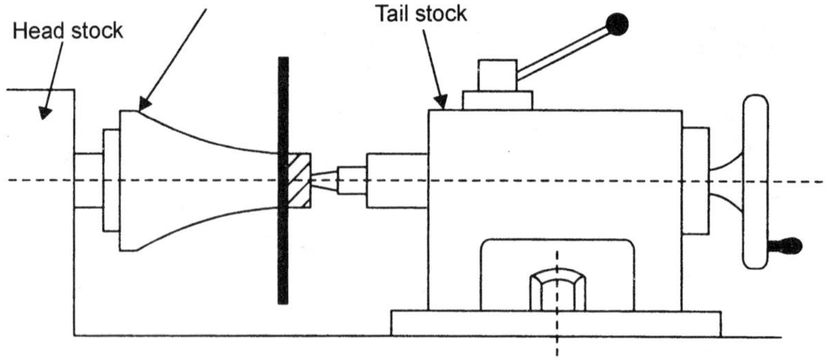

Fig. 2.45 Spinning Lathe

The bed supports headstock, tail stock and other accessories. The form block is fixed in the lathe spindle and turns with it. The work piece (metal disc) is bent over the form block to take its shape by applying pressure by means of to 01. The follower supports the work piece. Steps in spinning are shown in Fig. 2.46.

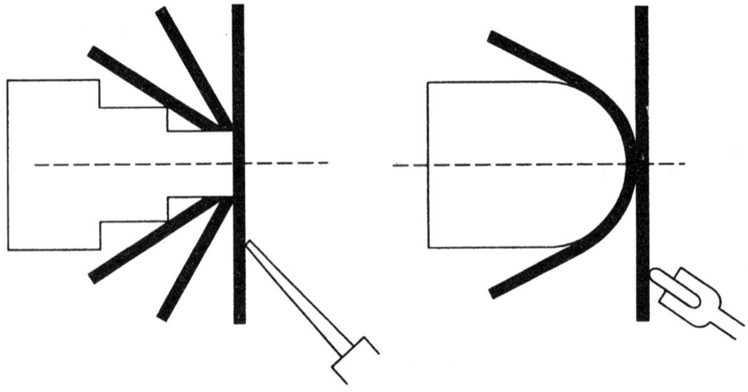

Fig. 2.46 Steps in Spinning

2.6.2 Applications

Aluminium and other soft metals are best suited for spinning. This is very suitable for making aluminium utensils, reflectors. Components used in chemical plants and stainless steel dairy utensils are produced by this process.

The three basic spin shapes are the cone, the hemisphere and cylinder. Of these the conical shape is the easiest to produce. Fig. 2.47 indicates the conical a group of products made by spinning.

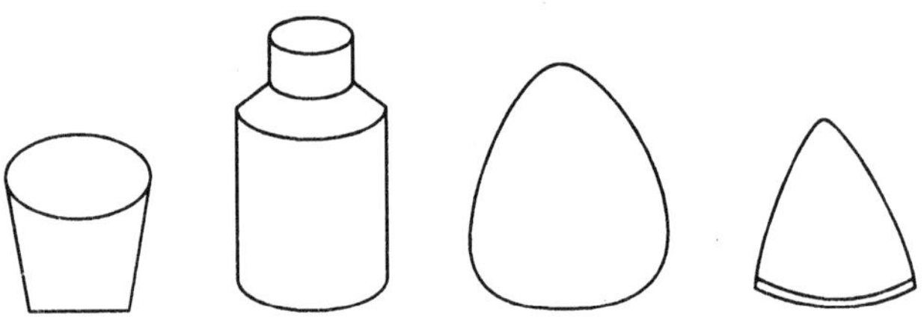

Fig. 2.47 Parts Produced by Spinning

2.7 WIRE DRAWING

Wire is made by cold-drawing hot rolled wire rod through one or more dies as shown in Fig. 2.48 to decrease its size and increase the physical properties.

The wire rod is rolled from a single billet and cleaned in an acid to remove scale, rust and immersed in a lime solution to neutralize the acid.

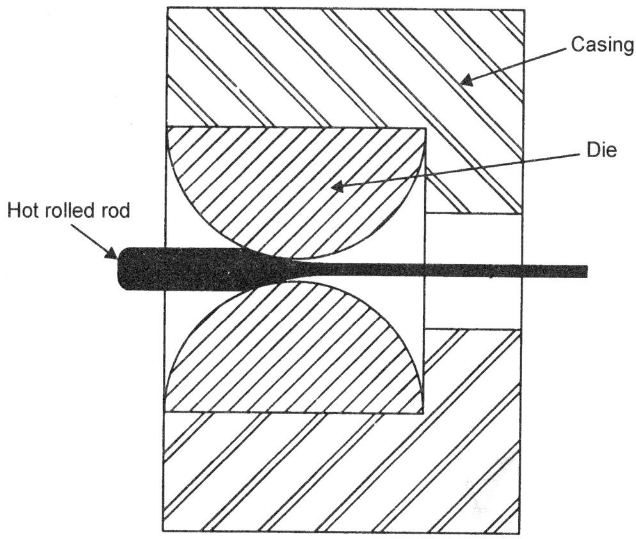

Fig. 2.48 Wire Drawing Process

Both single draft or continuous drawing processes may be used. In the first method, a coil is placed on the reel or frame and the end of the rod is pointed so that it will enter the die. The end is grasped by tongs on a draw bench and pulled out and wound around the reel. After the entire coil has passed through one die, the process of drawing wire through holes of small size is repeated until the desired diameter of the wire is obtained. A typical draw bench of this type with three sets of dies is shown in Fig. 2.49. After the wire has passed through several dies, it becomes brittle due to strain hardening. It should therefore be annealed.

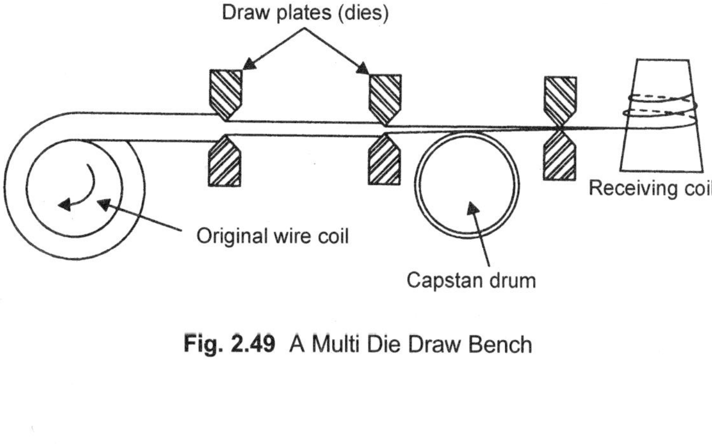

Fig. 2.49 A Multi Die Draw Bench

In continuous drawing, the wire is fed through several dies and draw blocks, which are arranged in series. The number of dies depends upon the reduction required and also on the kind of material being drawn.

2.8 TUBE DRAWING

Tube drawing is also similar to the other drawing processes. There are three basic types of tube drawing processes as shown in Fig. 2.50.

Figure 2.50(a) indicates the simplest form of tube drawing process. In this process, no internal mandrel is used hence it is called sinking process. The technique shown in Fig 2.50(b) reduces the tube diameter and controls its thickness. However, the imitation is length of the tube by length of the mandrel. To over come this problem, moving mandrel a shown in Fig. 2.50(c) is used. The tubes are also first pointed and then entered through the die and on the other side of the die, this end is gripped in tongs, which is connected to the draw bench. There may be more than one pass required to get the final size. The reduction in one pass is about 40 per cent. The metal is annealed after every pass in order to remove the effect of strain hardening. Hot drawn tubes are also cold drawn to provide good surface finish, better dimensional accuracy and improved physical properties.

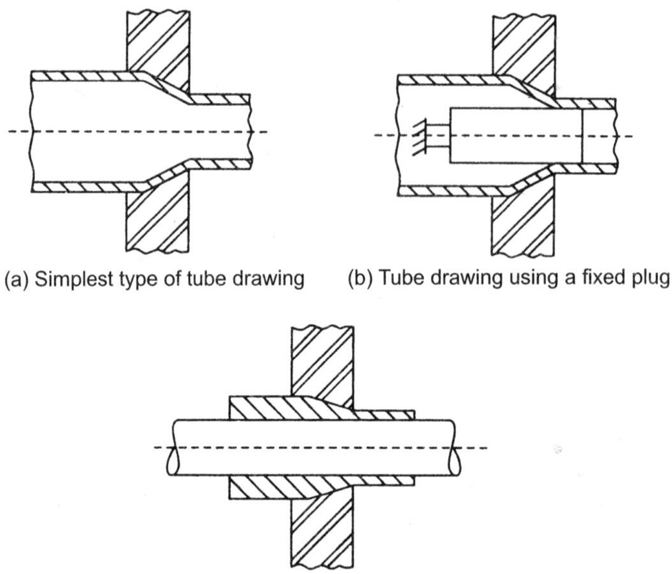

(a) Simplest type of tube drawing

(b) Tube drawing using a fixed plug

(c) Tube drawing using a removable

Fig. 2.50 Tube Drawing Process

2.9 STRETCH FORMING

This process consists of gripping a sheet of metal at each end in suitable jaws and stretching it over a form block of required contour until complete forming has been achieved. The material is stretched beyond its elastic limit causing permanent set. A simple design of stretch forming press is shown in Fig. 2.51. It consists of die or forming punch mounted on a ram placed between two slides, which grip the metal sheet.

Two types of hydraulic stretch forming presses are in common use. In the first type, the sheet is gripped by stationary jaws and subsequently stretched by moving the form block (die) vertically which is actuated by hydraulic system, until forming operation is completed. In the second type, the form block is fixed and jaws are moved horizontally by hydraulic. In this type, the sheet is pre stretched free of form block up to the yield point of the material, then wrapped until tension around the form block and give a final stretch to set the material to the die contour. The stretching of the blank takes place along the tangent to the die surface and thus friction forces developed between the material and die surface are reduced. In this process the spring back is completely eliminated. Side and top panels of car as well as aircraft wing components are manufactured by this process.

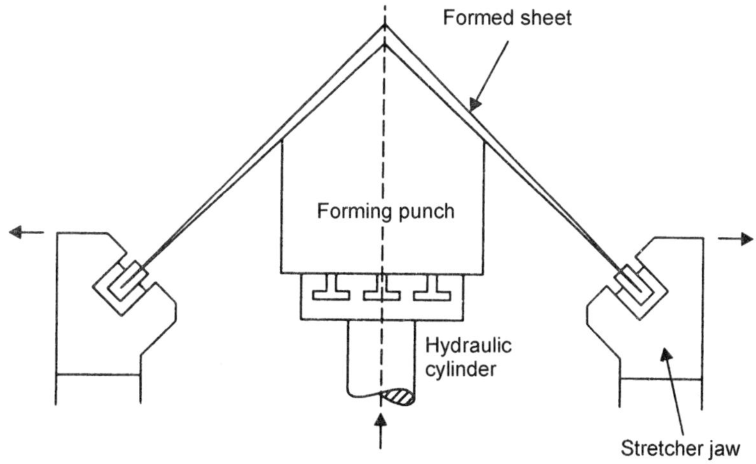

Fig. 2.51 Stretch Forming Press

QUESTIONS

1. Distinguish between hot working and cold working.
2. Explain the types of forging processes.
3. Explain upset forging process.
4. What are the defects in forges parts?
5. What are the advantages and limitations of forging?
6. Sketch and explain various types of Rolling Mills.
7. Explain the sequence of passes to get a round bar from a billet.
8. Discuss various types of Extrusion processes with neat sketches.
9. Explain Backward Extrusion process, what are its advantages and limitations?
10. Briefly describe the spinning process. What are its applications?
11. Explain wire drawing process.
12. With the help of neat sketch explain the tube drawing process.
13. Explain the stretch forming process.

3 Welding

3.1 INTRODUCTION

The art of joining metals by heating and then pressing together is a very old process called smith welding. With the developments and advancements in this field has given rise to a different processes and techniques for welding of metals. The use of welding in present day technology is extensive. Many common items that is automobile cars, aircrafts, electronic equipments, ships, bridges, boilers, household appliances etc. depend upon welding for their economical construction.

Welding

What it is welding may be defined as joining two pieces of metal by application of heat with or without application of pressure and addition of filler metal. It may be also defined as a metallurgical bond accomplished by the attracting forces between atoms.

3.2 CLASSIFICATION OF WELDING PROCESSES

The welding processes are classified as follows:

1. *Gas Welding*
   (a) Oxyacetylene Welding
   (b) Oxyhydrogen Welding
2. *Arc Welding*
   (a) Carbon Arc Welding
   (b) Metal Arc Welding
   (c) Submerged Arc Welding
   (d) Inert Gas Welding
   (i) TIG (ii) MIG
   (e) Plasma Arc Welding
   (f) Electric Slag Welding
3. *Resistance Welding*
   (a) Spot Welding
   (b) Seam Welding
   (c) Projection Welding
   (d) Butt Welding

4. *Solid State Welding*
   (a) Friction Welding (b) Ultrasonic Welding
   (c) Explosive Welding
5. *Thermo-Chemical Welding*
   (a) Thermit Welding (b) Atomic Hydrogen Welding
6. *Radiant Energy Welding Process*
   (a) Electron Beam Welding (b) Laser Beam Welding

3.2.1 Gas Welding

Gas welding is a fusion welding process. It joins metals using the heat of combustion of an oxygen/air and fuel gas (i.e. acetylene, hydrogen, butane) mixture. The intense heat (flame), thus produced melts and fuses together the edges of the parts to be welded, generally with the addition of a filler metal.

(a) Oxy-Acetylene Welding

Oxy-acetylene is used for welding almost all metals and alloys. When acetylene is mixed with oxygen in correct proportions in the welding torch and ignited, the flame resulting at the tip of the torch is sufficiently hot to melt and join the parent metal. The flame reaches a temperature of about 3000°C. A filler metal rod is generally added to the molten metal pool to built up the seam slightly for greater strength.

Oxygen is produced by either electrolysis or liquification of air. Electrolysis separates water into hydrogen and oxygen by passing an electric current through it. Most commercial oxygen is made by liquifying air and separating the oxygen from the nitrogen. It is stored in the steel cylinders. Acetylene gas (C_2H_2) is obtained from the chemical reaction of water and calcium carbide.

$$\underset{\text{Calcium Carbide}}{CaC_2} + \underset{\text{Water}}{2H_2O} \longrightarrow \underset{\text{Acetylene}}{C_2H_2} + \underset{\text{Hydrated Lime}}{Ca(OH)_2}$$

The reaction provides acetylene gas and hydrated lime as sludge. A special hopper of dropping the calcium carbide into a tank of water at controlled rate is referred as acetylene generator. Acetylene cylinders are also readily available.

Equipment for oxy-acetylene welding

Oxy-acetylene welding equipment consists of the following: (see Fig. 3.1)

(*i*) *Oxygen Cylinder*: Oxygen is filled in the cylinder at a pressure of 150 kg/cm^2. This cylinder is made of steel and it is in black colour.

(*ii*) *Acetylene Cylinder*: Acetylene is dissolved in acetone in a cylinder containing porous calcium silicate filler. These cylinders are usually filled to a pressure of 16 kg/cm^2. The cylinder is made of steel and it is in maroon colour.

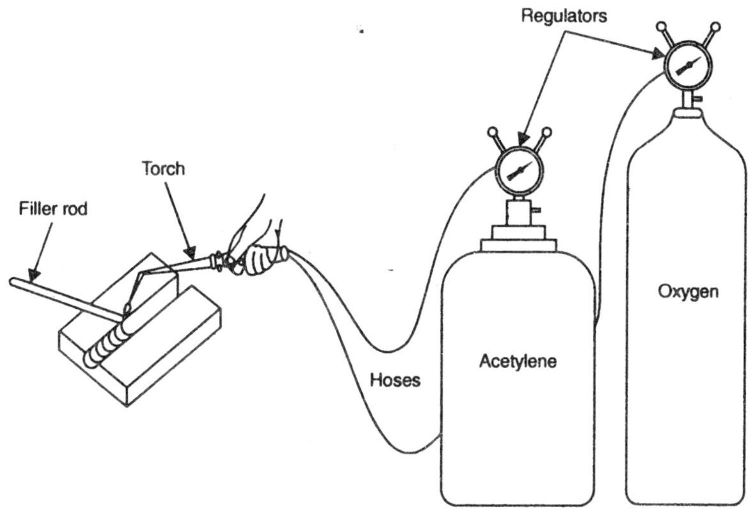

Fig. 3.1 Equipment for Oxy-Acetylene Welding

(*iii*) *Welding Torch*: It is used to mix the gases in the right proportions to control the volume of gases burned at the welding tip and to direct the flow. It has a handle to carry it and two inlet connections for gases at one end. Each inlet has a valve to control the volume of oxygen or other gases. The two gases mix up in a mixer and flame is produced by igniting the mixture at the tip of the torch.

(*iv*) *Pressure Regulator*: It is located on the top of the gas cylinder. Its function is to reduce the pressure from the cylinder and to maintain it at constant value. The pressure regulator located on the oxygen cylinder is called oxygen pressure regulator and the other one located on the top of the acetylene cylinder is called the acetylene pressure regulator.

(*v*) *Hose and Hose Fittings*: The hose is a rubber tube which permits the flow of the gas. Two hoses to carry oxygen and acetylene separately are required. They connect the regulator mounted on cylinders to the torch. Generally, green colour is adopted for oxygen hose and red colour for acetylene. The hose should be strong, durable, flexible and light in weight.

(*vi*) *Goggles*: Goggles fitted with coloured lenses should be provided to protect the eyes from harmful heat and ultraviolet and infrared rays.

(*vii*) *Gloves*: These are used to protect hands from heat and the metal splashes during welding.

(*viii*) *Spark Lighter*: It is used to provide a convenient and instant means for lighting the welding torch.

(*ix*) *Wire Brush*: Its function is to clean the surfaces of joints before and after welding.

Other equipments

Welding Rods: These are used for providing extra metal to the weld. These are also known as filler rods. The filler rod should have the same composition and properties as that of parent metal. The filler rods are available in 1, 1.25, 1.6, 2, 2.25, 3, 4.5, 6, 8 and 10 mm diameter. The selection of filler rod depends on the welding technique and thickness of the base metal. Steel rods are generally employed when welding ferrous metals. They have a higher carbon content and more manganese and silicon than the base metal. The last two components act as deoxidising agents and prevent the inclusions of oxide in the weld. Rods containing chromium and vanadium are used for welding alloy steels.

Flux: When the metal to be welded is heated by oxy-acetylene flame, the oxygen of the atmosphere combines with the heated metal and forms metal oxides. These metal oxides have higher melting point than the parent metal. Therefore, it is essential that these oxides are removed otherwise slag inclusions will result in poor quality of weld. These oxides can be removed from the weld location by the use of certain fluxes which react chemically with the oxides of most metals and from fusible slag and floats at the top of the molten puddle and do not interfere with the deposition of filler metal. Besides it also protects the molten puddle from atmospheric oxygen. Fluxes are available in several forms such as dry powder, paste or in the form of coating on the welding rod. For ferrous metals, borax, sodium carbonate, sodium bicarbonate are used as suitable fluxes. For copper and copper alloys, mixture of sodium and potassium borates, carbonates, chlorides and boric acid are suitable.

3.2.1.1 Types of Flames

The correct adjustment of the flame is important for the production of satisfactory welds. The flame must be of proper size, shape and condition in order to operate at maximum efficiency. The three types of oxy-acetylene flames, which are used in engineering works, are as follows (see Fig. 3.2).

(a) Neutral flame.
(b) Reducing or carburising flame.
(c) Oxidising flame.

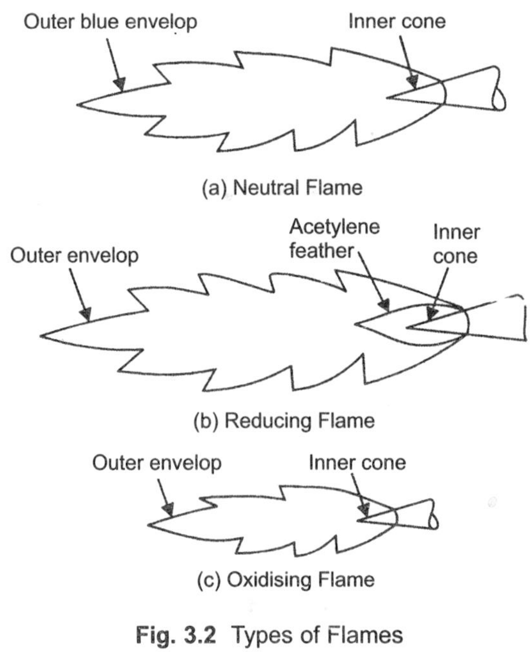

Fig. 3.2 Types of Flames

(*a*) *Neutral Flame*: A neutral flame is produced when approximately equal volumes of oxygen and acetylene are supplied to the torch. The temperature of the neutral flame is in order of about 3260°C.

The neutral flame consists of sharp brilliant inner cone extending a short distance from the tip of the torch and an outer cone or envelop. The first one develops heat and second protects the molten metal from oxidation, because the oxygen in the surrounding atmosphere is consumed by gases from flame.

The neutral flame is commonly used for welding most of the metals such as mild steel, stainless steel, cast iron, copper, aluminium etc.

(*b*) *Reducing or Carburising Flame*: If the volume of oxygen supplied to the neutral flame is reduced, the resulting flame will be reducing flame. The temperature of the reducing flame is of the order of 3038°C.

This flame has three zones (i) Inner cone (ii) An intermediate of whitish colour (iii) The bluish outer cone. The outer flame envelop is longer than the other two flames.

Being rich in carbon, this flame is suitable for welding steel. It is also used for surface hardening.

(*c*) *Oxidising Flame*: If the volume of oxygen to the neutral flame is increased, the result will be oxidising flame. The temperature of the oxidising flame is of the order of 13000°C. It is hotter than neutral flame.

The oxidising flame consists of 1 smaller inner cone which is more pointed than the neutral flame. The outer envelop is shorter. Oxidising flame is used in welding brass, copper base metals, zinc base metals and few ferrous metals such as manganese, steels and cast irons.

3.2.1.2 Types of Welded Joints (see Fig. 3.3 on p. 96)

Five basic types of joints are used in fusion welding. These are: (a) Butt joint (b) Lap joint (c) T-Joint (d) Corner joint (e) Edge joint. These joints are shown in Fig. 3.3.

(*a*) *Butt Joint*: Figure 3.3 (a); shows the butt-joint which is used to joint the ends of two plates or surfaces located approximately in the same plane.

(*b*) *Lap Joint*: Figure 3.3 (b); shows the lap-joint which is used to join two overlapping plates so that the edge of each plate is welded to the surface of the other.

(*c*) *T-Joint*: Figure 3.3 (c); shows the T-joint which is used to weld two plates or sections whose surfaces are at right angels to each other.

(*d*) *Corner Joint*: Figure 3.3 (d); shows the corner-joint which is used to join the edges of two sheets or plates whose surfaces are at 90° to each other.

(*e*) *Edge Joint*: Figure 3.3 (e); shows the edge joint which is used in sheet metal work.

3.2.1.3 Edge Preparation (Fig. 3.4, see p. 96)

To obtain sound welds, good edge preparation is essential. Different edge preparation for butt welds are: (a) square (b) single V (c) double V (d) single U (e) double U as shown in Fig. 3.4.

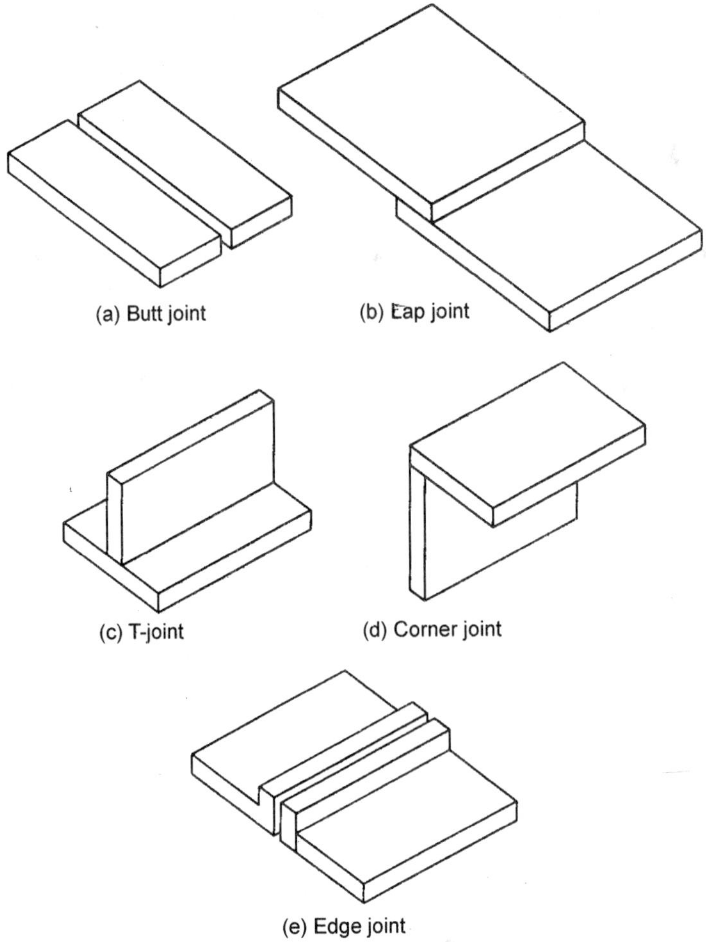

Fig. 3.3 Types of Welded Joints

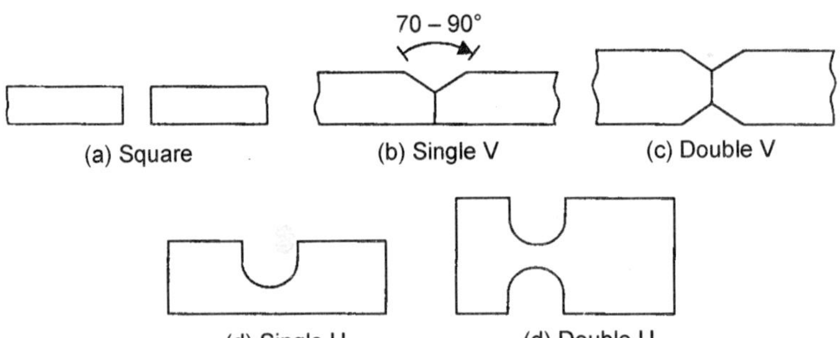

Fig. 3.4 Edge Preparation for Butt Welding

(*a*) *Square Butt Weld*: Square butt weld is used in welding the plates ranging from 3 mm to 5 mm. Before welding the edges are spaced about 3 mm.

(*b*) *Single V*: Single V edge preparation is used for plates of 8 mm and up to 16 mm thick.

(*c*) *Double V*: Double V is used for plates of over 16 mm thick.

(*d*) *Single U and Double U*: These are used for plates of over 20 mm thick.

3.2.1.4 Gas Welding Techniques

The selection of a proper technique will depend upon the metal to be welded, its thickness and the properties of the weld. The following methods are commonly used:

(*i*) *Position of Welding*: (a) Down hand welds (b) Vertical welds (c) Inclined welds (d) Horizontal welds (e) Over hand welds.

(*ii*) *Direction of Travel Welding Rod and Welding Torch*: (a) Leftwards or Forwards welding (b) Rightwards or backwards welding (c) Vertical welding.

(i) Position of welding (Fig. 3.5)

(*a*) *Down Hand Welds* (*flat*): These welds are deposited in any direction on a horizontal surface so that the flame is above the face of the weld (see Fig 3.5(a)).

(*b*) *Vertical Welds*: These welds are deposited on a vertical surface in a vertical direction as shown in Fig. 3.5(b).

(*c*) *Inclined Welds*: These welds are deposited on an inclined surface as shown in Fig. 3.5(c).

(*d*) *Horizontal Welds*: These welds are deposited on vertical surface in a horizontal direction as shown in Fig. 3.5(d).

(*e*) *Overhead Welds*: These welds are deposited on a horizontal surface in any direction so that the face of welds is above the flame as shown in Fig.3.5(e).

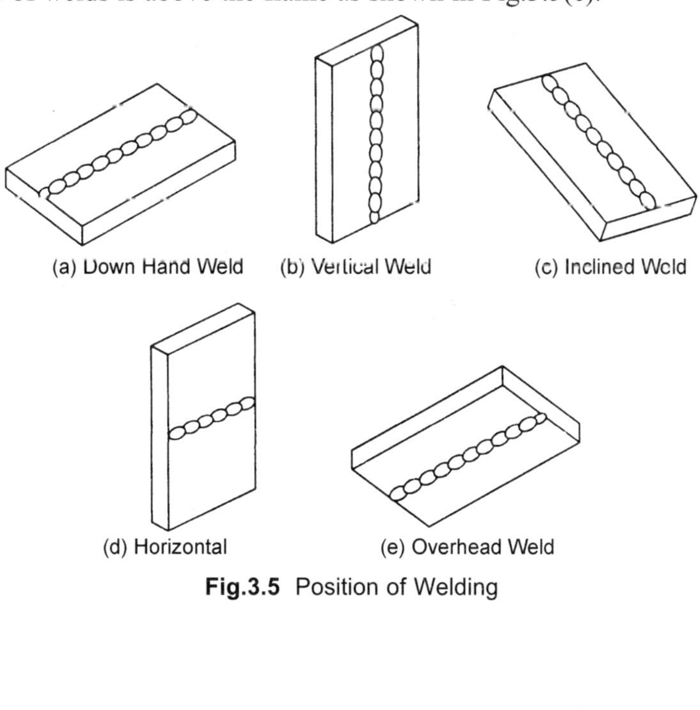

Fig.3.5 Position of Welding

(ii) Direction of travel welding rod and welding torch (Fig. 3.6)

(*a*) *Leftwards* (*or*) *Forward Welding*: The welder holds torch in the right hand and filler rod in the left hand. The weld is made working from right to left as shown in Fig. 3.6(a). Since, the flame is pointed in the direction of the welding, it preheat the edges of the joint. This method is suitable for mild steel, cast iron, aluminium, brass etc.

(*b*) *Rightwards* (*or*) *Backward Welding*: It is carried out from left to right as shown in Fig 3.6(b). Thicker materials can be welded by this method.

(*c*) *Vertical Welding*: It starts at the bottom of the weld joint and gives an oscillating movement to the welding torch which points slightly upwards. (see Fig. 3.6(c).

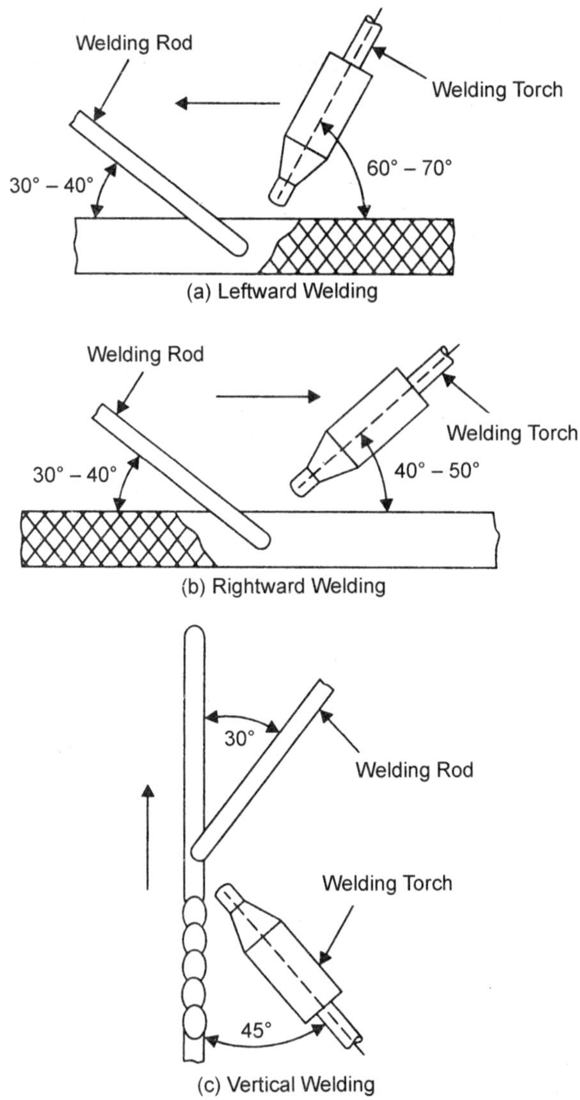

Fig. 3.6 Direction of Welding

3.2.1.5 Advantages of Oxy-Acetylene Welding

1. The equipment is comparatively in expensive.
2. Low maintenance cost.
3. The oxy-acetylene flame is generally more easily controlled and not as piercing as metallic arc welding. Therefore, it is used extensively for sheet metal fabrication and repair works.
4. The equipment is versatile. Besides gas welding, the equipment is used for preheating, brazing, metal cutting etc.
5. With proper technique, practically, all metals can be welded.
6. Since the source of heat and filler metal are separate, the welder has controlled over the filler material deposition rates.

3.2.1.6 Disadvantages of Oxy-Acetylene Welding

1. It takes considerable longer for the metal to heat up than in arc welding.
2. Prolonged heating of the joint in gas welding results in larger heat affected area. This often results in increased growth, more distortion.
3. These are safety problems involved in handling and storing of gases.
4. Flame temperature is less than the temperature of the arc.
5. Heavy sections cannot be joined economically.
6. Flux shielding in gas welding is not so effective as an inert gas shielding in TIG or MIG welding.

3.2.1.7 Applications of Gas Welding

1. For joining thin materials.
2. For joining most ferrous and non-ferrous metals.
3. In automatic and aircraft industries and sheet metal fabrications.

3.2.1.7(a) Oxy-Hydrogen Welding

Oxy-hydrogen welding is used for aluminium, magnesium, lead etc. In this process hydrogen is used in place of acetylene and the flame temperature is very low 2000°C. An advantage of this process is that no oxides are formed on the surface of the weld.

3.2.1.8 Oxy-Acetylene Cutting (Fig. 3.7)

It is a chemical process in the sense that the metal, at the portion where it is to be cut is actually made to oxidise under the action of flame with the following reaction.

$$\underset{\text{Iron}}{3Fe} + \underset{\text{Oxygen}}{2O_2} \longrightarrow \underset{\text{Black Iron Oxide}}{Fe_3O_4} \quad \underset{\text{Heat}}{+\ 26{,}691 \text{ cal}}$$

All ferrous metals can be cut by means of oxy-acetylene flame cutting. The oxy-acetylene flame cutting process makes use of a cutting torch. The torch mixes the acetylene and oxygen in the correct proportions to produce preheating flame and also the torch supplies a uniformly, concentrated stream of high purity oxygen to the reaction zone. The tip has a central hole for pure oxygen jet with surrounding holes for preheating flames as shown in Fig. 3.7.

To produce a cut, the steel is heated to ignition temperature (900°C) i.e., reddish yellow colour by preheating flame, keeping the torch 3 mm above the surface of material to be cut. A jet of pure oxygen is directed at this heated area. This forms the iron oxide there and the same melted immediately (burning the steel in its path). It is then blown off by the oxygen jet, thus providing a narrow slit along the cutting line.

Oxygen cutting can be accomplished manually or by machine (automatic)

Oxygen cutting machines are further divided in two classes:

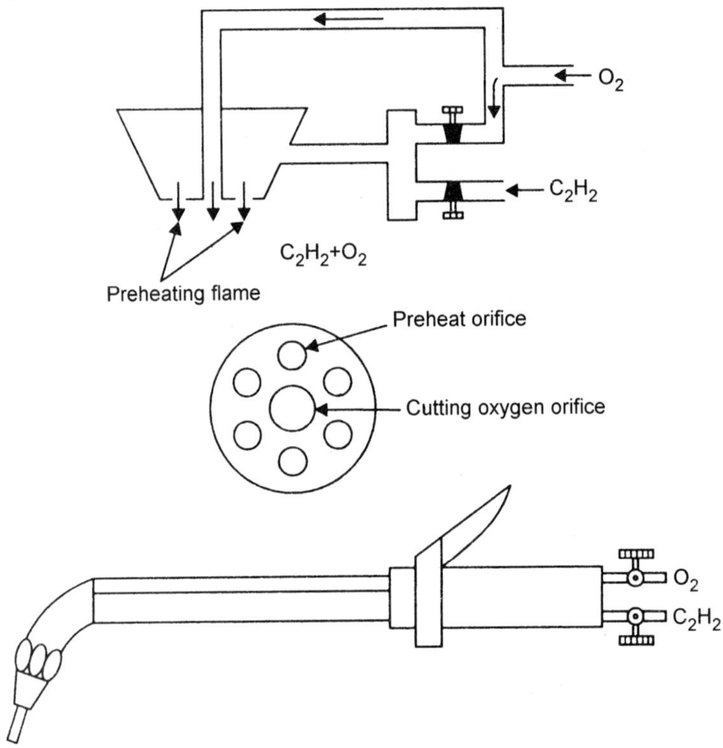

Fig. 3.7 Oxy-Acetylene Cutting Torch

1. Portable machine
2. Stationary machine

On a portable machine, the carriage supports the torch. It is usually run by an electric motor on a straight track. The speed of the motor is adjustable to the size of the metal being cut.

The stationary type of cutting machines are designed on two different mechanical principles for controlling the cutting torch. One is the pantograph design and the other uses a cross-carriage mechanism.

3.2.2 Arc Welding (Fig. 3.8)

In arc welding process, the welding temperature is produced by an electric arc, established between an electrode and the metal being welded. The temperature of the arc is 7000°C. The arc welding set up is shown in Fig. 3.8.

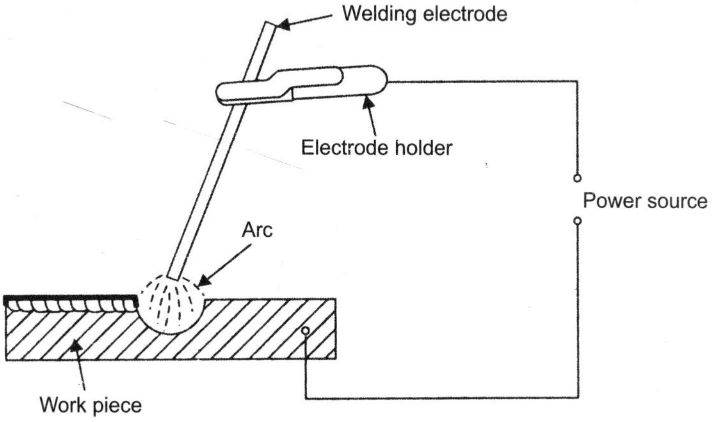

Fig. 3.8 Arc Welding Set up

3.2.2.1 Arc Welding Equipments

The equipments required for arc welding consists of:

(a) Arc welding power source	(f) Chipping hammer
(b) Electrode	(g) Helmet
(c) Electrode holder	(h) Safety goggles
(d) Cables, cable connectors	(i) Apron
(e) Earthing clamps	(j) Hand gloves

(*a*) *Arc Welding Power Source*: The power source required to maintain the arc between the electrode and base metal is available in (i) DC generator (ii) AC transformer with DC rectifier (iii) AC transformer.

(*i*) *DC Generator*: DC generator is run either by an electric motor or a diesel engine. These generator supplies voltage in the range of 15 to 50 volts and output current 200 to 600 Amps. These produce DC in either straight or reverse polarity. The heat generated is split into two parts in the ratio of 66 per cent at positive pole and 33 per cent at negative pole. For welding thin materials, the work is made negative and the electrode positive. This is called reverse polarity. For welding heavy sections the electrode is made negative and the work to be positive, this is called straight polarity. It can be used for welding ferrous and non-ferrous metals. The disadvantage of the generator is the high investment and maintenance cost. Its operation is noisy.

(*ii*) *AC Transformer*: AC transformer changes high voltage, low amperage to low voltage, high amperage. The main advantage of transformer over generator is low cost and ease of operation. Since there are no moving parts in the equipment, the operation is noiseless. The disadvantage of the transformer is that the polarity cannot be changed.

(*b*) *Electrodes for Arc Welding*: Electrodes *for* arc welding may be broadly classified as:

1. Non-consumable electrodes
2. Consumable electrodes

Non-consumable electrodes are usually made of carbon, graphite or tungsten. These electrodes do not get consumed during the arc welding. These are used in carbon arc welding. TIG welding, atomic hydrogen welding.

Consumable electrodes get consumed during the welding. These are made of various metals depending upon the purpose and chemical composition of parent metals being welded. These electrodes are further classified into (1) bare electrodes (2) coated electrodes.

Bare electrodes are used in submerged arc welding and Metal Inert Gas(MIG) welding.

Coated electrodes are again subdivided into (1) Light coated electrodes (2) Heavy coated electrodes.

Light coated electrodes are used for welding non-essential jobs. The primary purpose of light coated is to increase arc stability. These produce poor mechanical properties welds due to the lack of protection of the weld.

Heavy coated electrodes are used to produce high quality welds. *Functions of Coated Electrode*: The coating on electrodes perform the following functions:

1. Protects the weld from atmospheric oxygen and nitrogen by producing a shield of gas around the arc and weld pool.
2. Stabilize the arc.
3. Provide the slag so as to protect the weld from rapid cooling.
4. Remove oxides and impurities.
5. Add alloying elements to the weld metal.
6. Increase deposition efficiency.

3.2.2.2 Types of Arc Welding

(a) Carbon arc welding
(b) Metal arc welding
(c) Submerged arc welding
(d) Inert gas welding
    (i) TIG welding (ii) MIG welding
(e) Plasma arc welding
(f) Electro slag welding

(a) Carbon Arc Welding

In carbon arc welding process the arc is obtained between the carbon electrode and the work piece or between two carbon electrodes. This welding is suitably used in welding of steel sheets, copper alloys and brass etc.

(b) Metal Arc Welding

Figure 3.9 shows the metal arc welding. This is also called shield Metal Arc Welding (SMAW). Heat required for the welding is obtained from the arc struck between the coated electrode and the work-piece The material droplets are transferred from the electrode to the work piece through the arc and deposited along the joint to be welded. The coating produces a gaseous shield and slag to protect from atmosphere.

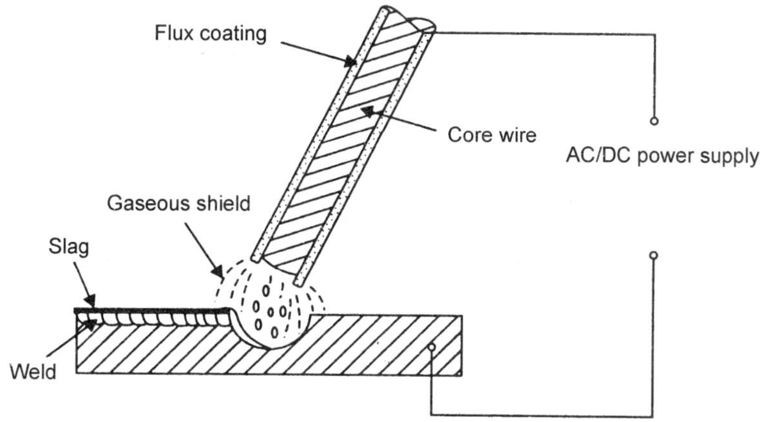

Fig. 3.9 Metal Arc Welding

Advantages

(i) It is the simplest of all the arc welding processes.

(ii) The equipment is portable and less cost.

(iii) Wide range of metals and their alloys can be welded.

Disadvantage

(i) Mechanization is difficult due to limited length of the electrode.

Applications

(i) All commonly used metals and their alloys can be welded.

(ii) This process finds applications in ship building, aircraft industries, automobile industries etc.

(c) *Submerged Arc Welding*

This process is so named because of metal arc is shielded by a blanket of flux as shown in Fig. 3.10. In this process instead of flux covered electrode, granular flux and a bare electrode is used. Flux is deposited continuously in front of the electrode and the flame feeder and the electrode feeder together move as the welding proceeds. The flux is sufficient depth to submerge completely the arc column so that there is no smoke or splatter and the weld is shielded from the effect of all atmospheric gases. As a result of this unique protection, the weld are exceptionally smooth.

The arc is started either by striking the electrode beneath the flux on the work or by placing the steel wool between the electrode and the work piece before switching on the welding current. The intense heat of the arc immediately, produces a pool of molten metal in the joint and at the same time the flux adjacent to the arc column melts and floats on top of the molten metal. This forms a blanket that eliminates spatter losses and protects the welded joint from oxidation. The current density is 300 to 400 amps which is 5 to 6 times than that of metal arc welding. Submerged arc welding is done manually or automatic and semi-automatic. The manual and the automatic submerged arc welding process are most suited to the flat welding position, or slightly vertical, down hill welding position. Backing strip of steel, copper or some refractory material is used under the joint to avoid loosing some of the molten metal.

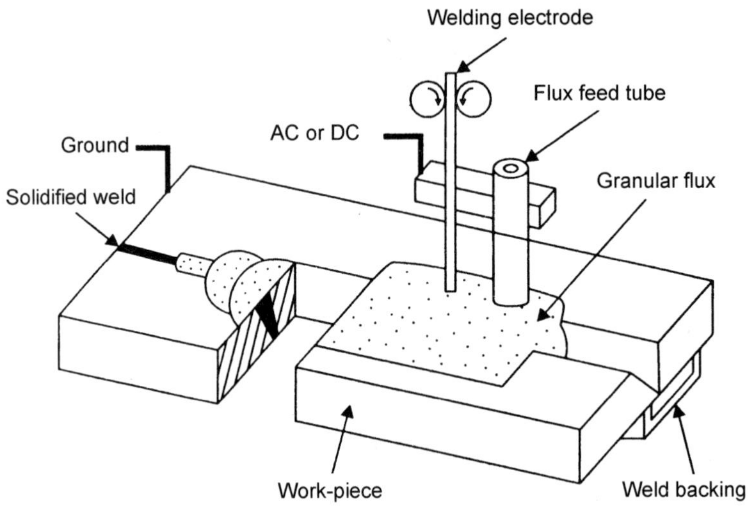

Fig. 3.10 Submerged Arc Welding

This process is used to weld low alloy, high tensile steels as well as mild steel, low carbon steels.

Advantages

(i) Deep penetration is obtained due to the high current, density which is 5 to 6 times than that of metal arc welding.
(ii) Welding is fast due to high melting rate of electrodes.
(iii) Minimum distortion due to high speed.
(iv) Quality of the weld is excellent and uniform.

Applications

The submerged arc welding process has many industrial applications. It is used for fabricating pipe, boiler vessels, structural shapes and practically any job where straight line welding is required.

(d) Inert Gas Welding

In conventional arc welding, the fluxes are used to shield the atmosphere around the molten metal. In inert gas welding, inert gases such as argon, helium, carbon dioxide are used for surrounding the electric arc and thus keeping the atmospheric air and other contaminations away from the molten metal pool. Two methods are employed.

(i) Tungsten-inert Gas (TIG) welding
(ii) Metal-inert gas (MIG) welding

(i) Tungsten-Inert Gas (TIG) Welding

A tungsten inert gas welding equipment is shown in Fig. 3.11. This process is also known as gas tungsten arc welding (GTAW). It uses a non-consumable tungsten electrode mounted at the centre of the torch. The inert gas is supplied to the welding zone through the angular path surrounding the tungsten electrode. Welding operation is done by striking the arc between the work piece and tungsten electrode in the atmosphere of inert gas.

Advantages

(i) No flux is required.

(ii) TIG welds are stronger, more ductile and more corrosion resistance than welds made with ordinary shield arc welding.

(iii) Welding is easily done in all the position.

Disadvantages

(i) Equipment is costier.

(ii) Separate filler rod is needed.

(iii) Decrease in welding speed.

Applications

(i) It is used for fusion welding of aluminium, magnesium alloyes, stainless steel, low alloy steel high alloy steel, brass, bronze, silver, molybdenum and a wide range of other metals.

(ii) It can also be used to weld many dissimilar metals.

(iii) The TIG process can be used to braze and to supply the heat source for braze welding.

(iv) It can also be used as heat source for the hard surfacing of the metals.

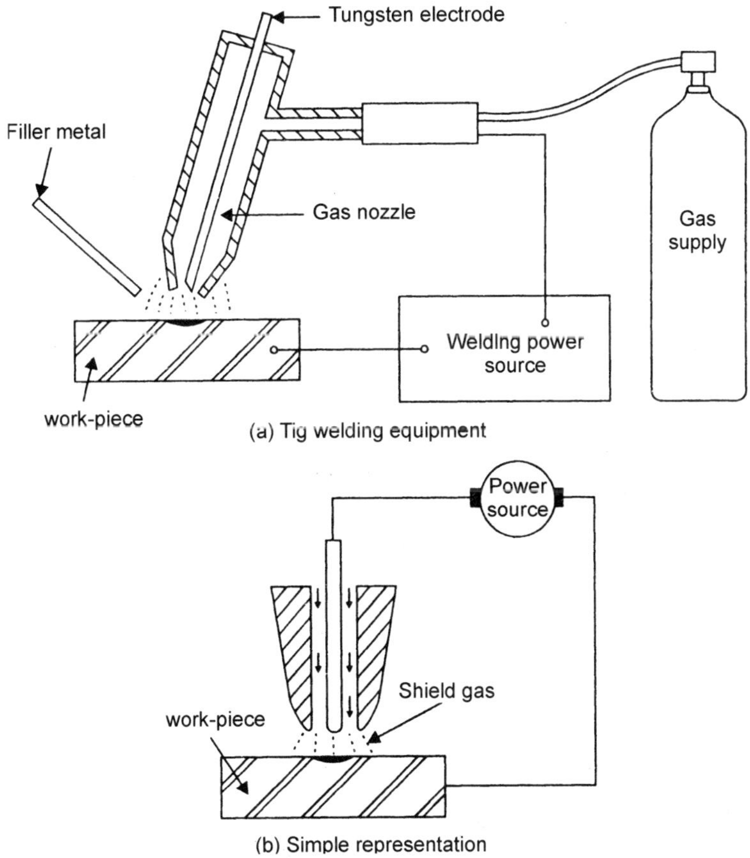

Fig. 3.11

(ii) Metal Inert Gas (MIG) Welding (see Fig. 3.12)

MIG welding stands for Metal Inert Gas Welding. In this process, the tungsten electrode is replaced with a consumable electrode. The electrode is continuously fed to the arc at the rate at which it is consumed and transferred to the base metal. Arc is shielded by an inert gas, which flows from the holder nozzle through which the electrode also passes. It is similar to submerged arc welding in feeding the bare electrode from a reel. It differs in the fact that the shielding is done by an inert gas and the arc is visible during the welding process.

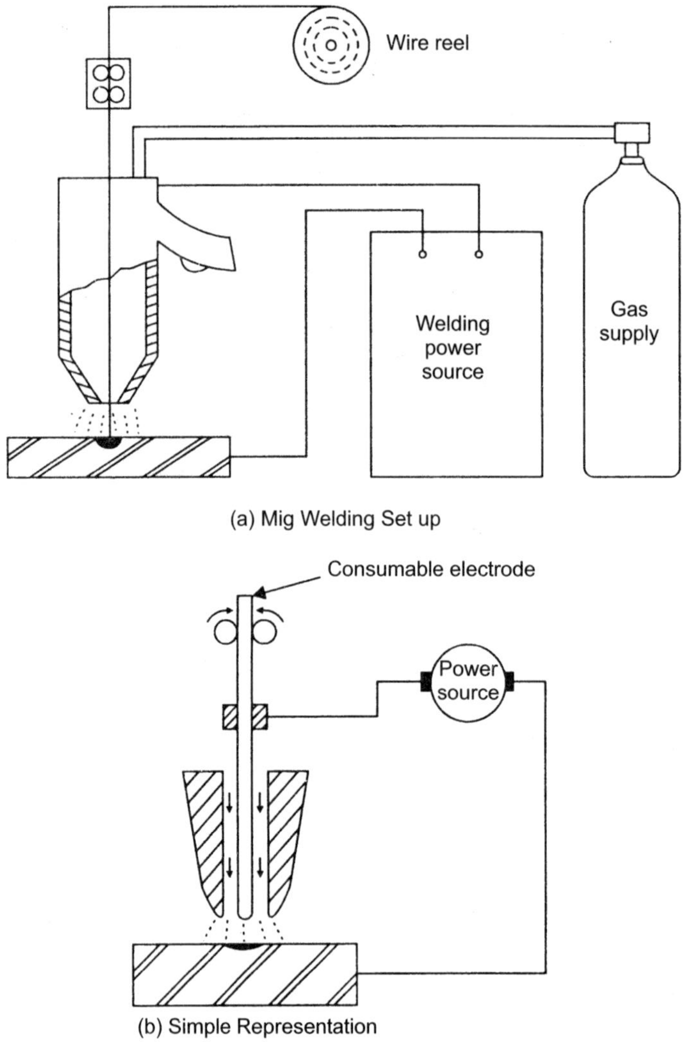

Fig. 3.12

Advantages

(i) No flux is required

(ii) High quality welds are produced

(iii) Less operator skill is required
(iv) High welding speed
(v) It is suitable for ferrous and non-ferrous metals.

Disadvantages

(i) Welding equipment is more complex and costly
(ii) It is difficult to weld in small corners.

Applications

(i) It can be done on most of the commercial metals.
(ii) It is used for welding carbon and low alloy steels, stainless steels nickel and its alloys, copper alloys.
(iii) MIG welding is used in aircraft and automobile industries.

(*e*) *Plasma Arc Welding*

It is an electric arc welding process which employs a high temperature constricted arc or plasma jet to obtain the melting and join of metals. In this process a gas (argon or hydrogen) is passed through an electric arc, where it gets ionised. This process uses two inert gases, one forms the plasma and second shield the arc plasma. Plasma arc welding can be divided into two basic types:

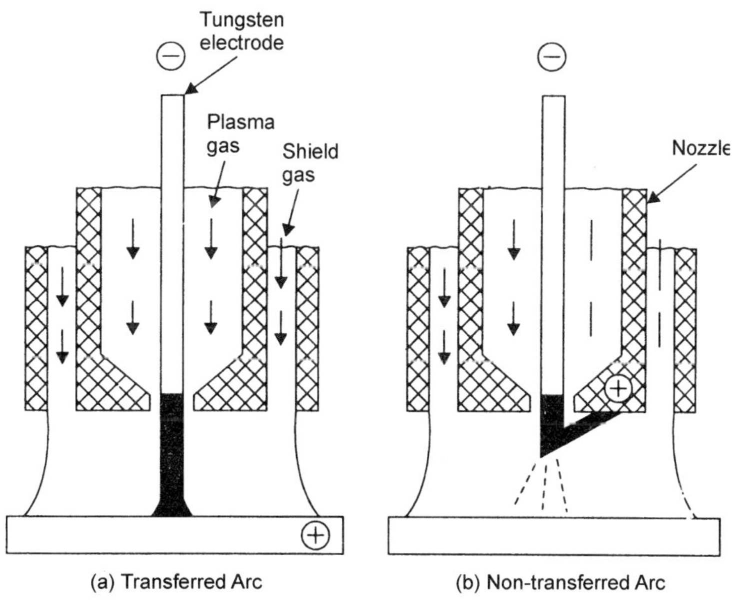

Fig. 3.13

(i) Transferred arc process
(ii) Non-transferred arc process
(*i*) *Transferred Arc Process* (Fig. 3.13(a)): The arc is formed between the electrode (–) and work piece (+). In other words, the arc is transferred from electrode to the work-piece. This possesses high energy density. For this reason, it is used to cut and melt the metals.

(*ii*) *Non-transferred Arc Process* (Fig. 3.13(b)): The arc is formed between the electrode (–) and water cooled constructing nozzle (+). Plasma arc comes out of the nozzle as a flame. This arc is independent of the work-piece and the work does not form a part of electric circuit. This arc is used only for welding.

(*f*) *Electro Slag Welding*

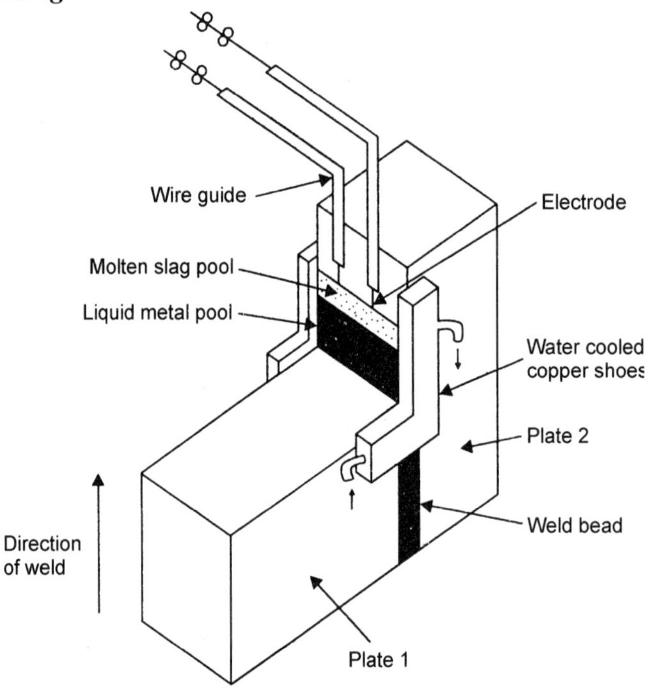

Fig 3.14 Electro Slag Welding

The Fig. 3.14 illustrates the principle of electro slag welding. The pieces to be welded are positioned vertically with necessary gap between them. Two copper shoes (water cooled) slides on either side of the gap form a well in which flux is deposited. An electric arc is struck between the electrode and the joint bottom with the help of a piece of steel wool. The arc melts the electrode and flux and forms the molten slag. When enough slag accumulate, the arc action stops and further requirement heat is provided by the resistance offered by the slag to the current flowing through it. The molten metal temperature is 2000°C. This heat is sufficient to fuse the edges of the work pieces and the welding electrode. The heated metal collects in the pool beneath the slag slowly solidifies thereby forming the weld bead joining the two work pieces.

Advantages

(i) Thicker plates can be welded in a single pass and economically.

(ii) High welding speed

(iii) Minimum joint preparation

(iv) Little distortion

(v) The weld metal is totally out of contact with atmosphere and hence the best quality of weld.

Applications

It is used particularly for welding thick (30 mm over) plates and structures for turbine shafts, boiler parts and heavy presses.

3.2.3 Resistance Welding

In resistance welding, a heavy electric arc current is passed through the metal pieces to be joint over a limited area, causing them to be locally heated to plastic state and the weld is completed by the application of pressure. In this process two copper electrodes are used. The metal pieces to be welded are pressed between electrodes and current is passed through the electrodes. A transformer in the welding machine reduces the voltage from either 120 or 240 volts to 4 to 12 volts and raises the amperage sufficiently to produce a good heat.

The amount of heat (H) generated is given by the following relation:

$$H = KIRT$$

Where, H = The heat generated in the work in Joules

I = Electric current in amperes

R = Resistance of the joint in ohms

T = Time of current flow in seconds

K = A constant to account for the heat loss from the welded joint.

For good resistance welding the following factors are properly controlled.

(*i*) *Welding Current*: Enough current is required to bring the work pieces to plastic state for welding. It is properly adjusted on the current control device on the machine.

(*ii*) *Welding Pressure*: Mechanical pressure is required to hold the work-pieces and squeeze the pieces to form the weld during plastic state.

(*iii*) *Cycle Time*: It is the combination of weld time and hold time. The duration of current flowing through the work pieces to raise the temperature is called welding time. After this the current is switched off while the pressure is still acting. The pressure is applied till the weld cools and regain sufficient strength. This period is known as hold time.

Types of resistance welding:

(a) Spot welding
(b) Seam welding
(c) Projection welding
(d) Butt welding

(*a*) *Spot Welding*

It is the simplest and most commonly used method of overlap welding of strips, sheets or plates of metal at small areas.

In this method, sheets of a metal to be welded are held between copper electrode (water cooled) by applying pressure through foot pedal lever. A current of low voltage and sufficient amperage is passed between electrode causing the parts to be brought to welding temperature. The metal under electrodes pressure is squeezed and welded. After this, the current is turned off while the pressure is still acting. The pressure is applied till the weld cools and produce a solid bond. Now, the pressure is released and the work is removed from the machine (see Fig. 3.15)

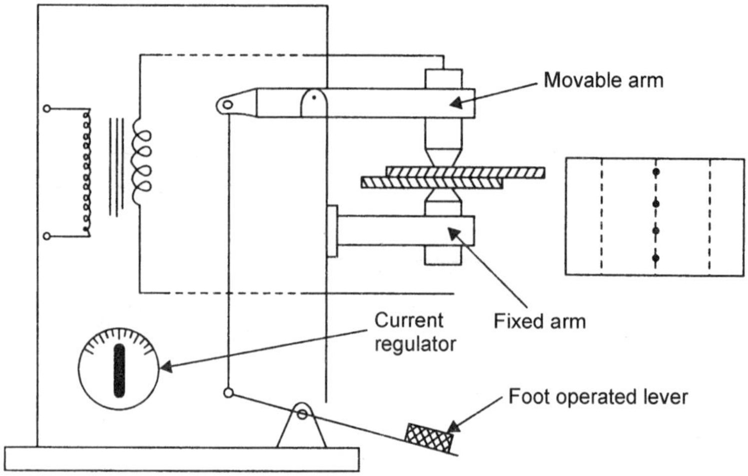

Fig. 3.15 Spot Welding Machine

Advantages

(i) No edge preparation is needed
(ii) Low cost
(iii) High speed of welding

Applications

(i) This technique is used mostly in thin sheetwork like making sheet metal boxes, containers such as receptacles.
(ii) Thicker metals up to 12.5 mm have been successfully spot welded.
(iii) It finds application in automobile and aircraft industries.

(*b*) *Seam Welding*

Seam welding is similar to spot welding, except that the electrodes in spot welding are replaced by copper rollers or wheels. The work pieces to be welded are passed between the rollers as shown in Fig. 3.16. A current impulse is applied through the rollers to the material in contact with them. The heat generated makes the metal plastic and the pressure from the rollers completes the weld.

In seam welding, there are two types of welds are obtained:

(i) Stitch welding
(ii) Roll welding

(*i*) *Stitch Welding*: Stitch weld is made by the current on the rollers off and on quickly enough, so that continuous fusion zone made of overlapping nugget is obtained (Fig. 3.17(a)).

(*ii*) *Roll Welding*: It is obtained by constant and regular timed interruptions of welding current, which causes individual nuggets to be formed.

Seam welding is used on many types of pressure tight or leakproof tanks for various purposes, exhaust systems, barrels etc.

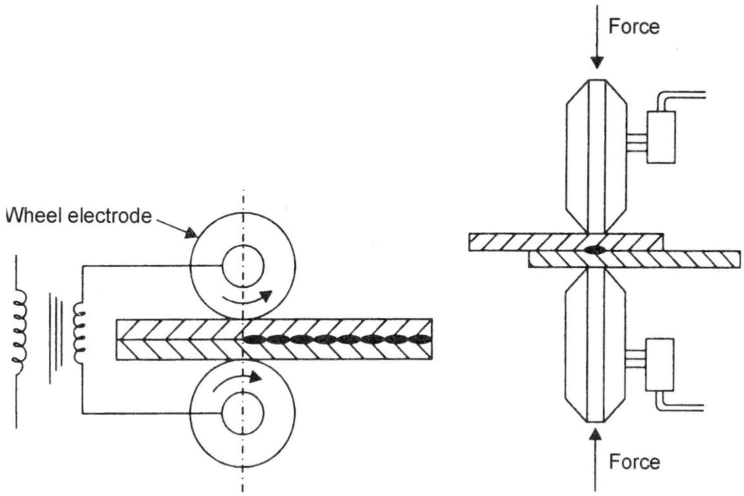

Fig. 3.16 Seam Welding

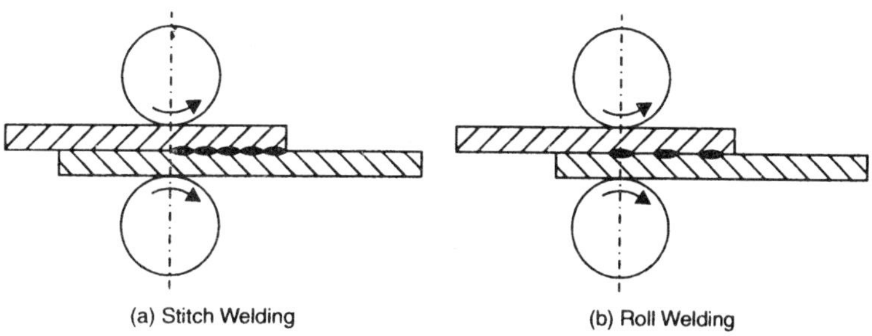

Fig. 3.17 Types of Seam Welding

(c) ***Projection Welding*** (Fig. 3.18)

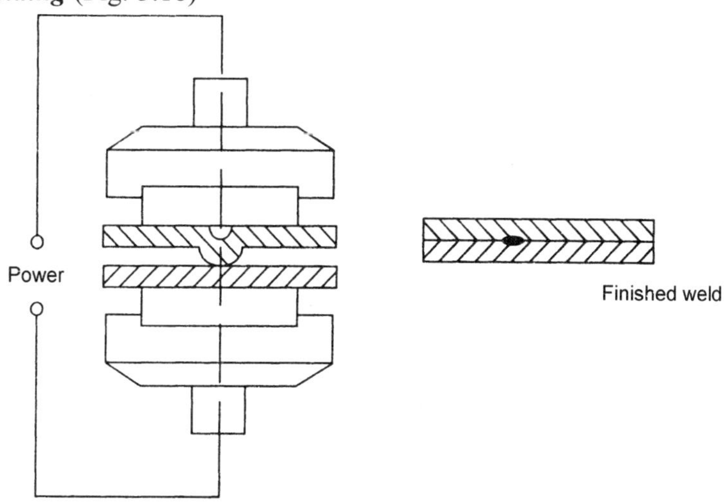

Fig. 3.18 Projection Welding

Projection welding is another variation of spot welding. Small projections are raised on one side of the sheet or plate where it is to be welded to another. The projections serve to concentrate (localize) the welding heat at these areas and facilitate fusion without the necessity of employing a large current. During the welding process, the heated and softened projections collapses under the pressure of the electrodes thereby forming the weld. The working principle of projection welding is shown in Fig. 3.18.

Advantages

(i) This method of welding gives longer electrode life.

(ii) Outer or top surfaces can be produced with no electrode marks.

Disadvantages

(i) All projections should be seated in one blow.

(ii) A prior operation is necessary to form the projection.

Applications

A common use of projection welding is attaching small fasteners, nuts, special blots, studs and similar parts to large components.

(*d*) *Butt Welding*

Butt welding is used to join the pieces end to end. This process is best suited to rods, pipes and many other parts of uniform cross section. Butt welding can be as follows:

(i) Upset welding (ii) Flash welding

(i) Upset Welding

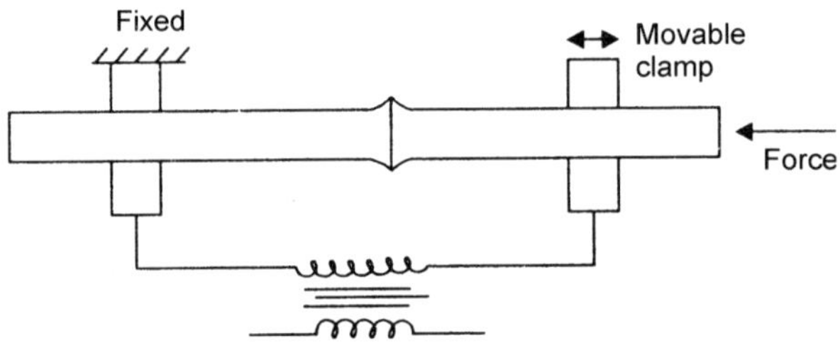

Fig. 3.19 Upset Butt Welding

In upset welding, the parts are clamped and brought in solid contact and current is applied so that the heat is generated through the contact area of the parts as illustrated in Fig. 3.19. At this point, the two parts are pressed together firmly. This action of pressing together is called upsetting. It is used on non-ferrous materials for welding bars, rods, tube formed parts etc.

(ii) Flash Welding

Flash welding is similar to upset welding except that the heat is obtained by means of an arc than the simple resistance heating. The two parts are brought together and the power supply is switched on. As the parts are moved closer, flashing or arcing raises the temperature of the parts to a welding temperature. Now power is switched off and the parts are forced together to form a weld (see Fig. 3.20).

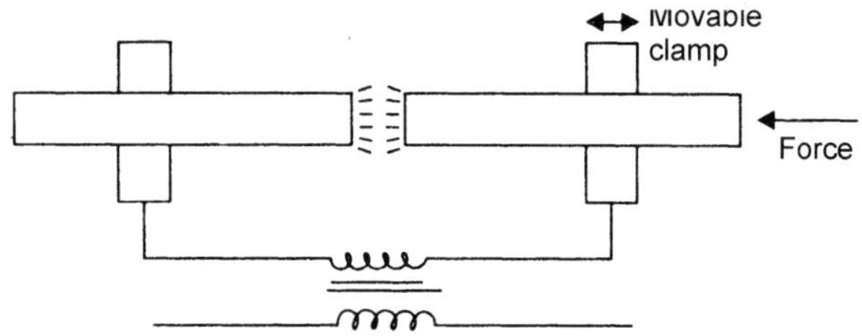

Fig. 3.20 Flash Butt Welding

3.2.4. Solid State Welding

The various solid state welding processes are as follows:

(a) Friction welding
(b) Ultrasonic welding
(c) Explosive welding

(a) Friction Welding

This is solid state welding process in which joining is made by conversion of frictional energy into heat and simultaneous application of axial force.

Figure 3.21 shows the sequence of friction welding process. The two parts are held axially aligned. One part is rotated at a predetermined speed while the other remains stationary.

The non-rotating part is gradually advanced towards the rotating part till contact is made. Axial pressure applied during rotation generates sufficient heat to facilitate fusion. When sufficient heat is produced, the rotation is stopped and the pressure may be increased until the parts are welded. Weld time is between 2 to 30 seconds.

Surface preparation for welding is not necessary in this process. Friction welding is mostly used for butt welding of rods and tubes.

Applications

In aerospace, Automobile industries.

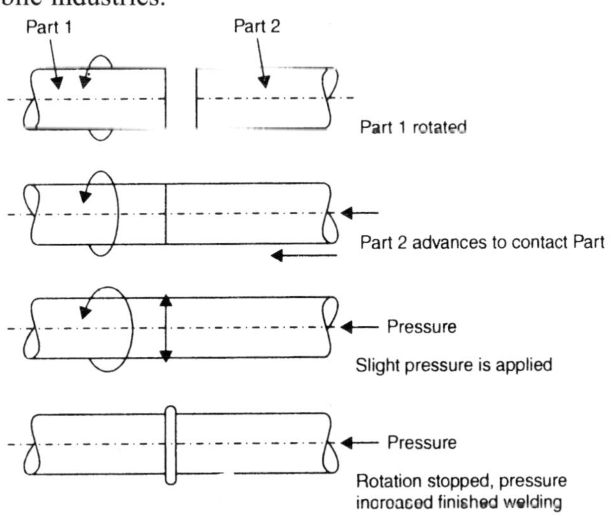

Fig. 3.21 Sequence of Friction Welding

(b) Ultrasonic Welding

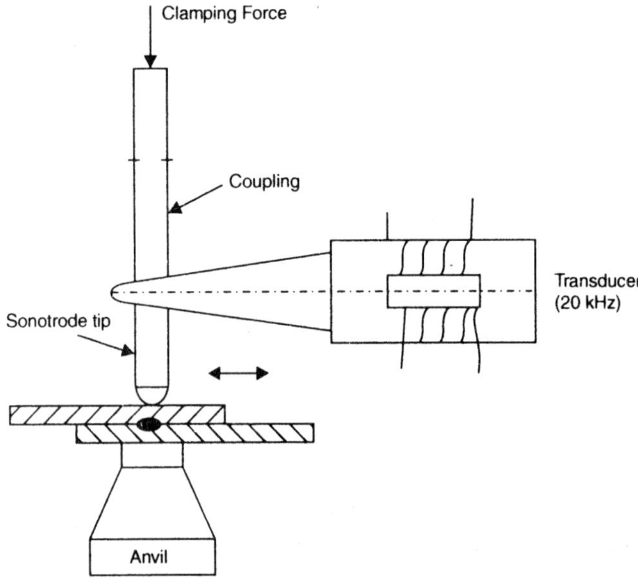

Fig. 3.22 Ultrasonic Welding

In ultrasonic welding, the weld joint is obtained by applying pressure and high frequency vibration motions (20kHz). Pieces to be welded are placed between sonotrode tip and anvil as shown in Fig. 3.22. The combined clamping pressure and vibratory forces introduce dynamic interfacial stresses between the pieces to be joined, then local deformation occurs at the interface. Due to pressure the work piece gets welded.

Advantages

(i) Very thin section can be welded
(ii) Surface preparations are not required
(iii) Dissimilar metals can be welded
(iv) Minimum surface deformation

Limitations

(i) It is not possible to weld heavy gauge metals.
(ii) It is not economical as compared to other processes.
(iii) The work pieces to be welded tend to get welded to the sonotrode or anvil.

(c) Explosive Welding

In this process, the weld joint is made with high relative velocity at a high pressure using high explosives. As the plate moves at high velocity and meets the other plate with massive impact, high stress waves created between the plates, which clears all the oxides and scales present in the interface and make a clean joint. Explosive welding eliminates the problems, associated with fusion welding methods such as the heat affected zone etc. Generally low detonation velocity explosives are used in explosive welding. The detonation velocity depends on the thickness of the plate being welded.

Figure 3.23 illustrates the two common setups used in explosive welding.
It contains four basic components.

1. Target plate
2. Flyer plate
3. Buffer plate
4. Explosive and a detonator

The target plate is fixed in an anvil of large mass. When the explosive is detonated. It thrusts the flyer plate towards the target plate. To protect the flyer plate from surface damage due to impact, a thin layer of rubber or PVC sheet is placed between the flyer plate and the explosive. The explosive may be in sheet form or granular form which is spread uniformly over the buffer plate. Welding is completed in microseconds.

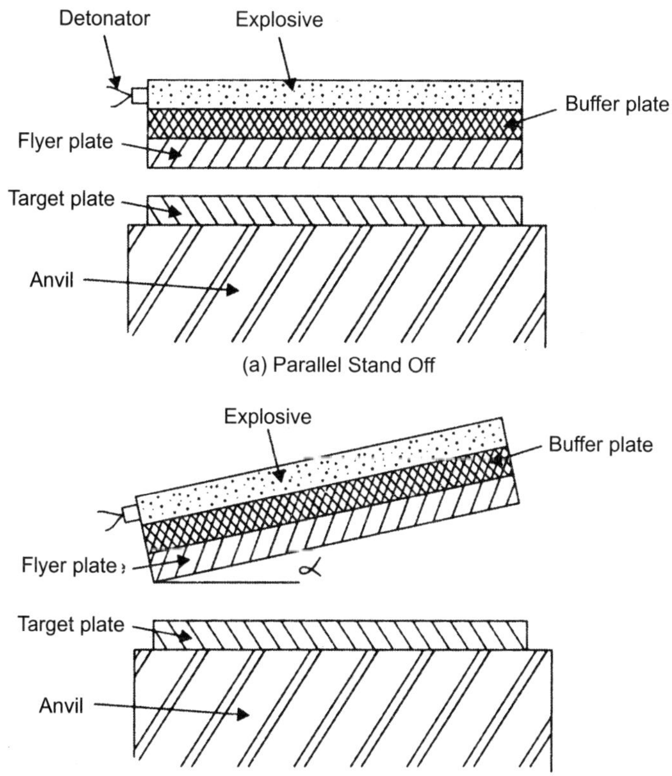

Fig. 3.23 Explosive Welding

Applications

It is used in lap joints. Aluminium and copper can be welded to stainless steel, aluminium to nickle alloys, tungsten to steel, stainless steel to nickel. Cladding of plates is one of the major commercial applications.

3.2.5 Thermo Chemical Welding

The various thermochemical welding process are:

(a) Thermit welding

(b) Atomic hydrogen welding

(a) *Thermit Welding*

This process is basically a fusion welding process in which welding is effected by pouring super heated steel arounds the parts to be welded. In this process, neither arc is produced to the parts nor flame is used. In this an exothermic chemical reaction is utilized for developing high temperature.

A mixture of finely divided aluminium and iron oxide called 'Thermit mixture' is kept in a crucible hanging over the mould. The thermit mixture is ignited using a magnesium ribbon or highly inflammable powder having barium peroxide. The reaction takes about 30 seconds only and heat is liberated which is twice the temperature of melting point of steel. The following reaction takes place as per equation:

$$8Al + 3F_3O_4 \longrightarrow 4Al_2O_3 + 9Fe + heat$$

The resultant is super heated molten iron. The molten iron is made to flow into the mould and fuse with the parts to be jointed.

The Fig. 3.24 shows the method of preparing the mould. The two pieces to be joined are cleaned and a gap is left between them. Then wax is poured on the joint and a wax pattern is formed. Moulding sand is rammed around the wax pattern and pouring, heating and risering gates are cut. A gas flame is used melt the wax pattern and at the same time it preheats the parts to be welded. Then the preheating gate is plugged with sand.

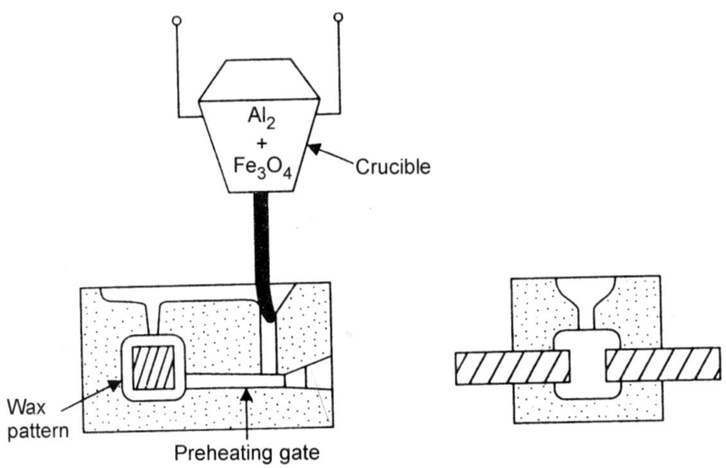

Fig. 3.24 Thermit Welding

Advantages

(i) The welds are sound and free internal residual stresses.

(ii) Broken parts can be welded on the site itself

(iii) The heat necessary for welding is obtained from a chemical reaction and thus no costly power-supply is required.

Limitations

Thermit welding is applicable only to ferrous metal parts of heavy sections.

Applications

It is applicable in the repair of heavy parts such as rail tracks, spokes of driving wheels, broken motor castings, connecting rod etc.

(*b*) *Atomic Hydrogen Welding*

In this process arc is maintained between two electrodes (non-consumable) and hydrogen is introduced into the arc. As hydrogen enters the arc, its molecules are broken up into atoms and again combine into molecules outside the arc. Due to this reaction heat is generated which is sufficient to join the work pieces. The hydrogen also serve as a shielding gas to the molten metal. Filler rod may be added during welding.

3.2.6 Radiant Energy Welding Processes

The various processes are:

(a) Electron beam welding
(b) Laser beam welding
(c) High frequency induction welding

(*a*) *Electron Beam Welding*

In electron beam welding, the heat required for the welding is obtained by bombarding high velocity electron beam on to the work pieces to be joined.

In this process, the electrons emitted from the cathode of electron gun accelerated towards anode and aligned by means of focus lenses and finally strikes the work piece. When the beam strikes the work piece, the kinetic energy of high velocity electrons is converted into heat. This heat is sufficient to melt and fuse the metal. It is carried in vacuum. (see Fig.: 3.25)

Advantages

(i) The heat effect zone is extremely narrow
(ii) No filler material is required
(iii) High penetration
(iv) Welding is not contaminated

Limitations

Costly equipment.

Applications

This welding is used in automobile, aeroplane industries.

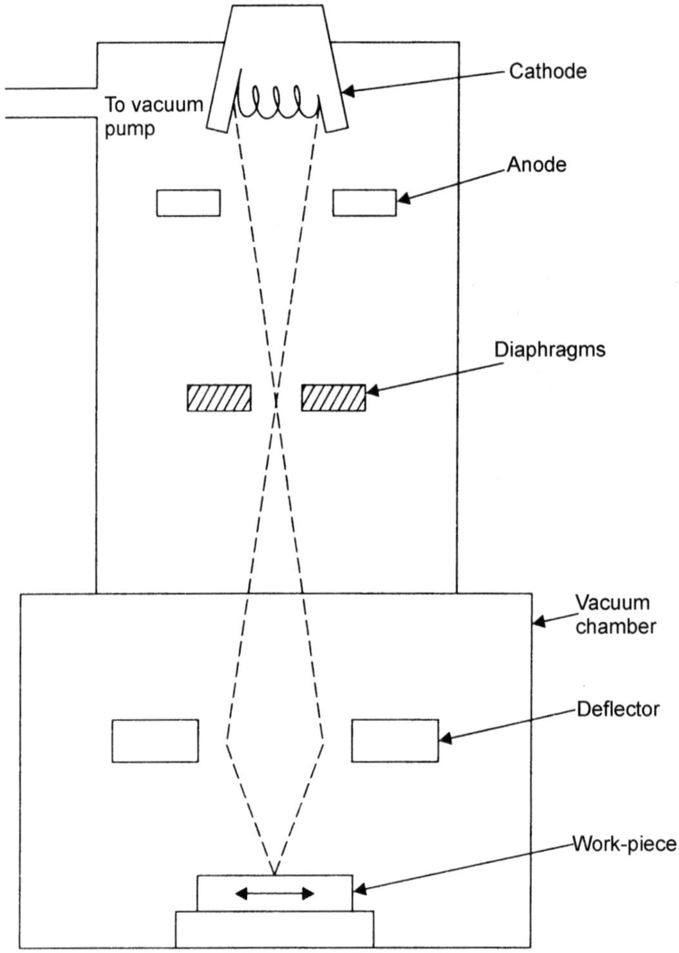

Fig. 3.25 Electron Beam Welding

(b) Laser Beam Welding: (See Fig. 3.26)

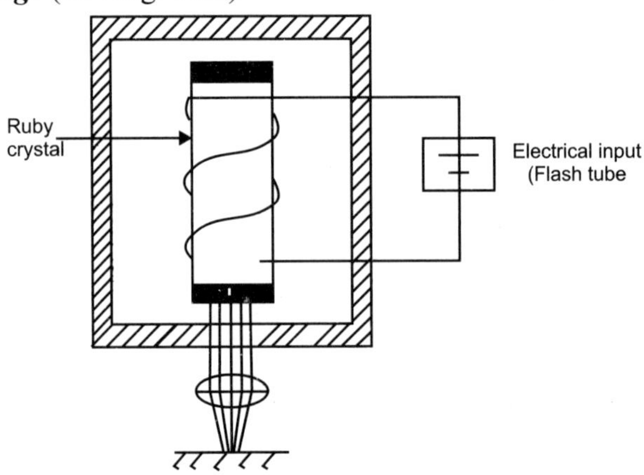

Fig. 3.26 Laser Beam Welding

The term laser stands for light amplified by stimulated emission of radiation. The laser beam is highly directional, strong, monochromatic and coherent.

The ruby crystal is illuminated by the flash tube where the chromium atoms are driven to an exited state. The photons are reflected repeatedly from one mirror to the other mirror at the two ends of the ruby crystal increasing the exitation of chromium atoms still further, to form a narrow beam of red light which leaves crystal through small hole in the mirror at one end of the crystal. By suitable focussing, the control of melting for welding can be done.

Advantages

(i) Heat affected zone is very less

(ii) Deeper penetration

(iii) No vacuum is required as in electron beam welding.

Disadvantages

(i) Low welding speed

(ii) Limited to the thickness of 1.5 mm.

Applications

(i) Dissimilar metals can be welded

(ii) It can be used for cutting as well as welding

(iii) Welds with high precision can be made.

(c) *High Frequency Induction Welding* (*See Fig. 3.27*)

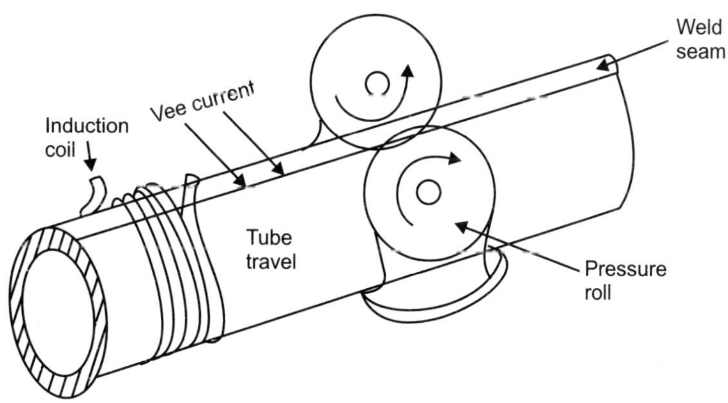

Fig. 3.27 HFIW Process

The high frequency current is introduced into the work piece at the surface to be welded. Due to the skin effect, at such high frequency the current does not flow through entire thickness of the strip, but tends to concentrate at the surface of the work piece. At the area between the squeezing rolls the work piece material is at plastic temperature. The rollers apply pressure to the joint and thus weld is made.

Advantages

(i) It is used to make tubing from coated materials.

(ii) It eliminates surface marking by electrical contacts.

Disadvantage

(i) Not suitable for high conductive materials.

3.3 SOLDERING AND BRAZING

These are used for joining similar or dissimilar metals by a non-ferrous filler metal having a melting point below that of a base metal. The filler is distributed by the capillary action.

3.3.1 Soldering

It is a method of joining two pieces of metal by means of a fusible alloy called solder, applied in the molten stage. The melting point of the filler metal is below 420°C. The solder is usually an alloy of Lead and Tin, Lead and Silver. A suitable flux is used in soldering to prevent oxidation of the joint. Fluxes are available in the form of powder, past or liquid.

Soldering is done in the following ways:

(*i*) *Hand soldering*: The soldering iron is heated by keeping in a furnace or by means of electrically. The joint is heated by soldering iron and solder is applied which melts and flows around the joint by capillary action.

(*ii*) *Dip soldering*: In dip soldering, the parts to be soldered are first cleaned and dipped in flux bath and finally dipped in the molten solder bath and lifted after the soldering is completed.

(*iii*) *Wave soldering*: In this method, parts are not dipped into the solder tank, but a wave is generated in the tank so that the solder comes up and makes necessary joint. This is used in electronic printed circuit board, PCB.

3.3.2 Brazing

It is a process of joining two pieces of metals in which a non-ferrous filler metal or alloy is introduced between the pieces to be joined. The melting point of the filler metal is above 420°C, but lower than the melting temperature of parent metal. The filler metal is distributed between surfaces by capillary action. The copper base alloys and silver base alloys are commonly used as filler metal in brazing. A suitable flux such as borax is used.

Brazing Methods

The selection of brazing method is based on the size and shape of the components to be joined, the base metal and the production rate.

(*a*) *Torch Brazing*: Torch brazing is the most versatile method. It is similar to oxy-acetylene welding. In this process, reducing flame is used to heat the joint area. A flux is applied and as soon as it melts, the filler metal is hand fed to the joint area. When the filler metal melts, it flows into the clearance between the base metal components by capillary action. This method finds applications in fabrication industry and repair work.

(*b*) *Furnace Brazing*: In this method the atmosphere of the furnace is controlled to prevent oxidation by hydorgen, dissociated ammonia, nitrogen or any gas, thus allowing the molten brazing metal to flow smoothly and uniformly around the joint.

(*c*) *Induction Brazing*: In this metals, the metals to be welded are surrounded by metallic coils through which high frequency current is passed. This induces eddy current which produces localized heating. The parts to be brazed are pre-fluxed and the brazing is placed in the joint before switching on the current.

(*d*) *Dip Brazing*: In dip brazing, the parts to be brazed are dipped into a bath of molten filler metal covered by a layer of molten flux. Surface not required to be coated with the brazing alloy must be protected by molasses or by lamp black. This process is used for small parts.

(*e*) *Salt Bath Brazing*: The source of heating in salt bath brazing is a molten bath of fluoride and chloride salts. This salt bath removes thin oxide films from the metals to be joined. The filler metal replaced in the joint area and is also sometimes cladded before dipped in the salt bath.

(*f*) *Resistance Brazing*: It is similar to spot welding. Electrical resistance is used for joining parts. The parts to be joined are placed between the electrodes of the welding machine with the filler metal and flux preloaded at the joint area. Current is then applied until the filler metal melts and flows around the joint.

This method is used in the manufacture of copper transformer leads.

Advantages

(i) It gives a stronger joint than soldering
(ii) Joint is clean
(iii) Any metal can be brazed
(iv) Less distortion and residual stress

Applications

Brazing is used for the assembly of pipe fittings, carbide tips to tool shank, radiators, heat exchangers and the repair of castings.

3.3.3 Braze Welding (or) Bronze Welding

Braze welding, also called Bronze welding due to the use of a bronze filler rod. This process is similar to brazing except that the capillary attraction is not used to distribute the filler metal into the joint. Instead, the filler metal is deposited by gravity since the joint gap is more. Braze welding is extensively used for repairing of ferrous casting and steel machine parts.

3.4 DEFECTS IN WELDING

A weld may have the following defects:

(a) Porosity
(b) Cracks
(c) Lack of fusion
(d) Over penetration
(e) Slag inclusion

(*a*) *Porosity*: Porosity in welding is caused by the presence of gases entraped during the solidification. The gas is released during the welding.

(*b*) *Cracks*: Welding cracks may be hot cracks and cold cracks. Hot cracks occur at elevated temperature just after the molten metal starts to solidify. Cold crack may be due to the formation of martansite the metal very hard as a result of rapid cooling.

(*c*) *Lack of Fusion*: This defect is due to insufficient temperature rise of the base metal, failure to remove oxide films.

(*d*) *Over Penetration*: This defects is due to excessive current in arc welding

(*e*) *Slag Inclusion*: Slag is formed by the reaction with fluxes and floats on the top of the weld pool. Due to the arc force, the slag goes into the weld pool and solidifies inside the fusion zone and forms slag inclusion. This defects occurs in multipass welding since the slag solidifies in the previous pass is not cleaned before depositing the next bead.

3.5 WELDING EQUATIONS

(a) The voltage length characteristic of a D.C.'s

$$V = A + BL$$

where,

V = Voltage drop across the arc

A = Electrode drop

B L = Column drop

(b) Power source characteristic equation is

$$V = OCV - \frac{OCV.I}{ISV}$$

where,

OCV = Open circuit voltage

ISC = Short circuit current

I = Arc current

V = Arc voltage

(c) Power = P = V.I

For maximum power

$$= \frac{\partial P}{\partial I} = 0$$

Example 3.1: The voltage length of arc characteristic of DC arc is given by V = 20 + 40 L, where L = arc length in cm. The power source characteristic can be approximately a straight line. Open circuit voltage is 80 and short circuit current is 1000 amps. Determine the optimum arc length. [*GATE 1991*]

Solution:

For welding arc $V = 20 + 40\ L$ ...(1)

For power source $V = 80 - (80/100)\ I$ ...(2)

For stable arc (1) = (2)

$$20 + 40L = 80 - (80/1000)\ I$$
$$I = 750 - 500\ L$$

Now Power $P = V.I.$

$$= (20 + 40L)\ (750 - 500L)$$
$$= (20 \times 50)\ (15+20L - 20L^2)$$

For optimum arc length $\dfrac{\partial P}{\partial I} = 0$

$$= 20 - 40L = 0$$

$\therefore$ $L = 0.5$ cm.

Arc Length $L = 0.5$ cm.

Example 3.2: The Voltage – arc length characteristics of a power source is V = 20 + 40 L, where V = operating voltage and L = arc length. Determine the open circuit voltage and short circuit current for arc length ranging from 3 to 5 mm and current from 400 to 500 amps during welding operations. [*GATE 1993*]

Solution: Voltage – arc length characteristic

$$V = 20 + 40\ L \quad \text{... (1)}$$

Power source characteristic

$$V = OCV - (OCV/ISC) \quad \text{... (2)}$$

OCV = open circuit voltage

ISC = short circuit current

L_1 = 3mm $\qquad L_2 = 5$ mm

1_1 = 400 amps $\qquad I_2 = 500$ amps

$$V_1 = 20 + 40I_1 = 20 + 40 \times 3 = 140\ V$$
$$V_2 = 20 + 40I_2 = 20 + 40 \times 5 = 220\ V$$
$$V_1 = OCV - (OCV/ISC)\ I_2$$
$$140 = OCV - (OCV/ISC).\ 500 \quad \text{...(3)}$$
$$V_2 - OCV - (OCV/ISC).\ I_1$$
$$220 = OCV - (OCV/ISC).\ 400 \quad \text{...(4)}$$

Eqs.(3)/(4)

$$(140/220) = (1 - 500/ISC)/(1 - 400/ISC)$$
$$140\ (ISC - 400) = 220\ (ISC - 500)$$
$$ISC = 675 \text{ amps}$$

From eqn. (3)

$$140 = OCV(1 - 500/675)$$
$$OCV = 540 \text{ volts}$$

Example 3.3: Two sheets of low carbon steel 1.5 mm thick (t) each are spot welded by passing a current of 10000 amps for 5Hz to 50Hz supply. The maximum indentation is 10% of sheet thickness and density of the spot weld nugget is 8 gm/mm^3. If 1380 joules are required to melt one gram of steel find the per cent of heat actually utilized in making the spot weld. Assume effective resistance is 200 micro ohms and d = 6Öt to determine nugget diameter also assume the nugget size to be equal to metal between the two electrodes. [*GATE 1992*]

Solution: Thickness of sheet t = 1.5 mm

Current I = 10000 amps

Resistance R = 200 μ ohms

Heat developed = $H = I^2 RT$

$= (10000)^2 \times 200 \times 10^{-6} \times 5/50$

$= 2000$ J

Diameter of the nugget $d_n = 6\sqrt{t} = 6\sqrt{1.5} = 7.35$ mm

Height of the nugget h = $2 \times t \times (1 - \text{indentation})$

$= 2 \times 1.5\ (1 - 0.1)$

$= 2.7$mm

Volume of fused metal = $(\pi/4)\ dn^2\ h$

$= (\pi/4)\ (7.5)^2\ (2.7)$

$= 114.56$mm^3

Mass = volume × density

$$= \frac{114.56 \times 8}{10000} = 0.916 \text{ gm}$$

Heat required for 1 gm melt is 1380 J

∴ for 0.916 = 0.916×1380

$= 1264.73$ J

Percentage of heat utilization for making nugget

$= (1264.73 \times 100)/2000$

$= 63.24\%$

3.6 HEAT AFFECTED ZONE (HAZ): (See Fig. 3.28)

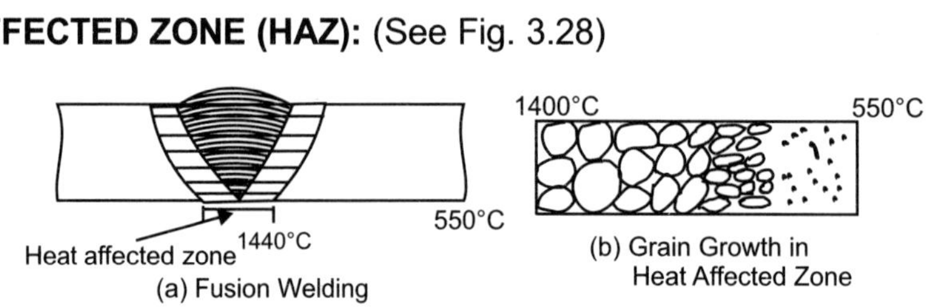

Fig. 3.28 Fusion Welding Showing Heat Afftected Zone

Heat Affected Zone (HAZ) is the zone where the base metal is affected metallurgically due to the heat of welding. It is the region closed to the weld, where large thermal fluctuations are encountered due to the fusion welding. This leads to changes in mechanical properties and structure. Heat Affected Zone contains three regions

(a) The grain growth zone (1150°C)

(b) Grain refined zone (1150°C to 950°C)

(c) The transition zone (950°C to 750°C)

(*a*) ***The grain growth zone*:** It is immediately adjacent to the fusion zone. In this zone, parent metal has been heated to a temperature above upper critical temperature. This resulted in grain growth.

(*b*) ***The grain refined zone*:** Adjacent to the grain growth zone is the grain refined zone. In this zone, parent metal has been heated just above the transition temperature where grain refinement is completed.

(*c*) ***The transition zone*:** In this zone, base metal temperature is below the transition temperature.

3.6.1 Economics of Welding

Accuracy of cost estimates for welding is essential for comparison as follows:

(a) To compare the economics of welding process with other process of fabrication or manufacturing like casting, machining etc.

(b) To determine the selling price of a product for a quotation so as to get reasonable profit to the company.

(c) To check the vendors quotations.

(d) To decide to make a part in the plant or purchased from outside.

3.6.2 Elements of Welding Cost

For estimating the welding cost, the following cost elements should be considered.

(a) Preparation cost

(b) Actual welding cost

(a) Preparation cost

It includes the cost of edge preparation before weld, positioning the parts in fixture for welding etc.

(b) Actual cost*: *This includes

(i) Material cost

(ii) Labour cost

(iii) Welding finishing cost

(iv) On-cost

(b) (i) Material cost

(a) *Oxyacetylene welding*: In this welding, cost of welding rods, flux and gases O_2 and C_2H_2 are calculated. Cost of welding rods = welding rods consumption × welding rod price

Cost of flux = Amount of flux used × flux price

Cost of fuel gases = Amount of gases used litres × Gas price

(b) *Arc welding*: In this welding, cost of welding rods, shield gate, power cost are calculated.

Cost of Electrodes = Electrode consumption × Electrode price

Shield gas or flux = Flux or shield gas used × price of gas or flux

Power cost = Total kwh used × Power charges.

(b) (ii) Labour cost

It will be obtained from wage sheets. This can also be calculated as the function of time. This is represented by operator factor or operator duty cycle and can be expressed as

$$(\text{K}) = \text{Operator factor} = \frac{\text{Actual welding time}}{\text{Total time}} \times 100$$

$$= \left(\frac{\text{Actual welding time}}{\text{Actual welding time} + \text{Preparation time}}\right) \times 100$$

(b) (iii) Welding Finishing Cost

This includes the expenditure made for finishing the weld joint after welding like grinding away the unwanted bead, cleaning the bead by brushing finally heat treatment cost etc.

(b) (iv) On-cost

All overhead costs such as equipment depreciation, taxes, technical supervision and other related costs.

3.6.3 Standard Time

The standard time is the sum of setup time(t_s), the base time (t_b), the auxiliary time (t_a), additional time (t_{ad}) and closing time (t_c)

$$T = t_s + t_b + t_a + t_{ad} + t_c$$

Setup time (t_s) = It refers to the time spent by the welder in getting the work order, reading specifications and instruction card and setting up equipment and fixture.

Base time (t_b) : It is the time during which the arc is burning.

Auxiliary time (t_a) : The time spent by welder to change the electrodes, clean and inspect the jointed welds.

Additional time (t_{ad}): The time spent to service the work place and for personal needs

Closing time (t_c) : The time spent to hand over the finished job.

(*a*) *Arc Welding*: In arc welding standard time (T) is quotient of the base time (t_b), by the operator factor (k) which is expresed as

$$T = \frac{t_b}{K}$$

where, t_b $= \frac{d.A.L}{\alpha.I}$ hours

where, d = The material density g/cm^3

A = cross-sectional area of weld cm^2

L = weld length cm

α = deposition ration gm/amp-hr.

I = welding current amp.

(*b*) *Gas Welding*: In oxy-acetylene welding the standard time is given by,

$$T = \frac{t_b}{K}$$

$$t_b = \frac{GL}{\alpha}$$

where, G = Mass of weld metal deposited per metre of weld length gm/m

L = Weld length m

(*c*) *Gas Cutting*:

$$T = \frac{Lt_b}{K}$$

where, L = Kerf length m

t_b = base time

When cutting low carbon steel, t_b may be taken as 2.5 min/m for 10 mm thick plate and 5 min/m for 60 mm thick plate.

Example 3.4: Calculate the standard time for shield metal Arc Welding of steel using 3 mm electrode with a welding current of 120 amps and deposition ratio 10 gm/A.H. The cross sectional area of the weld is 0.5 cm^2 and its length is 1m long. Density of steel as 7.85 g/cm^2 and operator factor is 0.25.

$$T = \frac{t_b}{K}$$

$$t_b = \frac{dAL}{\alpha I}$$

$$T = \frac{dAL}{\alpha IK}$$

$$= \frac{7.85 \times 0.5(1 \times 100)}{10 \times 120 \times 0.25} = 1.31\text{hours} = 78.5 \text{ minutes}$$

Example 3.5: Calculate the standard time for oxy-acetylene welding cutting (butt) for 6 mm thick plate, if mass of deposited metal is 35 gm/min, length of weld is 8m. Weld operation is down hand, vertical and overhead positions. Take operator factor as 0.25.

$$T = \frac{GL}{\alpha K}$$

$$= \frac{35 \times 8}{10 \times 0.25} = 112 \text{ minutes.}$$

Example 3.6: Calculate standard time for cutting strips 13 m long from plates of 10 mm and 60 mm thick, using manual oxy-acetylene cutting torch. Take operator factor as 0.3.

(JNTU-2005)

(a) For 10 mm thick plates

$$T = \frac{t_b \cdot L}{K}$$

for 10 mm thick plates $t_b = 2.5$ min/mm

$$T = \frac{2.5 \times 13}{0.3} = 108.3 \text{ minutes}$$

(b) For 60 mm thick

$$T = \frac{t_b L}{K}$$

for 60 mm thick plates, $t_b = 5$

$$T = \frac{5 \times 13}{0.3} = 216.6 \text{ minutes}$$

QUESTIONS

1. Discuss oxy-acetylene welding equipment with a neat sketch.
2. Explain the types of gas welding flames and under what conditions they are used.
3. Describe gas welding positions.
4. What are the advantages of gas welding.
5. Explain oxy-acetylene cutting process.
6. What are the function of coated electrodes.
7. Sketch and explain submerged arc welding.
8. Bring out the difference between TIG and MIG welding.
9. Draw a neat sketch and explain electro slag welding process.
10. Briefly explain the various resistance welding processes.
11. Explain the following:
    (a) Friction welding (b) Ultrasonic welding (c) Explosive welding.

12. With a neat sketch explain thermit welding process.
13. Explain the following:
    (a) Electron beam welding (b) Laser beam welding.
14. Differentiate between
    (a) Soldering and Brazing
    (b) Brazing and Braze Welding.
15. Explain Butt welding and Seam welding.
16. Describe the following metal joining techniques

(a) Dip Soldering

(b) Wave Soldering

(c) Furnace Blazing

(d) Induction Blazing

17. Write short notes on the following
    (a) Welding rods (b) Types of power supply for Arc Welding
18. Explain the term HAZ in welding

4

Powder Metallurgy

4.1 INTRODUCTION

Powder metallurgy is a process of making components from metallic powders. Initially, it was used to replace castings for metals which were difficult to melt because of high melting point. The development of technique made it possible to produce a product economically, and today it occupies an important place in the field of metal process. The number of material products made by powder metallurgy are increasing and include tungsten filaments of lamps, contact points. Self lubricating bearings and cemented carbides for cutting tools.

4.2 CHARACTERISTICS OF METAL POWDER

The performance of metal powders during processing and the properties of powder metallurgy are dependent upon the characteristics of the metal powders that are used. Following are the important characteristics of metal powders.

(a) Particle shape (b) Particle size
(c) Particle size distribution (d) Flow rate
(e) Compressibility (f) Apparent density
(g) Purity

(*a*) *Particle Shape*: The particle shape depends largely on the method of powder manufacture. The shape may be special nodular, irregular, angular, and dendritic. The particle shape influences the flow characteristics of powders. Special particles have excellent sintering properties. However, irregular shaped particles are good at green strength because they will interlook on computing.

(*b*) *Particle Size*: The particle size influences the control of porosity, compressibility and amount of shrinkage. It is determined by passing the powder through standard sieves or by microscopic measurement.

(*c*) *Particle Size Distribution*: It is specified in term of a sieve analysis, the amount of powder passing through 100, 200 etc., mess sieves. Particle size distribution influences the packing of powder and its behaviour during moulding and sintering.

(*d*) *Flow Rate*: It is the ability of powder to flow readily and confirm to the mould cavity. It determines the rate of production and economy.

(*e*) *Compressibility*: It is defined as volume of initial powder (powder loosely filled in cavity) to the volume of compact part. It depends on particle size, distribution and shape.

(*f*) *Apparent Density*: It depends on particle size and is defined as the ratio of volume to weight of loosely filled mixture.

(*g*) *Purity*: Metal powders should be free from impurities as the impurities reduces the life of dies and effect sintering process. The oxides and the gaseous impurities can be removed from the part during sintering by use of reducing atmosphere.

4.3 BASIC STEPS OF THE PROCESS

The manufacturing of parts by powder metallurgy process involves the following steps:

(a) Manufacturing of metal powders
(b) Blending and mixing of powders
(c) Compacting
(d) Sintering
(e) Finishing operations

(*a*) *Manufacturing of Metal Powders*

There are various methods available for the production of powders, depending upon the type and nature of metal. Some of the important processes are:

1. Atomization
2. Machining
3. Crushing and Milling
4. Reduction
5. Electrolytic Deposition
6. Shotting
7. Condensation

1. *Automization*: In this method as shown in Fig. 4.1 (a), molten metal is forced through a small orifice and is disintegrated by a powerful jet of compressed air, inert gas or water jet. These small particles are then allowed to solidify. These are generally spherical in shape. Automation is used mostly for low melting point metals/alloy such as brass, bronze, zinc, tin, lead and aluminium powders.

2. *Machining*: In this method first chips are produced by filing, turning etc. and subsequently pulverised by crushing and milling. The powders produced by this method are coarse in size and irregular in shape. Hence, this method is used for special cases such as production of magnesium powder.

3. *Crushing and Milling*: These methods are used for brittle materials. Jaw crushers, stamping mills, ball mills are used to breakdown the metals by crushing and impact. See Fig. 4.1 (b) and (c).

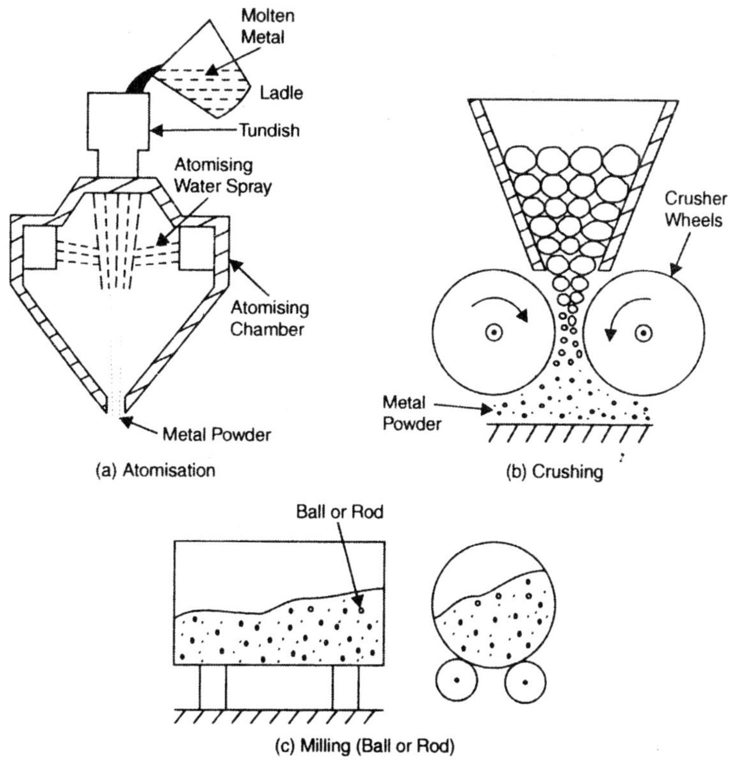

Fig. 4.1 Methods of Producing Metal Powders

In earlier stages of powder preparation gyratory crushers (Fig 4.1(b)) are used to crush brittle metals. For fine powder, the metal particles are fractured by impact. A ball mill (Fig. 4.1 (c)) is a horizontal barret shaped container holding a quantity of balls which arc free to tumble about as the container rotates, crushes and abrade the powder particles that are introduced into the container.

4. *Reduction*: Pure metal is obtained by reducing its oxide with a suitable reducing gas at an elevated temperature (below the melting point) in a controlled furnace. The reduced product is then crushed and milled to a powder.

Sponge iron powder is produced this way

$$Fe_3O_4 + 4C = 3Fe + 4CO$$

$$Fe_3O_4 + 4CO = 3Fe + 4CO_2$$

Copper powder by

$$Cu_2O + H_2 = 2Cu + H_2O$$

Tungsten, Molybdenum, Ni and Cobalt are made by the method.

5. *Electrolytic Deposition*: This method is commonly used for producing iron and copper powders. This process is similar to electroplating. For making copper powder, copper plates are placed as anodes in the tank of electrolyte, where as the aluminium plates are placed into electrolyte to act as anode. When D. C. current is passed through the electrolyte, the copper gets

deposited on cathode. The cathode plates are taken out from electrolyte tank and the deposited powder is scrapped off. The powder is washed, dried and pulverised to produce powder of the desired grain size. The powder is further subjected to heat treatment to remove work hardness effect. The cost of manufacturing is high.

6. *Shotting*: In this method, the molten metal is poured through a siever or orifice and is cooled by droping into water. This produces spherical particles of large size. This method is commonly used for metals of law melting points.

7. *Condensation*: In this method, metals are boiled to produce metal vapours and then condensed to obtain metal powders, This process is applied to volatile metals such as zinc, magnesium and cadmium.

(*b*) *Blending and Mixing of Powders*

Powder blending and mixing of the powders are essential for uniformity of the product. Lubricants are added to the blending of powders before mixing. The function of lubricant is to minimise the wear, to reduce friction. Different powder in correct proportions are thoroughly mixed either wet or in a ball mill.

(*c*) *Compacting*

The main purpose of compacting is converting loose powder into a green compact of accurate shape and size. The following methods are adopted for compacting:

1. Pressing
2. Centrifugal compacting
3. Slip casting
4. Extrusion
5. Gravity sintering
6. Rolling
7. Isostatic moulding
8. Explosive moulding

1. *Pressing*: The metal powders are placed in a die cavity and compressed to form a component shaped to the contour of the die as illustrated in Fig. 4.2. The pressure used for producing green compact of the component vary from 80 Mpa to 1400 Mpa, depending upon the material and the characteristics of the powder used. Mechanical presses are used for compacting objects at low pressure. Hydraulic presses are for compacting objects at high pressure. (See Fig. 4.2)

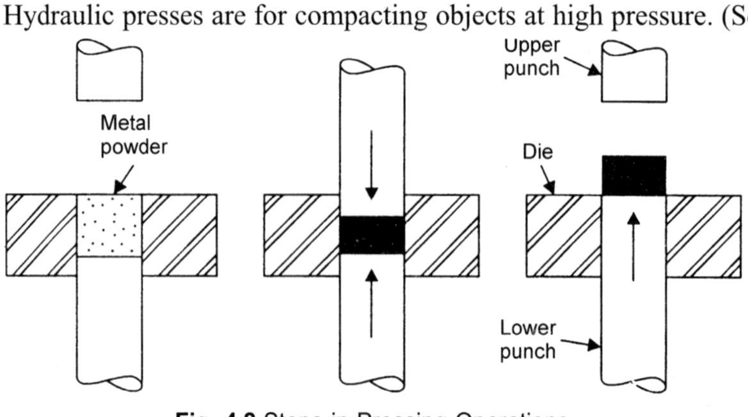

Fig. 4.2 Steps in Pressing Operations

2. *Centrifugal Compacting*: In this method, the moulder after it is filled with powder is centrifugal to get a compact of high and uniform density at a pressure of 3 Mpa. This method is employed for heavy metals such as tungsten carbide.

3. *Slip Casting*: In this method, the powder is converted into slurry with water and poured into the mould made of plaster of paris. The liquid in the slurry is gradually absorbed by the mould leaving the solid compact within the mould. The mould may be vibrated to increase the density of the compact.

This technique is used for materials that are relatively incompressible by conventional die compaction. The main drawback of this process is relatively slower process because it takes larger time for the fluid to be absorbed by the method.

4. *Extrusion*: This method is employed to produce the components with high density. Both cold and hot extrusion processes are for compacting specific materials. In cold extrusion, the metal powder is mixed with binder and this mixture is compressed into billet. The binder is removed before or during sintering. The billet is charged into a container and then forced through the die by means of ram. The cross-section of product depends on the opening of the die. Cold extrusion process is used for cemented carbide drills and cutters of ram. The cross-section of products depends on the opening of the die. Cold extrusion process is used for cemented carbide drills and cutters.

In the hot extrusion, the powder is compacted into billet and is heated to extruding temperature in non-oxidising atmosphere. The billet is placed in the container and extruded through a die. This method is used for refractive berium and nuclear solid materials.

5. *Gravity Sintering*: This process is used for making sheets for controlled porosity. In this process. the powder is poured on ceremic tray to form an uniform layer and is then sintered up to 48 hours in ammonia gas at high temperature. The sheets are then rolled to desired thickness. Porous sheet of stainless steel are made by this process and popularly used for fitters.

6. *Rolling*: This method is used for making continuous strips and rods having controlled porosity with uniform mechanical properties. In this method, the metal powder is fed between two rolls which compress and interlock the powder particles to form a sheet of sufficient strength as shown in Fig. 4.3. It then situated, rerolled and heat treated if necessary. The metals that can be rolled are Cu, Brass, Bronze, Ni, Stainless steel and Monel.

7. *Iscstatic Moulding*: In this method, metal powder is placed in an elastic mould which is subjected to gas pressure in the range of 65-650 Mpa from all sides. After pressing. the compact is removed from gas chamber. If the fluid is used as press medium then it is called as hydrostatic pressing. The advantages of this method are: uniform strength in all directions, higher green compact strength and low equipment cost. This method is used for tungsten, molybdenum, niobium etc.

8. *Explosive Compacting*: In this method, the pressure generated by an explosive is used to compact the metal powder. Metal powder is placed in water proof bags which are immersed in water container cylinder of high wall thickness. Due to sudden deterioration of the charge at the

end of the cylinder, the pressure of the cylinder increase. This pressure is used to press the metal powder to form green compact.

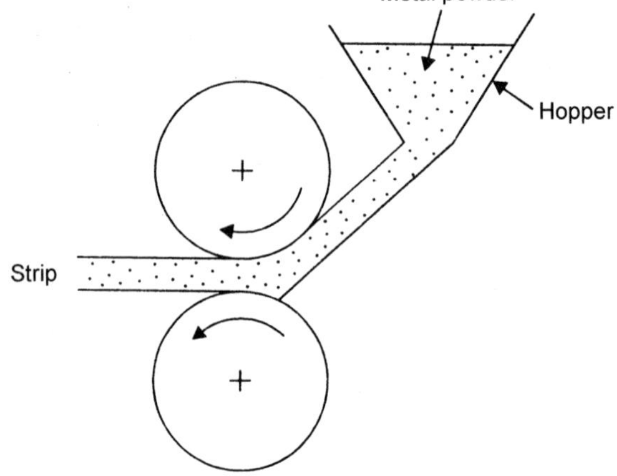

Fig. 4.3 Rolling

(*d*) *Sintering*

Sintering involves heating of the green compact at high temperatures in a controlled atmosphere [reducing atmosphere which protects oxidation of metal powders]. Sintering increases the bond between the particles and therefore strengthens the powder metal compact. Sintering temperature and time is usually 0.6 to 0.8 times the melting point of the powder. In case of mixed powders of different melting temperature, the sintering temperature will usually be above the melting point of one of the minor constituent [Ex : cobalt and cemented carbides] and other powders remain in soild state. The important factors governing sintering are temperature, time and atmosphere.

The sintering temperature and time of sintering for different metal powders are given below:

Type of powders	*Sintering temperature °C*	*Sintering Time*
Al & its alloys	370-520	24 hrs.
Cu, Brass & Bronze	700-900	30 min.
Iron	1025-1200	30 min.
Stainless Steel	1180	20-40 min
Tungsten Carbide	1480	20-40 min

Hot pressing

Hot pressing involves applying pressure and temperature simultaneously, so that the compacting and sintering of the powder takes place at the same time in a die. Its application is limited and can be used for compacting. Fe and Brass powders at much lower pressure than conventional pressing and sintering operations.

(*e*) *Finishing Operations*

These are secondary operations intended to provide dimensional tolerances, physical and better surface finish. They are:

1. Sizing
2. Coining
3. Machining
4. Impregnation
5. Infiltration
6. Heat treatment
7. Plating

1. *Sizing*: It is repressing the sintered component in the die to achieve the required accuracy.

2. *Coining*: It is repressing the sintered components in the die to increase density and to give additional strength.

3. *Machining*: Machining operation is carried out on sintered part to provide under cuts, holes, threads etc. which can not be removed on the part in the powder metallurgy process.

4. *Impregnation*: It is filling of oil, grease or other lubricants in a sintered component such as bearing.

5. *Infiltration*: It is filling of pores of sintered product with molten metal to improve physical properties.

6. *Heat Treatment*: The process of heating and cooling sintered parts are to improve

(i) Wear Resistance
(ii) Grain Structure
(iii) Strength

The following heat treatment process are used to the parts made by powder metallurgy:

1. Stress relieving
2. Carburising
3. Nitriding
4. Induction Hardening

7. *Plating*: Plating is carried out in order to:

1. Import a pleasing appearance (Cr plating)
2. Protect from corrosion (Ni plating)
3. Improve electrical conductivity (Cu and Ag plating)

4.4 DESIGN CONSIDERATIONS FOR POWDER METALLURGY PARTS

In designing of powder metallurgy parts, the following are the some of tooling and pressing considerations.

1. Side holes and side ways are very difficult to achieve during pressing and must be made by secondary machining operations.
2. Threads, kurling and other similar shapes should not be formed compacting. They should be produced by machining.
3. Abrupt changes in section thickness and narrow and deeper sections should be avoided as far as practicable.

4. It is recommended that sharp corners be avoided wherever possible. Fillets with generous radii are desirable.
5. Chambers can be made.
6. Under cuts that are perpendicular to the pressing direction can not be made, since they prevent the part ejection.

4.5 ADVANTAGES OF POWDER METALLURGY

1. Although the cost of making powder is high there is no loss of material. The components produced are clean, bright and ready for use.
2. The greatest advantage of this process is the control of the composition of the product.
3. Components can be produced with good surface finish and close tolerance.
4. High production rates.
5. Complex shapes can be produced.
6. Wide range of properties such as density, porosity and particle size can be obtained for particular applications.
7. There is usually no need for subsequent machining or finishing operations.
8. This process facilitates mixing of both metallic and non-metallic powders to give products of special characteristics.
9. Porous parts can be produced that could not be made any other way.
10. Impossible parts (cutting tool bits) can be produced.
11. Highly qualified or skilled labour is not required.

4.6 LIMITATION OF POWDER METALLURGY

1. The metal powders and the equipment used are very costly.
2. Storing of powders offer great difficulties because of possibility of fire and explosion hazards.
3. Parts manufactured by this process have poor ductility.
4. Sintering of low melting point powders like lead, zinc, tin etc., offer serious difficulties.

4.7 APPLICATIONS OF POWDER METALLURGY

Powder metallurgy techniques are used for making large number of components. Some of the application are as follows:

1. *Self-Lubricating Bearing and Filters*: Porous bronze bearings are made by mixing copper and tin powder in correct proportions, cold pressed to the desired shape and then sintered. These bearings soak up considerable quantity of oil. Hence during service, these bearings produce a constant supply of lubricant to the surface due to capillary action. These are used where lubricating is not possible. Porous filters can be manufactured and are used to remove, undesirable materials from liquids and gases.

2. *Friction Materials*: These are made by powder metallurgy. Clutch liners and Brake bands are the example of friction materials.

3. *Gears and Pump Rotors*: Gears and pump rotor for automobile oil pumps are manufactured by powder metallurgy. Iron powder is mixed with graphite, compacted under a pressure of 40 kg/cm and sintered in an electric furnace with an atmosphere and hydrocarbon gas. These are impregnated with oil.

4. *Refractor Materials*: Metals with high melting points are termed as refractory metals. These basically include four metals tungsten, molybdenum, tantalum and niobium. Refractory metals as well as their alloys are manufactured by powder metallurgy. The application are not limited to lamp filaments and heating elements, they also include space technology and the heavy metal used in radioactive shielding.

5. *Electrical Contacts and Electrodes*: Electrical contacts and resistance welding electrodes are made by powder metallurgy. A combination of copper, silver and a refractory metal like tungsten. molybdenum and nickle provides the required characteristics like wear resistant, refractory and electrical conductivity.

6. *Magnet Materials*: Soft and permanent magnets are manufactured by this process. Soft magnets are made of iron, iron-silicon and iron-nickle alloys. These are used in D.C. motors, or generators as armatures and in measuring instruments. Permanent magnets known as Alnico which is a mixture of nickle, aluminium, cobalt, copper and iron are manufactured by this technique.

7. *Cemented Carbides*: These are very important products of powder metallurgy and find wide applications as cutting tools, wire drawing dies and deep drawing dies. These consist of carbides of tungsten, tantalum, titanium and molybdenum. The actual proportions of various carbides depend upon its applications, either cobalt or nickle is used as the bonding agent while sintering.

Diamond Impregnated Tools

These are made from a mixture of iron powder and diamond dust. Diamond dust acts as a cutting medium and iron powder acts as the bond. These tools are used for cutting porcelain and glass. These bits are welded or brazed to a steel shank.

QUESTIONS

1. Discuss the characteristics of metal powders.
2. Describe various methods of producing metal powders.
3. Write shortly on (a) Centrifugal compacting (b) Extrusion (c) Slip casting.
4. Discuss advantages and limitations of powder metallurgy.
5. What are the various finishing operations used in powder metallurgy.
6. Discuss various applications of powder metallurgy.

5

Plastics

5.1 INTRODUCTION

The word plastics is from the Greek word plastikos, means is moulded and shaped. Plastics can be easily machined, formed and joined into required shapes. Hence, plastics find place in engineering materials and domestic applications. Plastics are available in rods, sheets, films and tubes.

5.2 TYPES OF PLASTICS

Plastics are classified as

(a) Thermosetting plastics

(b) Thermoplastics

(*a*)*Thermosetting plastics*: These are formed to shape with heat, with or without pressure, resulting in a product that is permanently hard. The heat first soften the material, but as additional heat or special chemicals are added, the plastic is hardened by chemical change known as "polymerization" and cannot be resoftened. Thermosetting plastics are Phenol-formaldehyde, Urea-formaldehyde, Epoxy resins etc. Products made by thermosetting plastics are T.V. cabinets, telephone receivers, Camera bodies and Automobile parts.

(*b*)*Thermoplastics*: Thermoplastics undergo no chemical change in moulding. They remain soft at elevated temperatures until they are hardened by cooling. These plastics can be reused or recycled by melting and remoulding. Most commonly used thermoplastics are polystyrene, polytene, P.V.C.(Polyvinyl Chloride) Nylon, Teflon etc. Products made by thermoplastics are photographic films, insulating tapes, hose pipes etc.

5.3 COMPARISON BETWEEN THERMOSETTING PLASTICS AND THERMOPLASTICS

Thermosetting Plastics

1. Once hardened and set, they do not soften with the application of heat
2. These are more stronger and harder
3. Objects made by these plastics can be used at comparatively high temperatures
4. These are supplied in monomeric or partially polymerized form in which these are either liquids or semi solids.
5. T.V. cabinets, Automobile parts are made by these plastics.

Thermoplastics

1. They can be repeatedly softened by heat and hardened by cooling
2. They are comparatively softer and less stronger
3. Objects made by thermoplastics cannot be used at higher temperatures as these tend to soften under heat
4. These are usually supplied as granular material
5. Insulating tapes, photographic films are made by these plastics.

5.4 ADVANTAGES OF PLASTICS

1. Light is weight compared to metals
2. Excellent surface finish
3. Close dimensional tolerances
4. Moisture and corrosion resistance
5. Easy to shape and mould

5.5 DISADVANTAGES

1. Low strength
2. Low heat resistance
3. Deteriorate in sunlight

5.6 APPLICATIONS OF PLASTICS

Plastics find applications in manufacturing of:

1. Photofilms in film industry
2. Insulating tapes
3. Electrical parts like plugs, switches etc.
4. Radio, T.V. cabinets
5. Furniture like chairs, tubs
6. Telephone receivers
7. Camera bodies

8. Gears and Bearings
9. Toys, bottles, buckets etc.
10. Hose pipes
11. Automobile parts

5.7 METHODS OF PROCESSING

(a) Moulding of thermoplastics

1. Injection Moulding
2. Blow moulding
3. Extrusion
4. Thermoforming (Vacuum forming)
5. Calendering

(b) Moulding of thermosetting plastics

1. Compression moulding
2. Transfer moulding

***a.1 Injection Moulding*:** (See Fig 5.1)

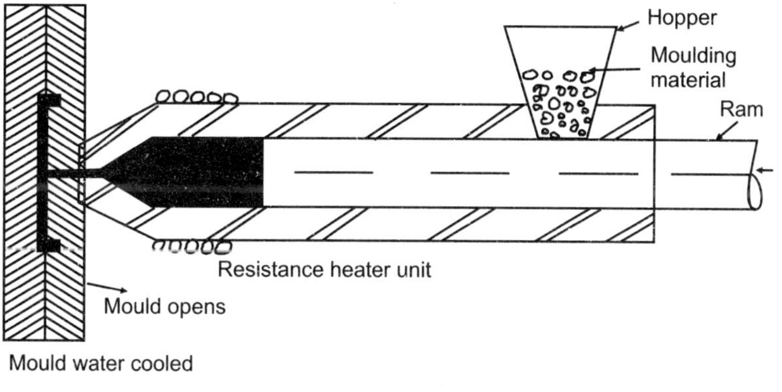

Fig. 5.1 Injection Moulding

Injection moulding machines are somewhat similar to those used for die casting. In this method, the moulding material in the form of granules or pellets is fed through the hopper into the cold end of the injection cylinder. Then the injection ram forces the powder into the heating section of the cylinder where its temperature is raised to 300°C. Then the ram is moved forward by applying hydraulic pressure to inject the soften material through die into water cooled mould. After the mould is filled, it is allowed to cool and harden. Then the ram is retracted, the mould is opened and the product is ejected.

Injection mould products are cups, containers, electrical and communication components.

a.2 Blow Moulding (See Fig. 5.2)

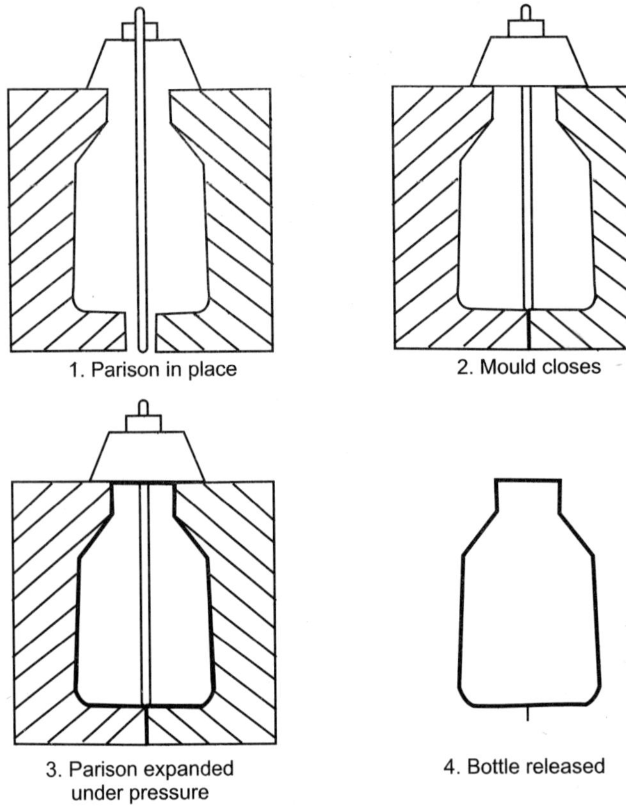

Fig. 5.2 Blow Moulding Process

The blow moulding commences with the extrusion of (heated) tubular piece of plastic known as parison which is transferred to the two-piece mould. The parison is gripped in the two-piece mould and its bottom end is sealed. Compressed air is blown into the parison to force the plastic against the walls of water cooled mould. Air pressure ranges from 0.7 to 10 kg/cm^2. The mould is allowed to cool and then opened to remove the article.

Blow moulding is used for making plastic bottles, toys, doll bodies and many other items.

a.3 Extrusion Moulding

Polymer sheets and films can be produced using a flat extrusion die. These are advantages of extrusion process. The tooling cost is low compared to injection moulding. Material thickness can be controlled. In addition production rates are high and intricate profiles can be produced.

a.4 Thermoforming (See Fig. 5.3)

It consists of heating a thermoplastic sheet until it softens and then forcing it to confirm to some mould either by vacuum or air pressure.

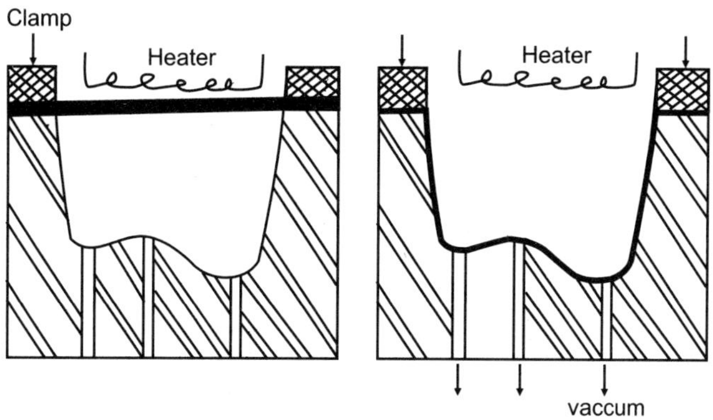

(a) Heating of thermoplastic sheet (b) Vacuum is applied to the mould

Fig. 5.3 Thermoforming Process

The products made by thermoforming are jelly containers used in restaurants, refrigerator inner panels, packing containers etc.

a.5 Calendering (See Fig. 5.4)

This is an intermediate process in which extruded thermoplastic sections are reduced to sheets of films.

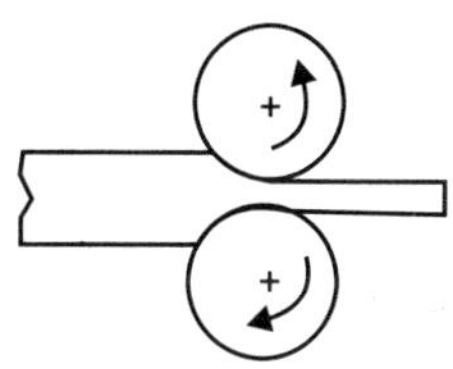

Fig. 5.4 Calendering of Thermoplastic

b.1 Compression Moulding (See Fig. 5.5)

It is usually used for thermosetting plastics.

In compression moulding thermosetting material is placed in heated mould (female die). The upper part of thc mould is brought down to compress the material into the required shape and density. When the mould is closed, the material undergoes a chemical change or polymerization, that harden it.

The moulding temperature ranges from 150–180°C. The moulding pressure ranges from 150–500 kg/cm^2. The time required to harden the product ranges from 1 to 1.5 minutes. It also depends on the thickness of the product.

The products made by this process include dishes, container caps etc.

b.2 Transfer Moulding (See Fig. 5.6)

Transfer moulding is a variation of compression moulding in which heat and pressure is applied to the moulding materials outside the mould until they become fluid. The fluid material is then forced through a series of channels from external chamber to the mould cavity where final cure takes place.

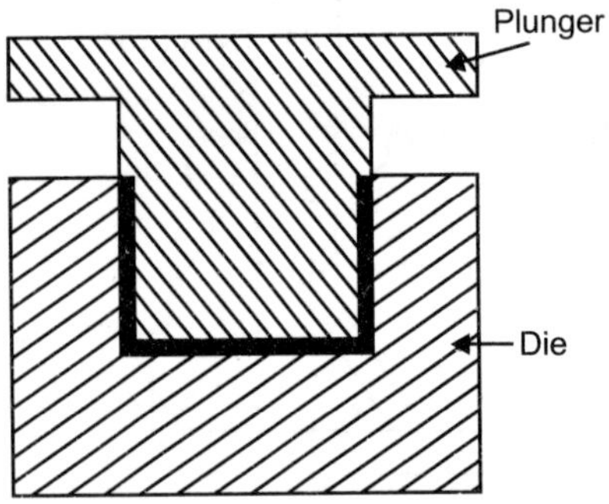

Fig. 5.5 Compression Moulding

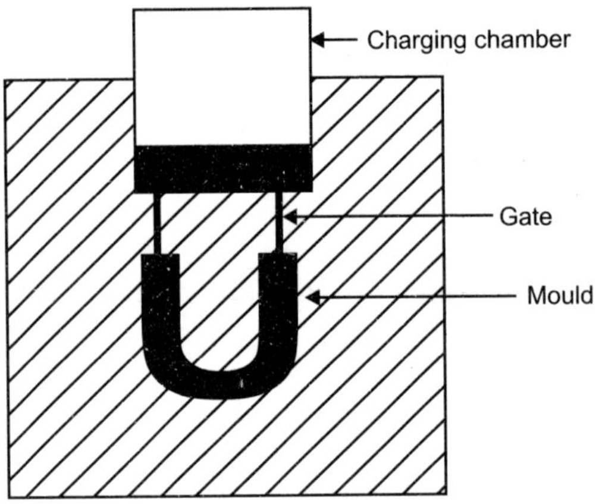

Fig. 5.6 Transfer Moulding

5.8 WELDING OF PLASTICS

Welding is superior to cementing and riveting in many aspects for plastics.

5.8.1 Types of Welding

(a) Hot gas welding
(b) Heated tool welding
(c) Hot platen welding
(d) Friction welding
(e) Ultrasonic welding

(a) Hot Gas Welding (See Fig. 5.7)

In this method, instead of the oxy-acetylene torch, an electrically heated gun is used. Compressed air passes through the heating element and strikes the joint area at about 200°C. Filler rod also can be used.

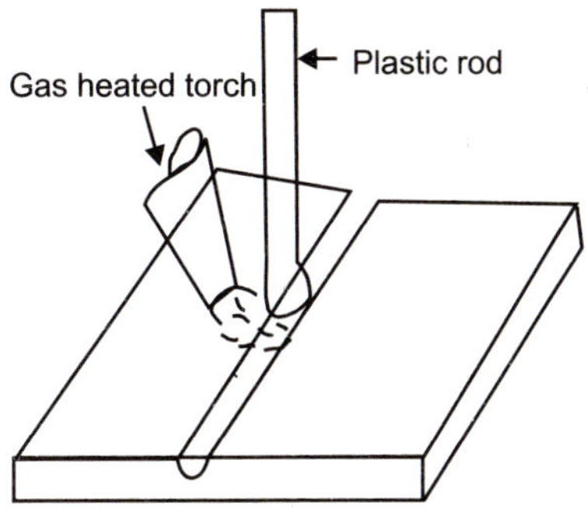

Fig. 5.7 Hot Gas Welding

(b) Heated Tool Welding (See Fig. 5.8)

In this method, the thermoplastic is first softened contacting it with a heated tool and then press together.

With films or sheet, heating wedge is placed between the surfaces to be joined and is moved along the line of welding as the edges are softened. Roller applied pressure to the top sheet and this welded to the bottom sheet.

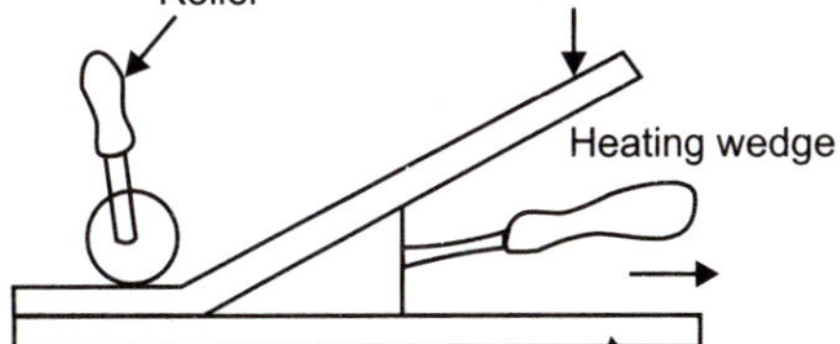

Fig. 5.8 Heated Tool Welding

(c) Hot Platen Welding (See Fig. 5.9)

This method is used for welding large, irregular shaped plastic parts.

The parts to be joined are placed into fixtures. A heated platen is kept between the parts to be welded. Now edges to be joined are soften due to the heat of heated platen. Then remove the heated platen, push the parts against each other and kept for some time till they cooled. After the welding, the fixture is opened and the complete assembly is removed.

(d) Friction Welding (See Fig. 5.10)

The two plastic rods may be fusion welded together by frictional heat.

The stationary part is bring against the spinning part. After the desired temperature is reached due to friction, the spinning rod is brought to stop and pressure is axially applied to complete the joint. Then the welded rods are allowed to cool.

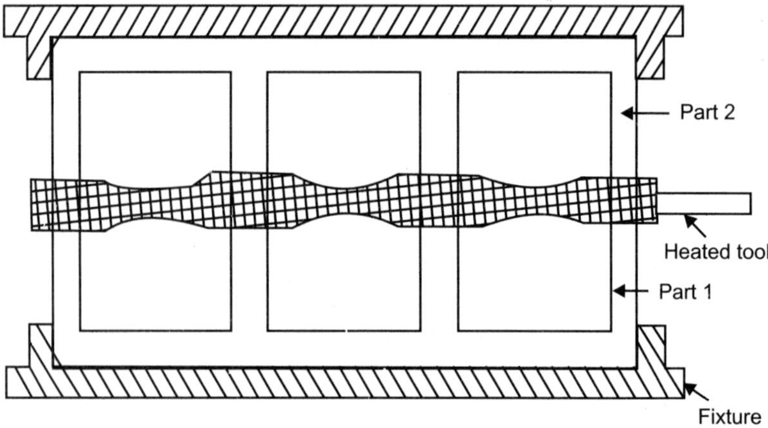

Fig. 5.9 Hot Platen Welding

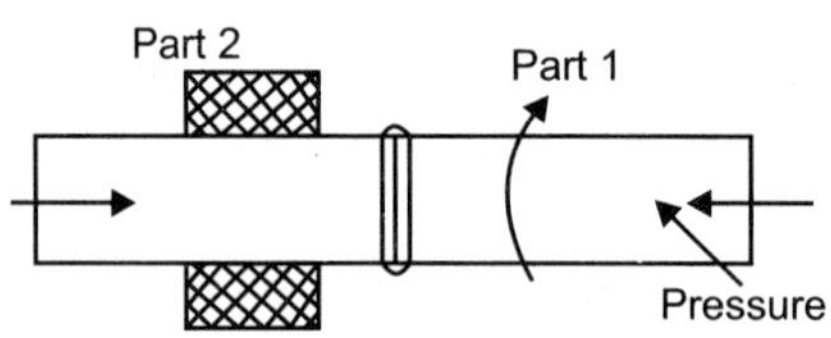

Fig. 5.10 Friction Welding

(*e*) *Ultrasonic Welding* (See Fig. 5.11)

In this method, two parts to be joined are placed together and vibrations are transmitted to the parts by a vibrating tool. Parts are soften due to friction between parts and then pressure is applied to get welded together.

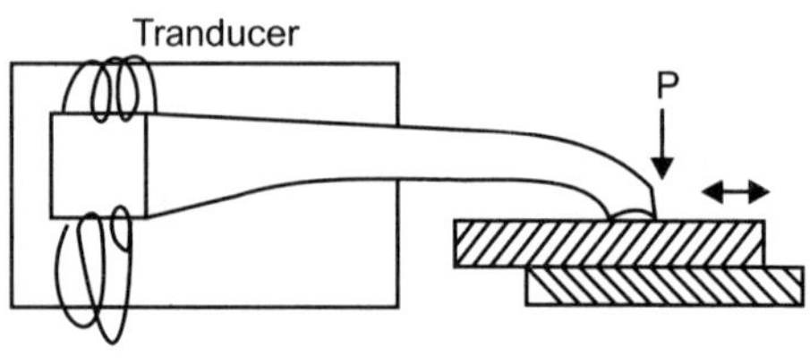

Fig. 5.11 Ultrasonic Welding

5.9 MACHINING OF PLASTICS

Machining of plastics can be satisfactorily performed with conventional machine tools used for machining of metal. However, there are some principles of machining that apply to plastics alone because the properties of plastics are different from metals. Further more plastics have a greater thermal expansion rate, soften and distort at elevated temperatures. This behaviour of plastics imposes certain precautions which include keeping the work cool to avoid sticking to the tool and excessive expansion and deflection during machining or after machining. Air, water and oil coolants are commonly used during machining of plastics.

(*a*) ***Turning*** (See Fig. 5.12)

Sharp tools avoid too much heat. Hence, the basic consideration in turning plastics on a lathe is the back rake angle. The back rake of 20° will be found suitable for virtually all thermoplastics. A zero rake angle suits for thermosets. Small depth of cut and feed should be given while machining. Proper support to the work-piece should be given because of lack of stiffness.

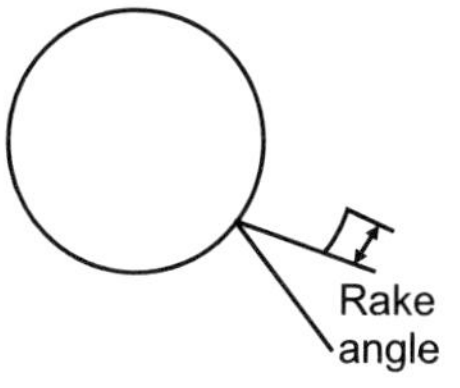

Fig. 5.12 Turning of Plastics

(*b*) ***Drilling*** (Fig. 5.13)

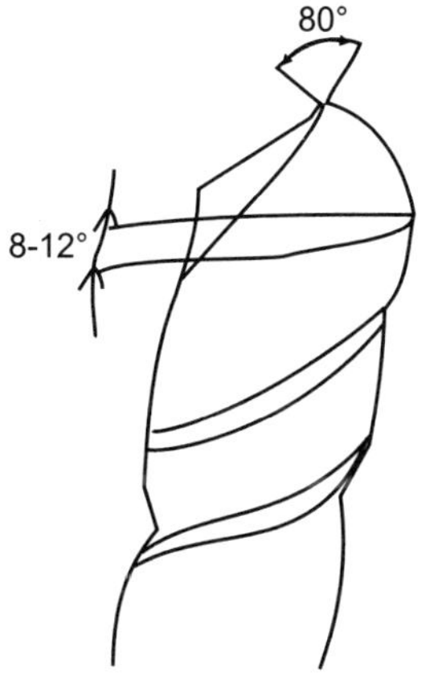

Fig. 5.13 Drilling of Plastic

If the chips of hot plastics adhere to the flutes of the drill, then the chips cannot escape from the hole and temperature will rise rapidly. A point angle of 80° is recommended for drilling thermoplastics and thermosetting plastics. This is very sharp point compared to drill bit used for drilling metals. This sharp point angle gives less end thrust in drilling operation.

(*c*) ***Bending***

Straight bends in thermoplastics are made by heating along the line of bend.

QUESTIONS

1. Distinguish between thermosetting plastics thermoplastics.
2. What are the advantages limitations of plastics?

3. Explain Injection Moulding Process.
4. With the help of suitable figures explain the blow moulding process. Mention its applications.
5. Explain various methods of welding of plastics.
6. Explain compression moulding process.

6

Presses

6.1 INTRODUCTION

A press is a machine tool used to shape or cut metal by applying force. These are used for mass production.

6.2 TYPES OF PRESSES

Classified as follows:

(a) Based on source of power
   - (i) Fly Press or Hand Press
   - (ii) Power Press

(b) Based on design of frame
   - (i) Gap frame
   - (ii) Inclinablc
   - (iii) Straight side
   - (iv) Horn Press

(c) Method of actuation of press
   - (i) Single action
   - (ii) Double action
   - (iii) Triple action

(d) Mechanism used for applying power to Ram
   - (i) Crank
   - (ii) Eccentric

(iii) Cam
(iv) Toggle
(v) Knuckle
(vi) Rack of Pinion
(vii) Hydraulic

(a)(i) Fly Press or Hand Press (See Fig. 6.1)

The fly press is the most simple of all presses and hand operated press. It carries a robust cast iron frame. The rotary motion of heavy balls converted into vertical motion of a ram by screw. The top portion of the frame forms a sort of nut. Some portion of the frame below nut guide the ram. Punch is attached to ram.

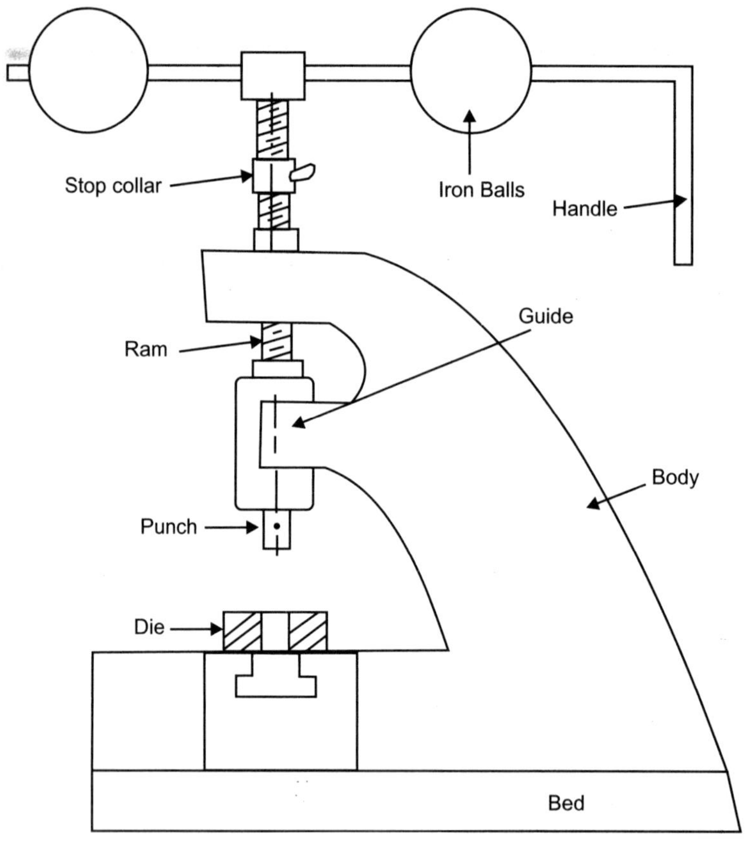

Fig. 6.1 Fly Press

(a)(ii) Power Press

The power press may be mechanical or hydraulic to transmit power to the ram. In a mechanical press, the rotary motion of electric motor is converted into reciprocating motion of the ram by using different mechanisms like crank mechanism, eccentric etc. In hydraulic press the fluid under high pressure is pumped on one side of the piston and then on the other side in a hydraulic cylinder to get the reciprocating motion of the ram. Fig. 6.2 shows power press driven by crank and connecting rod mechanism. Fig. 6.3 shows hydraulic press.

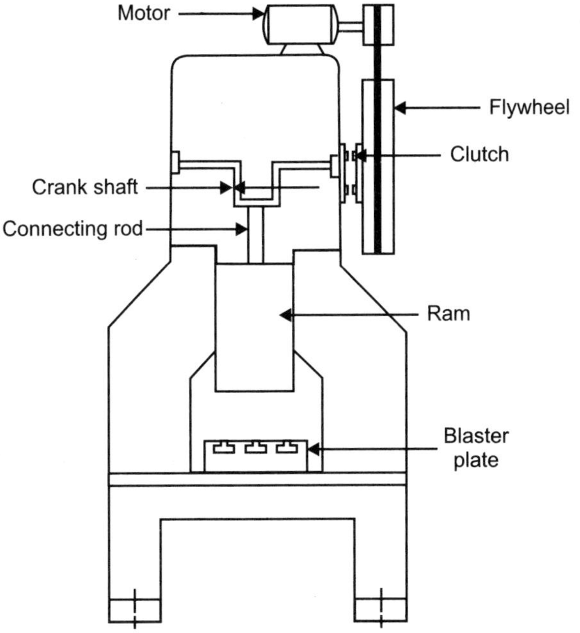

Fig. 6.2 Power Press (Mechanical)

(b)(i) Gap Frame Press (See Fig. 6.4 (a))

The Gap press has a gap like opening in the frame for feeding the sheet metal from one side of the press. This will permit the use of long and wide work-pieces.

(b)(ii) Inclinable Frame Press (See Fig. 6.4 (b))

This is the most common type of press used in industry. It has an ability to tilt back on its base, so that, it permits the scrap and finished products to fall down from the die by gravity without the help of any type of handling mechanism.

(b)(iii) Straight Side Press (See Fig. 6.4 (c))

The straight side frame is preferred for larger presses which provides larger bed area. This is used for press forging, coining, deep drawing etc.

(b)(iv) Horn Press (See Fig. 6.4 (d))

In horn press, the front face of the frame is fitted with a cylindrical post called horn. This horn is used as a support for die and used for tubular work.

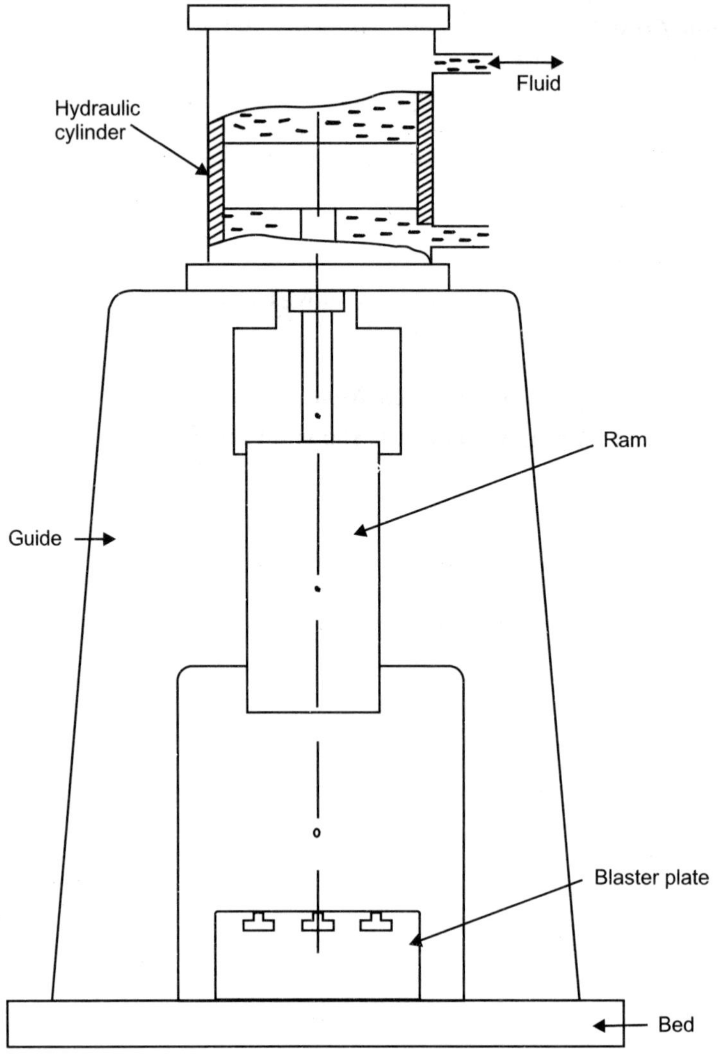

Fig. 6.3 Hydraulic Press

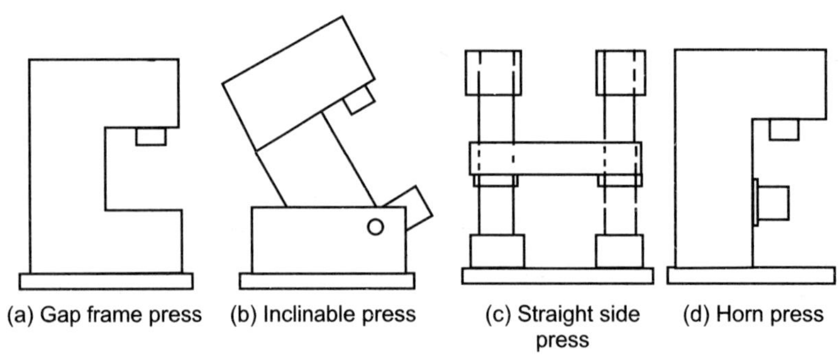

Fig. 6.4 Type of Pressed based on Design of Frame

(c)(i) Single Action Press

This press has only one ram or slide.

(c)(ii) Double Action Press

This press has two slides moving in the same direction against a bed. This type of press is used for deep drawing operation.

(c)(iii) Triple Action Press

This press has three slides in which two slides moves in the same direction and the third slide moves upward through fixed bed.

(d) Mechanism Used for Applying Power to Ram

Various mechanisms used for ram movement is shown in Fig. 6.5.

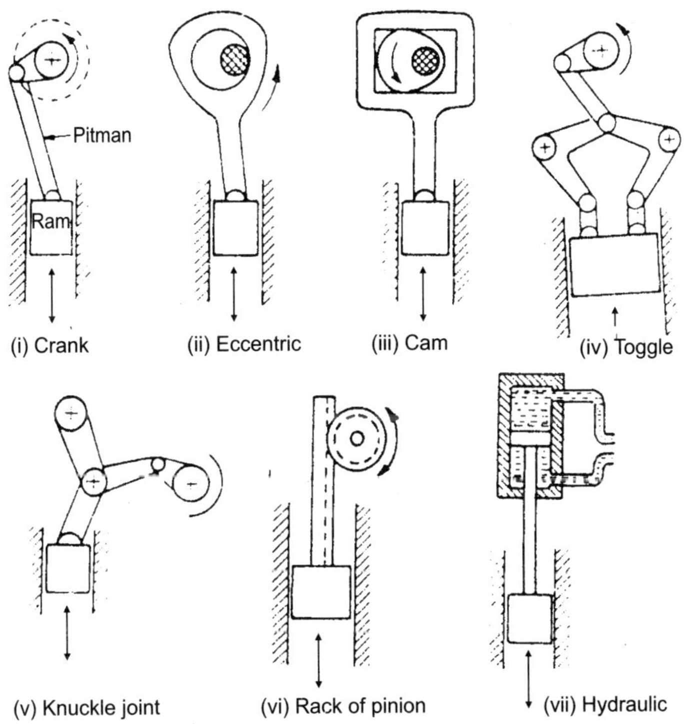

Fig. 6.5 Various Types of Power Transmitting Mechanisms

6.3 SELECTION OF PRESS

The following are taken into consideration for selecting a press.

(a) Thickness of the stock
(b) Material of the stock
(c) Force required to cut the metal

(d) Type of operation to be done
(e) Speed of the operation
(f) Type of the drive required

6.4 COMPONENTS OF SIMPLE DIE (See Fig. 6.6)

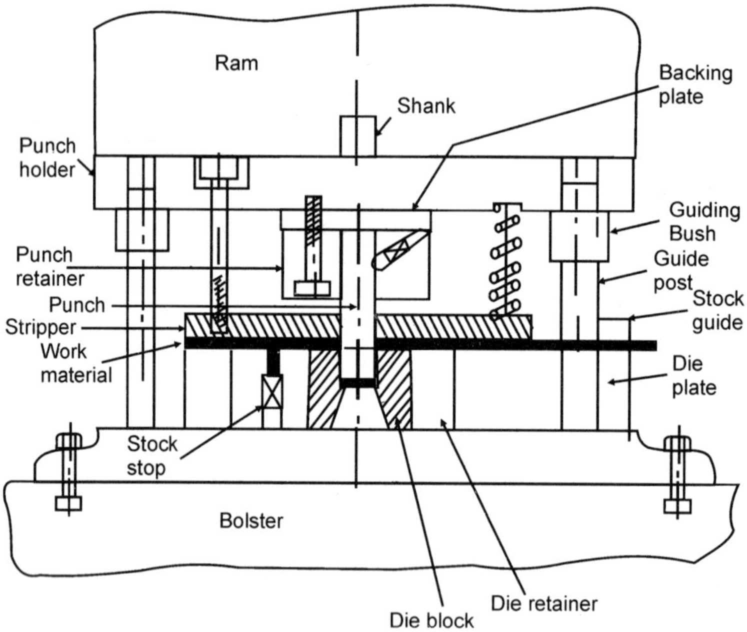

Fig. 6.6 Components of Die Assembly

BOLSTER PLATE: It is used for locating and supporting die assembly.

DIE BLOCK: The female part of a complete tool for producing work in a press.

PUNCH: This is the male component of the die assembly.

PUNCH PLATE: It fits closely over the body of the punch and holds it in proper relative position.

STRIPPER PLATE: When the punch goes down for operation and starts returning, the scrapstrip tries to go up along with it. The stripper plate prevents the upper movement of the scrap strip and frees the punch for next operation (stroke).

6.5 TYPES OF PRESS TOOLS OR TYPES OF DIES

(a) Simple Dies
(b) Compound Dies
(c) Combination Dies
(d) Progressive Dies

(a) *Simple Dies*

These dies are designed to perform only one operation such as blanking, piercing, notching etc.

(b) *Compound Dies* (See Fig. 6.7)

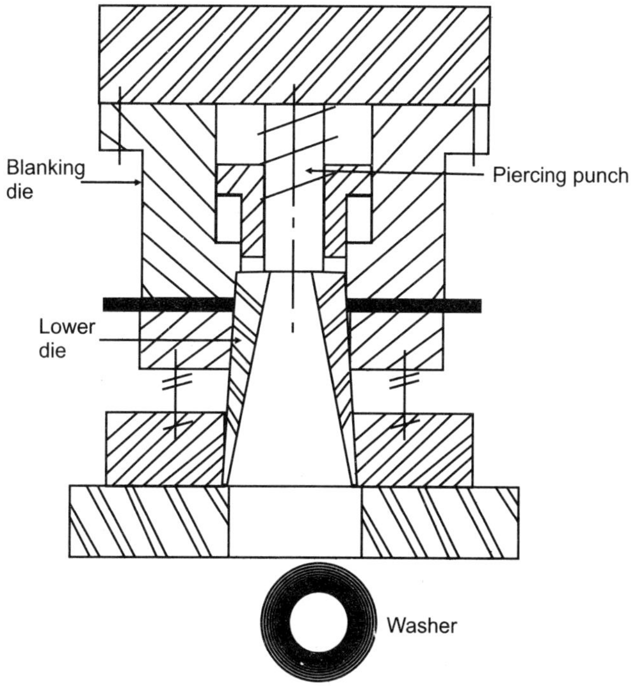

Fig. 6.7 Compound Die

They perform two or more cutting operations such as blanking and piercing. They are usually single action dies where all operations are completed in one ram stroke at the same station.

The lower die body has cutting edges both on its outward and inward surfaces. The outside cutting edges serve as a punch for blanking operation and inside cutting edges acts as a die for the piercing operation. This die is costly compared to progressive die.

(c) *Combination Dies* (See Fig. 6.8)

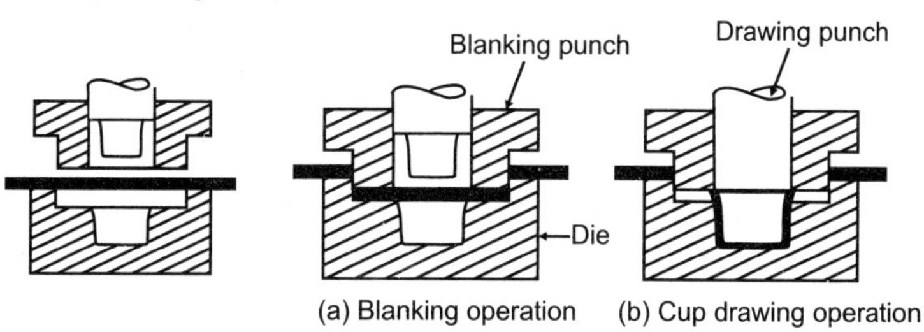

Fig. 6.8 Combination Die

In combination dies, the cutting and forming or drawing operations are combined and carried out in single operation.

(d) Progressive Dies (See Fig. 6.9)

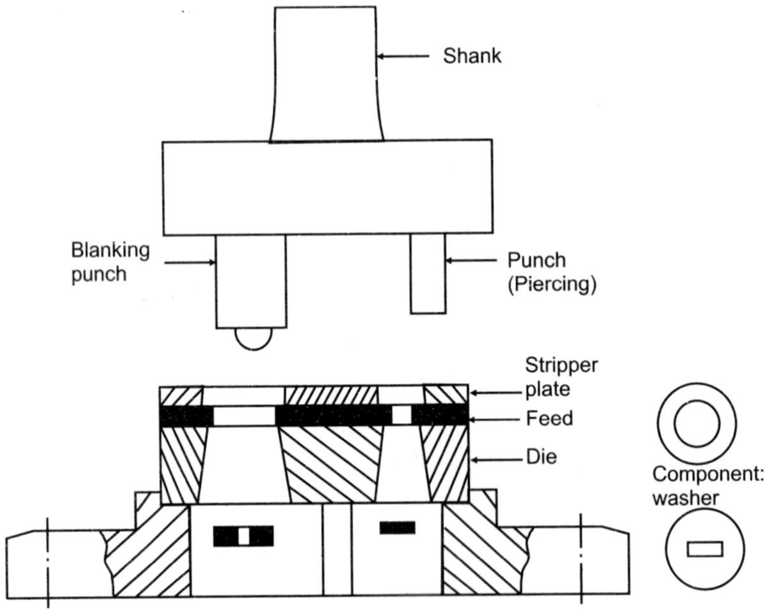

Fig. 6.9 Progressive Die

It consists of multiple stations aligned in a row. The part is moved from one station to the other. The part remains attached to the scrap, till the last station.

Washers are made by progressive dies as show about figure. At the first station, a hole is pierced by piercing punch on the sheet. The sheet is then advanced to the next station. The correct position is obtained by the stop. In the second operation, the pilot enters the pierced hole and locate it and then blanking punch shears the plate. In this way, a washer is made.

6.6 CUTTING ACTION IN A DIE AND PUNCH OPERATIONS (SHEARING ACTION) (See Fig. 6.10)

As the punch moves down, it contacts the work material supported by the die and pressure is build up gradually increases. When the elastic limit of the work material beyond the pressure of the punch, the material begins to flow plastically and displaced into the die cavity. A radius is formed at the top edge and bottom edge of the blank as shown in Fig (a); compression of the blank material against the walls of the die opening burnishes a portion of edges of the blank as shown in Fig. (b); further advancement of the punch, causes to start fractures at the cutting edge of the punch and die as shown in Fig. (c); finally the fracture will meet and broken away as shown in Fig. (d).

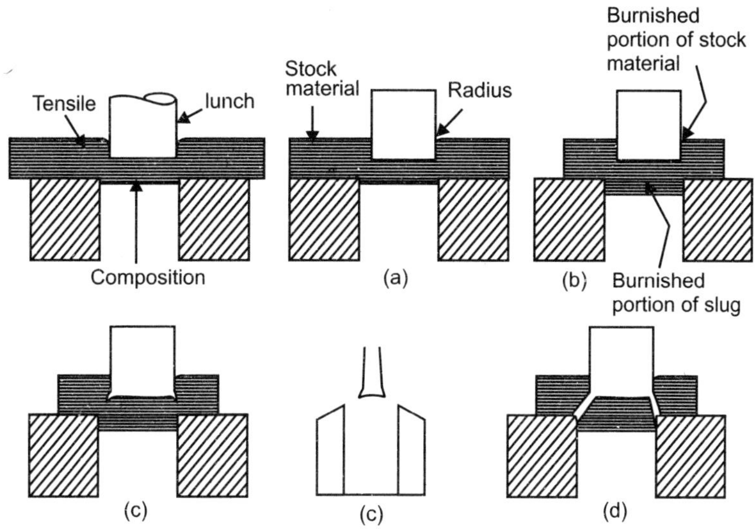

Fig. 6.10 Different Stages in Blanking

6.7 PUNCH FORCE

The punch force is a function of the area of the cut edge being sheared at any instant and the shearing strength of the material. The cut edge area is the product of the length of the cut and thickness of the material being cut.

cutting force P is given by the equation as given below:

$P = \pi D f_s t$ — For circular blank

$P = 2(l+b) f_s t$ — For rectangular blank

$P = 4 l f_s t$ — For square blank

where, fs = Shear strength of material, kgf/mm^2

D = Diameter of circular blank, mm

t = Thickness of the blank, mm

l = Length of the blank

b = Width of the blank

P = Cutting force, kgf

Example 6.1: A 25 mm square hole is to be cut in sheet metal 0.75 mm thick. The shear strength of the material is 2860 kg/cm^2. Calculate the cutting force.

$$P = 4 l f_s t$$

$$= 4 \times 2.5 \times 2860 \times \frac{0.75}{10}$$

$$= 2145 \text{ kgs.}$$

6.8 CONTROL OF HOLE AND BLANK SIZES BY CLEARANCE LOCATION (See Fig. 6.11)

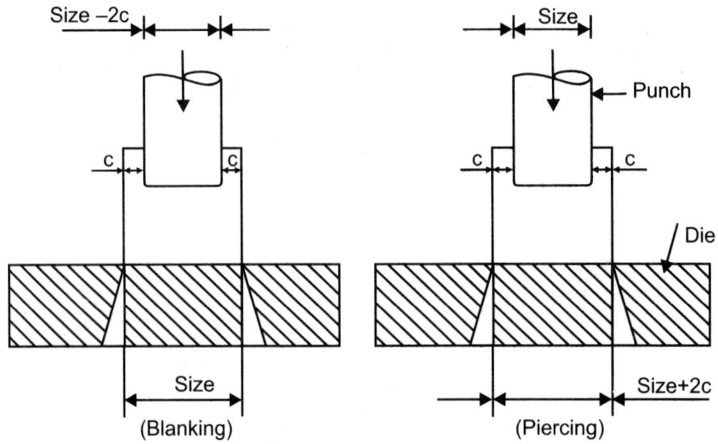

Fig. 6.11 Provision of Clearance

If a blank is to be made by the given size, then make the die to the size and punch smaller by total clearance. If hole is to be made to the exact size, punch is made the size and die larger by total clearance is die size controls the blank size, punch size controls hole size. It is shown in Fig. 6.11.

For Blanking Operation:

Die size = Blank size

Punch size = Blank size – 2c

For Punching operation:

Die size = Hole size + 2c

Punch size = Hole size

Clearance c= $0.0032 \times t \times T$

where, t = sheet thickness

T = Material shear stress

Example 6.2: Determine the die and punch sizes for blanking a circular disc of 20 mm diameter from steel having 294 kg/mm^2 shear strength and thickness of the sheet is 1.5mm.

The clearance to be provided is given by

$$C = 0.0032 \times t \times T$$
$$= 0.0032 \times 1.5 \times 294$$
$$= 1.41 \text{ mm}$$

For Blanking Operation

Die size = Blank size = 20 mm

Punch size = Blank size – 2c

$$= 20 - 2 \times 1.4$$
$$= 17.2 \text{ mm}$$

6.9 ANGULAR CLEARANCE (See Fig. 6.12)

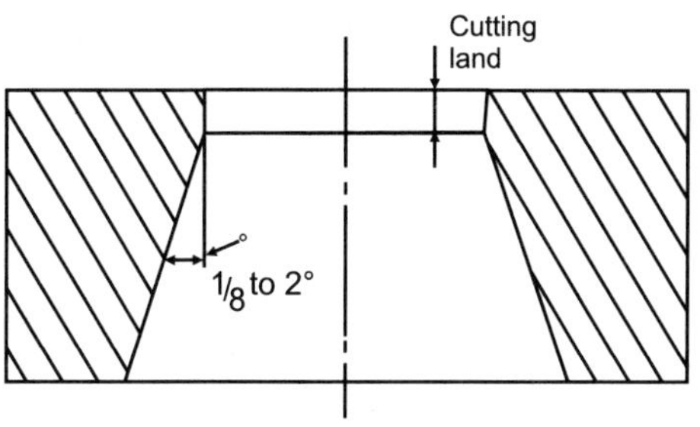

Fig. 6.12 Angular Clearance on Die

In the shearing operation, first the material is elastically deformed and then plastically and finally removed from the stock strip. After the final breaking, the slug will spring back due to release of stored elastic energy. This will make the blank to cling to the die face, unless the die opening is enlarged. This is referred as angular clearance or draft. The drat provided depends on the material, thickness and shape. For thicker and soft materials, higher angular clearances are provided. It is generally 1/8° to 2°.

6.10 SHEET METAL OPERATIONS

(a) Sheet Metal Cutting Operations
(b) Sheet Metal Forming Operations
(c) Sheet Metal Drawing Operations

(a) Sheet Metal Cutting Operations

(a)(i) Blanking (See Fig. 6.13)

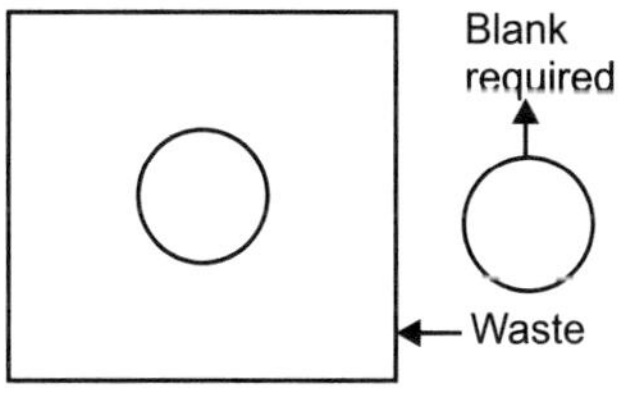

Fig. 6.13 Blanking

The article punched out is called "Blank" is required product. The plate with hole left goes as waste.

(a)(ii) Punching (See Fig. 6.14)

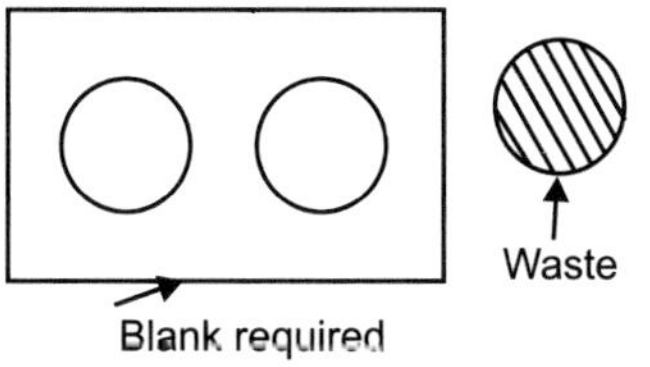

Fig. 6.14 Punching

It is the cutting operation by which various shaped holes are made in sheet metal. Punching is similar to blanking except that the hole is desired product, the material punched out to form hole being waste.

(*a*)(*iii*) *Notching* (See Fig. 6.15)

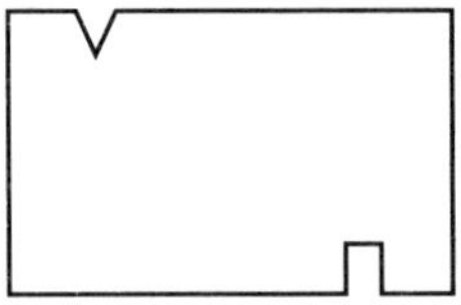

Fig. 6.15 Notching

Notching is a method to cut a specified small portion of metal towards the edge of the stock.

(*a*)(*iv*) *Perforating* (See Fig. 6.16)

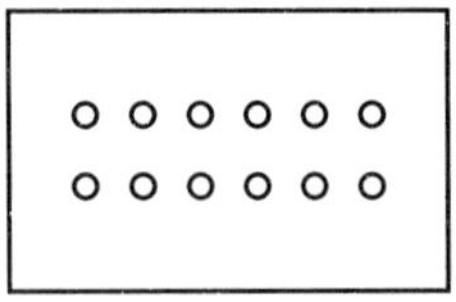

Fig. 6.16 Perforating

This is a process by which multiple holes of very small and arranged in a regular pattern are cut in the workpiece.

(*a*)(*v*) *Piercing*

Piercing is a punching operation. In piercing, a pointed, bullet shaped punch is forced through the sheet metal to produce hole with a rough flange around the hole.

(*a*)(*vi*) *Nibbling* (See Fig. 6.17)

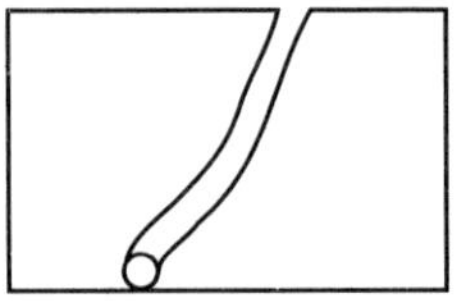

Fig. 6.17 Nibbling

Nibbling is removing metal in small increments. When a specified contour is to be cut in a sheet metal, a small punch is used to punch repeatedly along the necessary contour, generating required profile.

(*a*)(*vii*) *Slitting* (See Fig. 6.18)

It is the operation of cutting a sheet metal in a straight line along the length.

Fig. 6.18 Slitting

(a)(viii) Lancing (See Fig. 6.19)

In this operation, a cut is made part way across the metal and then one side is bent down to form a sort of tab or lower.

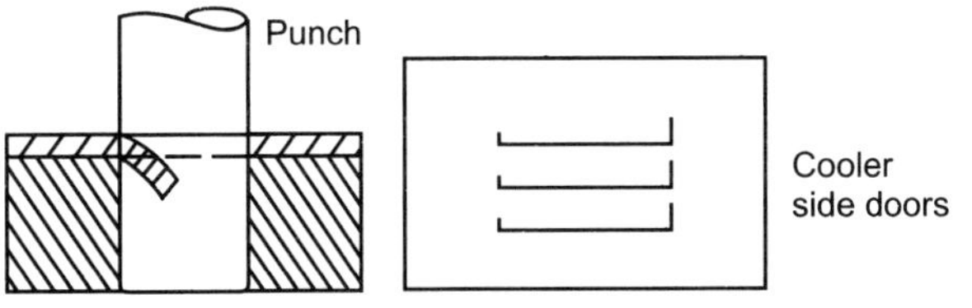

Fig. 6.19 Lancing

(b) Sheet Metal Forming Operations

(*i*) ***Bending***: In bending, the metal is stressed in both tension and compression at the two sides of neutral axis beyond the elastic limit, but below the ultimate strength of the metal. It may be V-bending and edge bending as shown in Fig. 6.20a and 6.20b:

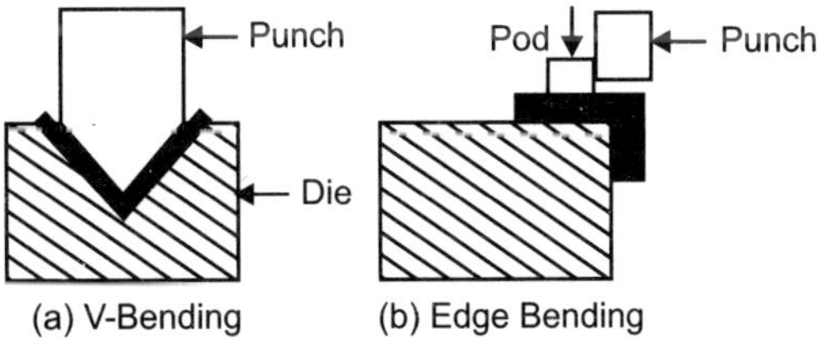

Fig. 6.20 Bending

(*ii*) ***Flanging*** (See Fig. 6.21)

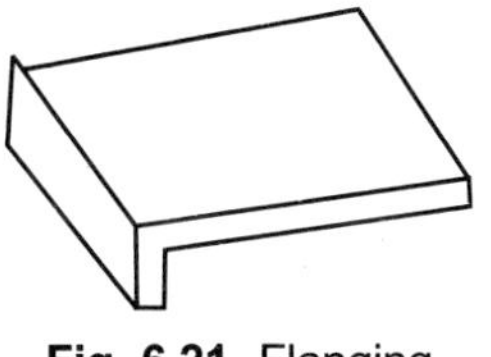

Fig. 6.21 Flanging

This operation is done to make the edges of sheet metal more rigid.

(*iii*) *Curling* (See Fig. 6.22)

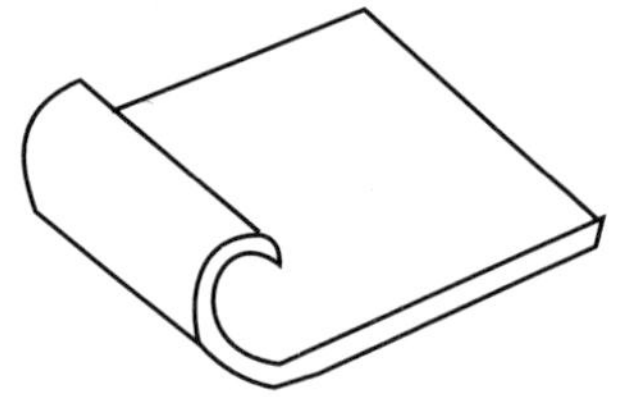

Fig. 6.22 Curling

It is the operation in which the edges of a component are formed into a roll or a curl by bending the sheet metal. This is done to reinforce or stiffen the edge or to provide smoothness to the surface.

(*iv*) *Coining* (See Fig. 6.23)

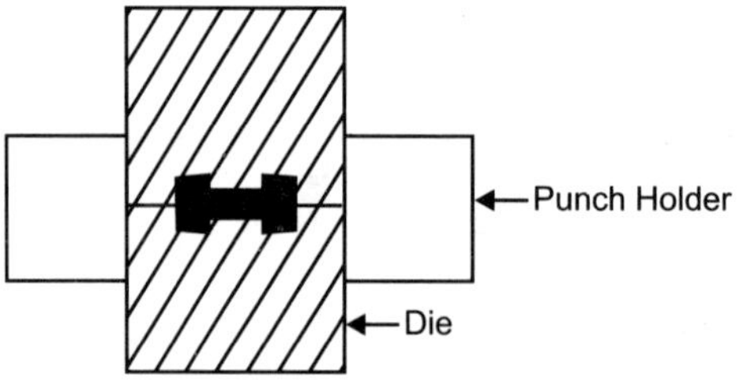

Fig. 6.23 Coining

It is the operation of production of coins, medals or other ornaments by squeezing operation.

Tremendous pressure is applied on the blank from both ends. Under severe compressive load, the metal flows in the cold state and fills up the cavity of the punch and die. The component thus produced gets a sharp impression on its surfaces corresponding to the engravings on the punch and die.

(*v*) *Embossing* (See Fig. 6.24)

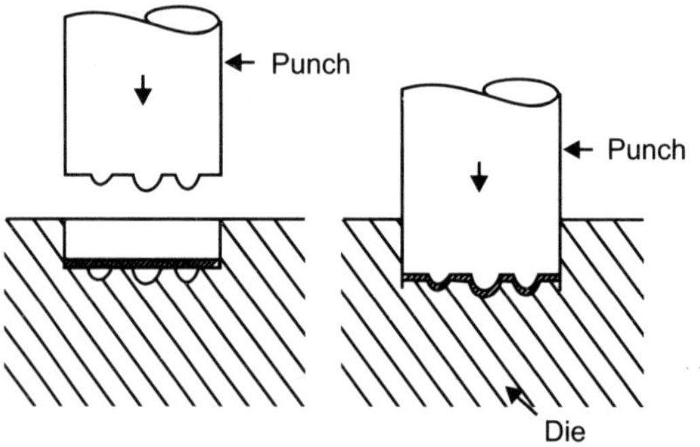

Fig. 6.24 Embossing

It is the operation of giving impressions of figures, letters or designs on sheet metal.

Embossing differs from coining in that the work-piece will have the same design on both sides, one side being depressed while the other is raised. Where in coining is usually used to create a different design on each side as in coins or medals.

(*vi*) *Crimping*

This is used for assembly purposes. The bottle cap is crimped on the circumference of cold drink bottle. The end of stove pin is crimped.

(c) Sheet Metal Drawing Operations (See Fig. 6.25)

Drawing involves shaping flat blanks into hollow components.

In drawing operation, the blank is forced by punch into the hole of die and is thus given the required shape.

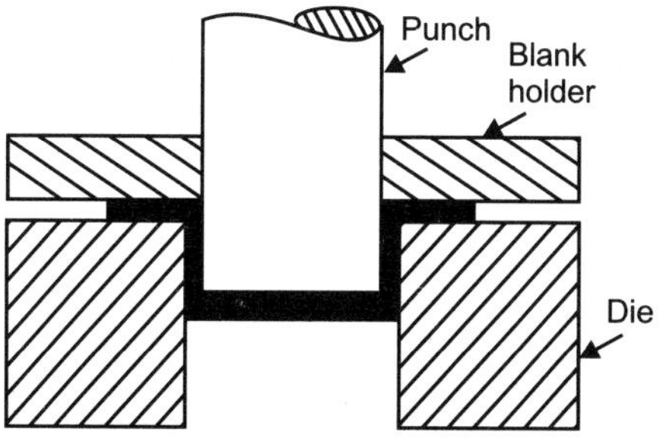

Fig. 6.25 Drawing Operation

6.11 SCRAP STRIP LAYOUT (Fig. 6.26 (a) and (b))

For an economical utilization of the stock, it is necessary to make a layout to show how the blanks can be best produced from the given stock.

In single row blank layout, material will be wasted. Hence, double row blank layout as shown in Fig. 6.26 (b) will be adopted. Double row blanks may bc a singlc pass or double pass design. Single pass design utilizes two punches where as double pass design utilizes single punch.

In a double pass design, the strip is to be overturned for the second pass.

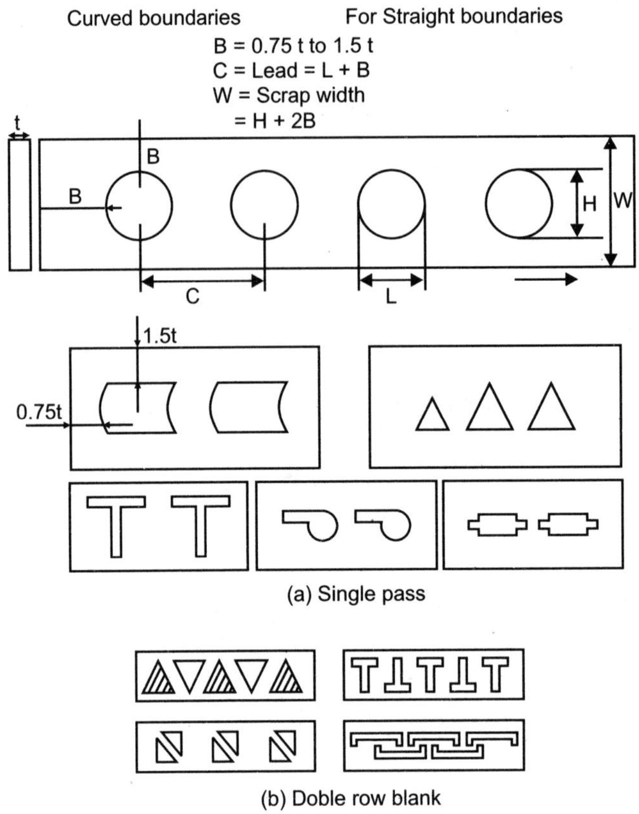

Fig. 6.26 Scrap Strip Layout

QUESTIONS

1. Explain with a neat sketch the working of fly press.
2. Explain the hydraulic press with neat sketch.
3. Explain the following dies
   (a) Simple Die
   (b) Compound Die
   (c) Combination Die
   (d) Progressive Die
4. Sketch and explain blanking die.
5. Explain various sheet metal cutting operations.
6. What is coining? Explain the process with neat sketch.

Appendix I

Objective Type Questions

1. Which of the following material can be used for making pattern?
   (a) Aluminium (b) Wax
   (c) Wood (d) All of these
2. Metal Patterns are used for
   (a) Small castings
   (b) Large scale production of castings
   (c) Large castings
   (d) Complicated castings.
3. When a pattern is made three parts, the middle part is known as
   (a) Drag (b) Cheek
   (c) Cope (d) Nonc of these
4. A taper provided on the pattern for its easy withdrawal from the mould is known as
   (a) Machining allowance
   (b) Shrinkage allowance
   (c) Draft allowance
   (d) Distortion allowance
5. Shrinkage allowance is made by adding to external dimensions and subtracting from internal dimension.
   (a) Agree
   (b) Disagree
   (c) None
   (d) None of the above
6. Shrinkage allowance for Cast Iron pattern is
   (a) 10 mm/m (b) 20 mm/m
   (c) 30 mm/m (d) 25 mm/m
7. Match plate pattern is used in
   (a) Green sand moulding
   (b) Bench moulding
   (c) Machine moulding
   (d) Pit moulding
8. The material of pattern in case of investment casting is
   (a) Thermosetting resin
   (b) Special plastic
   (c) Synthetic sand
   (d) Wax
9. Which of the following provide an additional projection on the pattern and forms a seat to support and locate the core in the mould?
   (a) Mould print (b) Core print
   (c) Drag (d) Cope

10. Loose piece pattern are
    (a) a sort of split pattern
    (b) Used when pattern cannot be drawn from the mould
    (c) Similar to core print
    (d) Never used in foundry
11. The following pattern is used for very large castings
    (a) Solid pattern (b) Cope and Drag
    (c) Skeleton Pattern (d) Split Pattern
12. A bell cast by using
    (a) Split pattern
    (b) Segmental pattern
    (c) Skeleton Pattern
    (d) Sweep Pattern
13. Tolerance produced by investment casting
    (a) + 0.05 mm (b) + 0.2 mm
    (c) 0.5 mm (d) 1 mm
14. In die casting, machining allowance is
    (a) Small (b) Large
    (c) Very large (d) Not provided
15. The surface to be machined are marked on pattern by
    (a) Red (b) Yellow
    (c) Black (d) Green
16. The surface to be unmachined is marked on pattern by
    (a) Red colour (b) Yellow colour
    (c) Black colour (d) Blue colour
17. The property of sand due to which it evolves a great amount of steam and gases is called
    (a) Collapsibility (b) Permeability
    (c) Cohesiveness (d) Adhesiveness
18. The property of sand due to which the sand grains stick together is called
    (a) Collapsibility (b) Permeability
    (c) Cohesivenss (d) Adhesiveness
19. Green sand is a mixture of
    (a) 30% sand and 70% clay
    (b) 50% sand and 50% clay
    (c) 70% sand and 30% clay
    (d) 90% sand and 10% clay
20. The purpose of gate
    (a) deliver molten metal into mould cavity
    (b) acts as a reservoir for molten metal
    (c) feed molten metal to compensate shrinkage
    (d) deliver molten from pouring basin to gate
21. Cast Iron pipes are produced by
    (a) Slash casting
    (b) Investment casting
    (c) True centrifugal casting
    (d) Die casting
22. In centrifugal casting method, impurities are
    (a) Forced outside the surface
    (b) Collected at the center of the casting
    (c) Uniformly distributed
    (d) None of the above
23. Core is used to
    (a) Make desired recess in casting
    (b) Strengthen moulding sand
    (c) Support loose pieces
    (d) To remove pattern easily
24. In a hot chamber die casting machine
    (a) Melting pot is separate from the machine
    (b) Melting pot is an integral part of machine
    (c) Melting pot may have any location
25. In a centrifugal casting method
    (a) Core is made of sand
    (b) Core is made of ferrous metal
    (c) Core is made of non-ferrous metal
    (d) No core is used
26. Shift is a casting defect which

(a) Results in a mis matching of top of bottom parts of mould
(b) Resulting in enlargement of casting
(c) Occurs as sand patches on upper surface of casting

27. A casting defect which results in enlargement of casting is known as
(a) Shift (b) Sandwash
(c) Swell (d) Scab

28. A casting defect which occur due to improper venting of sand is
(a) Cold shut (b) Blow holes
(c) Shift (d) Swell

29. Linseed oil is used in the core sand as
(a) Catalyst (b) Pasting agent
(c) Flux (d) Binder

30. Fettling is an operation performed
(a) Before casting
(b) During casting
(c) After casting
(d) After heat treatment

31. In gas welding, flame temperature of oxyacetylene gas is
(a) 1200°C (b) 1800°C
(c) 2400°C (d) 3200°C

32. Acetylene gas is stored in cylinder in
(a) Solid form (b) Liquid form
(c) Gaseous form (d) All the above

33. In gas welding, maximum thickness of material which can be welded with 30 mm diameter welding rod is
(a) 3 mm (b) 6 mm
(c) 15 mm (d) 300 mm

34. Black colour is generally painted on
(a) Oxygen cylinder
(b) Acetylene cylinder
(c) Hydrogen cylinder
(d) None of the above

35. Maroon colour is generally painted on
(a) Oxygen cylinder
(b) Acetylene cylinder
(c) Hydrogen cylinder
(d) None of the above

36. If the weld is made from right to left, it is known as
(a) Fore-hand welding
(b) Back-hand welding
(c) Vertical welding
(d) None of the above

37. For gas welding, the pressure desired by torch for oxygen is
(a) 7 to 130 kN/m^2 (b) 70–280 kN/m^2
(c) 280–560 kN/m^2 (d) 560–840 kN/m^2

38. For gas welding, the pressure desired by torch for acetylene is
(a) 7 to 130 kN/m^2 (b) 70–280 kN/m^2
(c) 280–560 kN/m^2 (d) 560–840 kN/m^2

39. The maximum flame temperature occurs
(a) at the outer cone
(b) at the inner cone
(c) between outer & inner cone
(d) at the torch tip

40. In the welding eyes need to be protected against
(a) Intense glare
(b) Sparks
(c) Infrared rays
(d) Infrared and ultra violet rays

41. The main criterion for selection of electrode diameter in arc welding
(a) Material to be welded
(b) Types of welding process
(c) Thickness of material
(d) Voltage used

42. Which of the following is preferred for welding non-ferrors metals by Arc Welding?
    (a) A.C. low frequency
    (b) A.C. high frequency
    (c) D.C.
    (d) All the above
43. In arc welding, arc is created between the electrode and work by
    (a) Flow of current
    (b) Voltage
    (c) Material characteristics
    (d) Contact resistance
44. A single V and single U butt welds are for sheets of approximate thickness
    (a) 1 to 5 mm
    (b) 5 to 15 mm
    (c) 15 to 25 mm
    (d) More than 25 mm
45. For welding plates of thickness less than 5 mm beveling of its edges
    (a) is done to single V or U-groove
    (b) is done to double V or U-groove on one side
    (c) is done to a double V or U groove on both sides
    (d) is not required
46. Open circuit voltage for arc welding is order of
    (a) 10–40 volts (b) 40–95 volts
    (c) 100–125 volts (d) 130–200 volts
47. Welding process in which two pieces to be joined are overlapped and placed between two electrodes is known as
    (a) Percusion welding
    (b) Projection welding
    (c) Spot welding
    (d) Seam welding
48. In resistance welding, voltage used for heating is
    (a) 1 V (b) 10 V
    (c) 100 V (d) 500 V
49. In resistance welding
    (a) Voltage is high and low current
    (b) Voltage is low and high current
    (c) Both voltage and current are high
    (d) Both voltage and current are low
50. Grey cast iron is usually welded by
    (a) Gas welding
    (b) Resistance welding
    (c) Arc welding
    (d) TIG welding
51. In resistance welding, two copper electrodes used are cooled by
    (a) Water
    (b) Air
    (c) Both (a) and (b)
    (d) None of the above
52. The temperature of the inner core of neutral flame is order of
    (a) 1,000°C (b) 2,000°C
    (c) 2,500°C (d) 3,500°C
53. The most commonly used flame in gas welding is
    (a) Neutral flame
    (b) Oxidising flame
    (c) Carbon flame
    (d) None of the above
54. Spot welding, projection of seam welding are classification of
    (a) Thermit welding
    (b) Arc welding
    (c) Electric Resistance welding
    (d) Forge welding

55. Spot welding is used for welding plates having thickness
    (a) 0.25 mm to 1.25 mm
    (b) 1.25 mm to 2.5 mm
    (c) 3 mm
    (d) 3 mm to 6 mm
56. In spot welding, the tip of electrode is of
    (a) Stainless steel (b) Aluminium
    (c) Copper (d) Brass
57. In spot welding electrode tip's diameter (d) should be
    (a) $\sqrt{t}$ (b) 1.5 t
    (c) 3 t (d) 4.5 t
58. In which cutting process using high velocity jet of Ionised hot gas?
    (a) Oxy-acetylene cutting
    (b) Plasma arc cutting
    (c) Metallic arc welding
    (d) Oxy-arc cutting
59. Welding of steel structures on site of a building is done by
    (a) Spot welding
    (b) Projection welding
    (c) Arc welding
    (d) Seam Welding
60. In arc welding, the temperature of heat of arc is in the range
    (a) 1000°C to 2,000°C
    (b) 2000°C to 4,000°C
    (c) 4000°C to 6,000°C
    (d) 7000°C to 8,000°C
61. Welding Process in which flux is used in the form of granules is
    (a) Gas welding
    (b) Submerged arc welding
    (c) Arc welding
    (d) Thermite welding
62. Which of the following welding process uses non-consumable electrode?
    (a) Laser welding
    (b) TIG welding
    (c) MIG welding
    (d) Ion beam welding
63. Laser beam welding finds widest application in
    (a) Heavy Industry
    (b) Structural work
    (c) Process Industry
    (d) Electronic Industry
64. Gases used in TIG welding are
    (a) Hydrogen and Oxygen
    (b) CO_2 and H_2
    (c) Argon and Neon
    (d) Argon and Helium
65. The following welding process uses consumable electrode
    (a) MIG (b) TIG
    (c) Thermite (d) Spot welding
66. TIG welding is preferred for
    (a) Mild steel (b) Aluminium
    (c) Silver (d) All the above
67. A collimated light beam is used for producing heat in
    (a) TIG welding (b) MIG welding
    (c) Laser welding (d) Plasma welding
68. Temperature of Plasma torch is order of
    (a) 1000°C (b) 5000°C
    (c) 10,000°C (d) 33,000°C
69. A soldering Iron bit is made of
    (a) Brass (b) Tin
    (c) Copper (d) Steel
70. Welding spatter is
    (a) Flux
    (b) Electrode coating
    (c) Welding defect
    (d) Welding test

71. Distortion in welding occurs due to
(a) Use of excessive current
(b) Improper clamping
(c) Use of wrong electrode
(d) Oxidation of weld
72. The melting point of the filler metal in brazing should be
(a) 420°C (b) 600°C
(c) 800°C (d) 900°C
73. Straight polarity is better for
(a) Thick materials
(b) Thin materials
(c) Any material
(d) None of the above
74. Reverse polarity is used for
(a) Thick materials
(b) Thin materials
(c) Any material
(d) None of the above
75. Electrode used in TIG is
(a) C.I. (b) Tungsten
(c) Al (d) Cu
76. In MIG welding, helium is used in order to
(a) Provide cooling effect
(b) Acts as flux
(c) Acts as shield medium
(d) Facilitate welding
77. Current range in SMAW is
(a) 10–500 Amps. (b) 10–50 Amps.
(c) 100–200 Amps. (d) 10–100 Amps.
78. Which of the following is correct in SAW
(a) Non-consumable electrode
(b) No electrode
(c) Consumable electrode
(d) None of the above
79. Oxy-acetylene flame having more amount of O_2 is called
(a) Oxidizing flame (b) Neutral flame
(c) Carburising flame (d) Red flame
80. Oxy-acetylene flame having more amount of C_2H_2 is called
(a) Oxidizing flame (b) Neutral flame
(c) Carburising flame (d) Red flame
81. Fuel gas used in gas welding is
(a) N_2 (b) C_2H_2
(c) CO_2 (d) He
82. Forge welding is an example of
(a) Arc welding
(b) Solid state welding
(c) Resistance welding
(d) Seam welding
83. Flux used in Brazing is
(a) Aluminium (b) Borax
(c) Zinc chloride (d) Calcium carbide
84. Flux used in soldering
(a) Borax (b) Sodium Silicate
(c) Calcium Carbide (d) Zinc Chloride
85. Soldering Iron bit is made of
(a) Brass (b) Copper
(c) Tin (d) Steel
86. Ultrasonic welding is best suited for
(a) Non-ferrous metals
(b) Brittle materials
(c) Conductive materials
(d) Metals
87. In ultrasonic welding, the frequency range
(a) 10–40 cps (b) 50–100 cps
(c) 4000–20000 cps (d) 200–500 cps
88. Products made from plastic materials are
(a) Light weight
(b) Corrosive resistance
(c) Excellent surface finish
(d) All of the above
89. The example of thermosetting material is
(a) Cellphone
(b) Phenol formaldehyde
(c) Synthetic rubber
(d) Resin

90. Bottles and floatable objects of thermo plastics are made by the process
    (a) Blow moulding
    (b) Transfer moulding
    (c) Extrusion
    (d) Slush moulding
91. Tooth paste tubes are manufactured by
    (a) Direct extrusion
    (b) Piercing
    (c) Impact extrusion
    (d) Indirect extrusion
92. In which type of extrusion, the frictional force is high?
    (a) Direct extrusion
    (b) Indirect extrusion
    (c) Impact extrusion
    (d) Hydrostatic
93. In Backward Extrusion
    (a) Metal flows in the direction of ram
    (b) Metal flows in opposite direction of ram
    (c) Metal flow in circular motion
    (d) None of the above
94. Upsetting is the process of
    (a) Increasing the cross-section at the expense of length
    (b) Increasing the length at the expense of cross-section
    (c) Some kind of welding
    (d) Some kind of bending
95. Hot press forging
    (a) causes a steadily applied pressure instead of impact force
    (b) is used to force the end of a heated bar into a desired shape
    (c) is forging operation in which two halves of rotary die openly and closely while impact the end of heated tube
    (d) is a forging method for reducing the diameter of a bar and in the process making it longer
96. Forging of plain carbon steel is carried out at
    (a) 750°C (b) 900°C
    (c) 1110°C (d) 1300°C
97. In forward extrusion, the ram and the metal moves in
    (a) Opposite direction
    (b) Same direction
    (c) Circular
    (d) None of the above
98. Which of the following material can not be forged?
    (a) Wrought iron
    (b) Mild steel
    (c) Cast iron
    (d) High carbon steel
99. Internal stress setup during forging can be removed by
    (a) Annealing (b) Normalizing
    (c) Tempering (d) Both (a) and (b)
100. What is the mechanical property of a material is desired so that it forged?
    (a) Ductility (b) Malleability
    (c) Elasticity (d) Machinability
101. In which metal forming process, the material is shaped by intermitten blows?
    (a) Drawing process (b) Extrusion
    (c) Wire drawing (d) Forging process
102. What is the advantage of forging?
    (a) Good surface finish
    (b) Low tooling cost
    (c) Improved physical properties
    (d) Close tolerances
103. Drawing is the process of
    (a) Increase in length
    (b) Increase in cross section
    (c) Reduce in length
    (d) Reduce in area

104. Extruded product takes the shape depending on
   (a) Press capacity
   (b) Deforming force
   (c) Shape of the hole in die plate
   (d) Depends on material

105. In hot working, the temperature of material used is
   (a) 200°C
   (b) Above the melting point of material
   (c) Room temperature
   (d) Above the recrystallization temperature

106. In cold working, the temperature of material is
   (a) Below the recrystallization temperature
   (b) Room temperature
   (c) Below room temperature
   (d) Above recrystallization temperature

107. Which rolling process gives better surface finish elongated grains?
   (a) Hot rolling (b) Cold rolling
   (c) Extrusion (d) Hot working

108. Product made by rolling
   (a) I-Section (b) Tubes
   (c) Rollers (d) Metal rolls

109. Machinery used in production of channel, I sections, angles etc.
   (a) Continuous casting method
   (b) Rolling mills
   (c) Forging Plant
   (d) Spinning machines

110. In four high rolling mill, bigger rollers are called
   (a) Guide rolls (b) Backup rolls
   (c) Main rolls (d) Support rolls

111. In four rolling mill, the diameter of backup rolls in comparison with the diameter of working rolls is
   (a) Same
   (b) Larger
   (c) Smaller
   (d) Depend upon capacity

112. Spinning operation is carried out on
   (a) Hydraulic press
   (b) Mechanical press
   (c) Lathe
   (d) Milling machine

113. Strech forming is
   (a) In which the edges of sheet metal are turned to provide stiffness
   (b) Producing contours in sheet metal permanently
   (c) Employed to expand a tubular or cylindrical part
   (d) None of these

114. Forces needed in cold working
   (a) Less
   (b) More
   (c) Equal
   (d) Sometimes more or less

115. Work hardening occurs in the following forming process:
   (a) Cold working (b) Hot working
   (c) Extrusion (d) None of these

116. The operation of cutting of a flat sheet to the desired shape is called
   (a) Shearing (b) Piercing
   (c) Punching (d) Blanking

117. Notching is the operation of
   (a) Cutting sheet metal in a straight line along the length
   (b) Removal of metal to the desired shape from the edge of plate
   (c) Cutting a sheet of metal through a part of length and bend the cut portion
   (d) Bending the sheet

118. Cutting a sheet of metal in a straight line along the length known as
(a) Plunging (b) Notching
(c) Slitting (d) Forming

119. In sheet metal blanking, shear is provided on punch and die so that
(a) Press load is reduced
(b) Good cut edge
(c) Working of sheet minimized
(d) None

120. The operation of bending a sheet metal along curved axis is known as
(a) Plunging (b) Notching
(c) Forming (d) Slitting

121. The operation of producing cup shaped parts from sheet metal
(a) Drawing (b) Coining
(c) Lancing (d) Squeezing

122. The operation of straightening a curved sheet metal is known as
(a) Drawing (b) Squeezing
(c) Lancing (d) Planishing

123. In compound die
(a) Only one operation is performed in single stroke
(b) Two or more operations are performed simultaneously in single stroke of ram
(c) Two or more cutting operations are performed at one station in every stroke of the ram
(d) Both cutting and non-cutting operations at one station in every stroke of ram

124. Cutting and forming operations are performed in single operation in
(a) Simple die
(b) Combination die
(c) Progressive die
(d) Compound die

125. Blanking and piercing operations can be performed simultaneously in
(a) Compound die (b) Simple die
(c) Progressive die (d) Combination die

126. In progressive dies
(a) Two or more cutting operations can be performed simultaneously
(b) Cutting and forming operations can be combined and carried in simple operation
(c) Work-piece moves from one station to other with separate operations performed at each station
(d) All of the above

127. The cutting force in punching and blanking operations depends on
(a) The modulus of elasticity
(b) shear strength
(c) Bulk modulus
(d) Yield strength

128. Angular clearance provide on dies is of order of
(a) 5 to 10°C (b) 3 to 5°C
(c) ½ to 2°C (d) 10 to 20°C

129. In bending operation, the metal takes the shape of
(a) Die
(b) Punch
(c) Average of two
(d) Could take any shape

130. In Blanking operation, the clearance is provided on
(a) Die
(b) Punch
(c) Half on die and half on punch
(d) Provided on any member

ANSWERS

1. (d)	2. (b)	3. (b)	4. (c)	5. (a)	6. (a)	7. (c)	8. (d)
9. (b)	10. (b)	11. (c)	12. (d)	13. (a)	14. (d)	15. (a)	16. (c)
17. (b)	18. (c)	19. (c)	20. (a)	21. (c)	22. (b)	23. (a)	24. (b)
25. (d)	26. (a)	27. (c)	28. (b)	29. (d)	30. (c)	31. (d)	32. (b)
33. (a)	34. (a)	35. (b)	36. (a)	37. (b)	38. (a)	39. (b)	40. (d)
41. (c)	42. (c)	43. (d)	44. (b)	45. (d)	46. (b)	47. (c)	48. (b)
49. (b)	50. (a)	51. (a)	52. (d)	53. (a)	54. (c)	55. (a)	56. (c)
57. (a)	58. (b)	59. (c)	60. (c)	61. (b)	62. (b)	63. (d)	64. (d)
65. (a)	66. (b)	67. (c)	68. (d)	69. (c)	70. (c)	71. (b)	72. (a)
73. (a)	74. (b)	75. (b)	76. (c)	77. (a)	78. (c)	79. (a)	80. (c)
81. (b)	82. (b)	83. (a)	84. (d)	85. (b)	86. (b)	87. (c)	88. (d)
89. (b)	90. (a)	91. (c)	92. (a)	93. (b)	94. (a)	95. (a)	96. (d)
97. (b)	98. (c)	99. (d)	100. (b)	101. (d)	102. (c)	103. (a)	104. (c)
105. (d)	106. (a)	107. (b)	108. (a)	109. (b)	110. (b)	111. (b)	112. (c)
113. (b)	114. (b)	115. (a)	116. (d)	117. (b)	118. (c)	119. (a)	120. (c)
121. (a)	122. (d)	123. (c)	124. (b)	125. (a)	126. (c)	127. (b)	128. (c)
129. (b)	130. (b)						

PART B
MACHINE TOOLS

7

Fundamentals of Metal Cutting

7.1 INTRODUCTION

Metal cutting is the process of producing required dimensional work-piece by removing the unwanted material in the form of chips.

7.2 CLASSIFICATION OF CUTTING TOOLS

Cutting tools are classified into two groups

1. Single point tools
2. Multi point tools

7.2.1 Single Point Tools (See Fig 7.1)

Single point tools have one cutting edge. These tools are classified under the following groups.

(a) According to the method of manufacturing tools
  (i) Solid tool
  (ii) Forged tool
  (iii) Tipped type tool
  (iv) Bit type tool

(b) According to the method of cutting edge (feed)
  (i) Right hand tool
  (ii) Left hand tool

(c) According to method of using the tool.

(*a*) (*i*) *Solid tool*: The cutting edge is formed by grinding the end of a piece of tool steel stock. (See Fig. 7.1(a)) shows a solid tool.

(*ii*) *Forged tool* (See Fig. 7.1(b)): Forged tools are manufactured from high carbon steel or high speed steel. The required shape of the tool is made by forging before hardening and grinding.

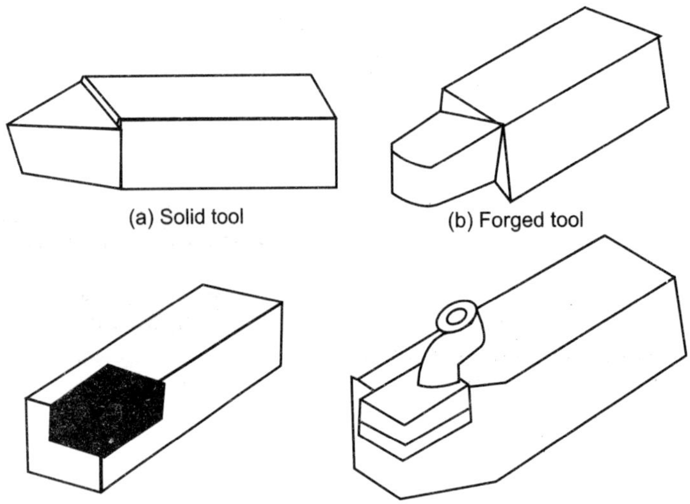

(a) Solid tool (b) Forged tool

(c) Tipped tool (Blazed tool) (d) Bit type tool

Fig. 7.1 Single Point Tools

(*iii*) *Tipped type tool* (See Fig. 7.1(c)): In tipped type tool, the cutting edge is in the form of small tip made of high grade material which is blazed to a shank of low grade material.

(*iv*) *Bit type tool* (See Fig. 7.1.(d)): In bit type, a high grade material like carbide or ceramic of a square or rectangular shape is held mechanically in a tool holder.

(*b*) (*i*) *Right hand tool* (See Fig. 7.2(a))

(b) Right hand tool (a) Left hand tool

Fig. 7.2 Left and Right Cutting Tool

In a right cutting tool, the side of cutting edge is on the side of the right thumb when the right hand is placed on the tool with the palms downwards and the fingers pointed towards the tool nose.

(*b*) (*ii*) *Left hand tool* (See Fig. 7.2(b)): A left cutting tool is one which has the side cutting edge is on the thumb side when the left hand is applied.

(*c*) *According to method of using tool*

(i) Turning tool
(ii) Facing tool
(iii) Parting off tool
(iv) Thread cutting tool
(v) Forming tool
(vi) Boring tool.

7.2.2 Multi Point Cutting Tools

They have more than one effective cutting edge to remove the excess material from the work-piece. Milling cutters, reamers, broaches and grinding wheels are multipoint tools. These tools may have rotary travel, in case of drilling or milling operations or may have linear travel, in case of Broaching operations.

7.3 ELEMENTS OF SINGLE POINT TOOL (See Fig. 7.3)

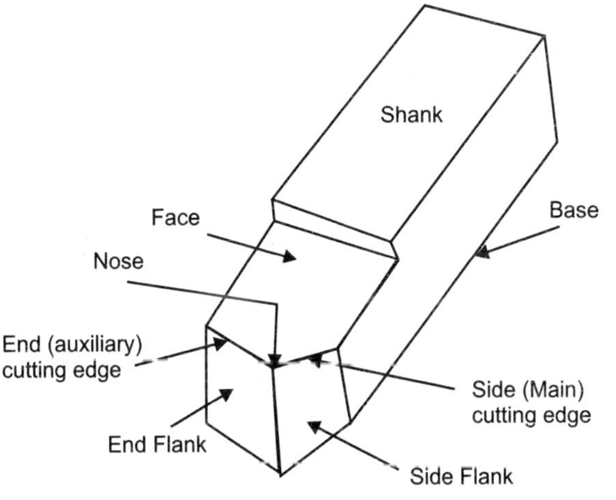

Fig. 7.3 Elements of Single Point Tool

Face: It is the surface over which the chips flow

Flank: It is the surface below the cutting edge

Nose: It is the junction of the side and end cutting edges

Side cutting edge: It does the main work in the cutting process. It is the intersection of the face and side flank.

End or auxiliary cutting edge: It is the intersection of face and end flank.

7.4 GEOMETRY OF A SINGLE POINT TOOL (TOOL ANGLES) (See Fig. 7.4)

In a single point tool, these are various angles, each of them has a definite purpose.

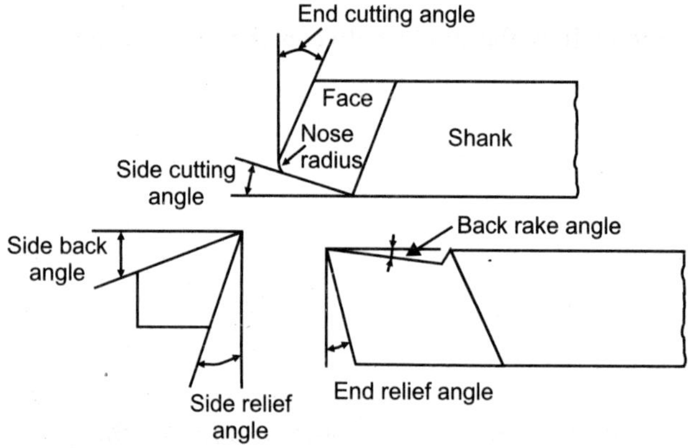

Fig. 7.4 Geometry of a Single Point Tool (Tool Angles)

Back rake angle (*Top rake angle*): It is also called as Top rake angle. It measures the downward slope of the top surface of the tool from nose to the rear along the longitudinal axis. Its purpose is to guide the direction of the chip flow. The size of the angle depends upon the material to be machined. The softer material requires greater positive rake angle. Aluminium requires more back rake angle than C.I. or steel. Back rake angle may be positive, neutral, negative rake angle. Positive rake angle is used to cut low tensile strength and non-ferrous metals. Negative rake angle is used for high tensile strength materials.

Side rake angle: It measures the slope of the top surface of the tool to the side in a direction perpendicular to the longitudinal axis. It also guides the direction of the chip away from the job. The amount that a chip is bent depends upon this angle. (See Fig. 7.5)

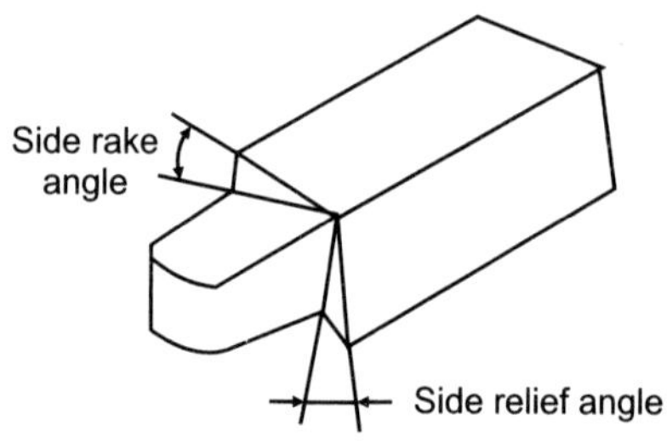

Fig. 7.5 Side Rake Angle of Side Relief Angle

Side relief angle: It is the angle, made by the flank of the tool and a plane perpendicular to the base just under the side cutting edge. This angle permits the tool to be fed side-ways into the job so that it can cut without rubbing.

End relief angle: It is the angle between a plane perpendicular to the base and the end flank. This angle prevents the cutting tool from rubbing against the job.

End cutting edge angle: It is the angle between the face of the tool and a plane perpendicular to the side of the shank. It acts as a relief angle that allows only a small section of the end cutting edge to contact the machined surface.

Side cutting edge angle: It is the angle between the side cutting edge and the longitudinal axis of the tool. It avoids the formation of built up edges controls the direction of chip flow and distributes the cutting force and heat produced over large cutting edge.

Nose Radius: It is provided to increase finish and strength of the cutting tip of the tool.

7.5 TOOL SIGNATURE

The tool angles have been standardized by American Standards Association (ASA). Seven important elements comprise the signature of the cutting tool and stated in the following order.

Back rake angle, side Rake angle, end relief angle, side relief angle, end cutting edge angle, side cutting edge angle and nose radius. It is usual to omit symbols for degrees, and mm, but simply stating the numerical values of each element.

For example a tool signature as 10, 10, 6, 6, 8, 8, 2

Back rake angle = 10°
Side rake angle = 10°
End relief angle = 6°
Side relief angle = 6°
End cutting edge angle = 8°
Side cutting edge angle = 8°
Nose radius = 2 mm.

7.6 TOOL NOMENCLATURE SYSTEMS (Tool Angle Specification Systems)

The most important nomenclature systems are:

(a) British system (maximum normal rake system)
(b) American system (ASA system)
(c) German system (DIN system)
(d) Orthogonal rake system (ORS).

(a) British System (MRS) (See Fig. 7.6)

According to this system, the rake angle is specified as the steepest slope of the rake face. It is equal to the angle between the rake face and the tool face measured in a plane perpendicular to both the rake face and the tool base.

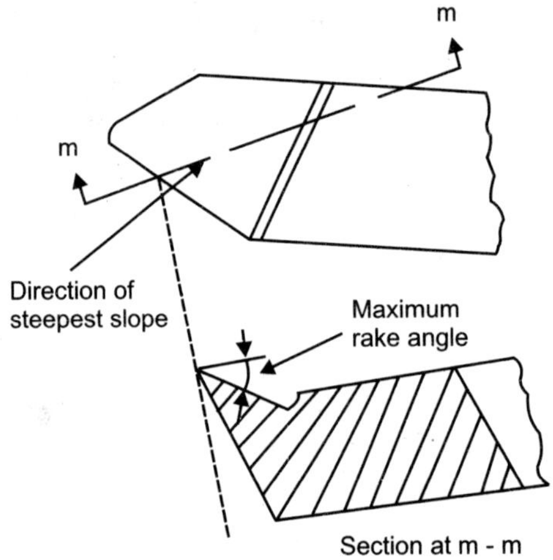

Fig. 7.6 British System of Tool Nomenclature

(b) American System (ASA System) (See Fig. 7.7)

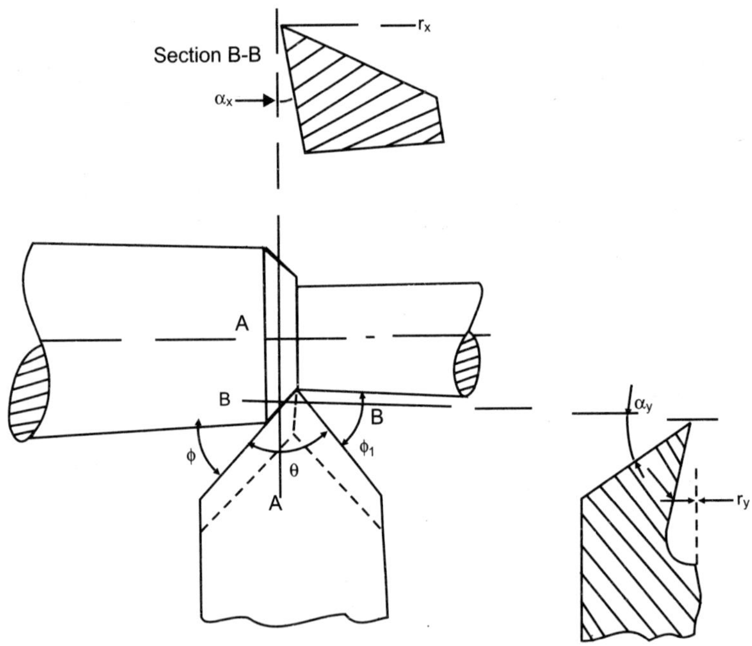

Fig. 7.7 American System of Tool Nomenclature (x-y-z planes)

r_y = Top rake angle ; α_y = End relief angle

r_x = Side rake angle ; α_x = Side relief angle

θ = Nose angle ; ϕ = Side cutting edge angle

ϕ_1 = End cutting edge angle

In this system, the geometry of the rake face is expressed in terms of the side rake angle and the back rake angle. The back rake angle is the angle between the rake face and the base of the tool measured in a plane perpendicular to the base of the tool and parallel to the longitudinal axis of the tool. The side rake angle is defined as the angle between the rake face and the tool base measured in a plane normal to that in which the back rake angle is measured.

(c) German System (DIN System) (See Fig. 7.8)

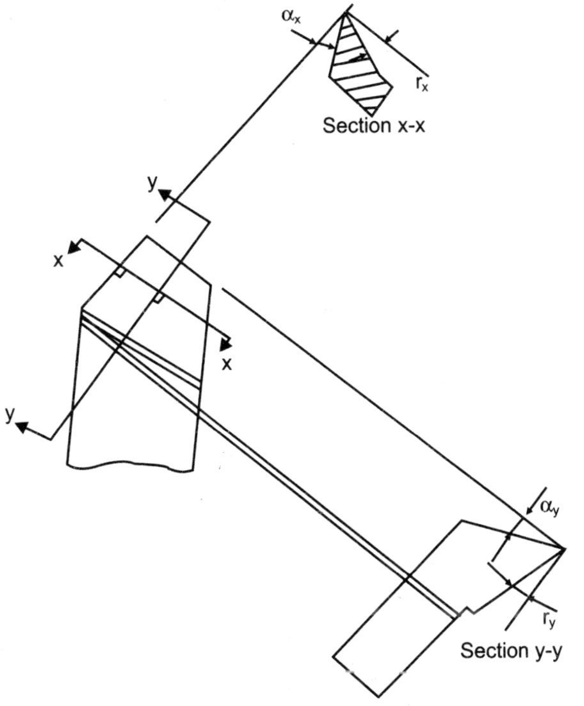

Fig. 7.8 German System of Tool Nomenclature

r_y = Back rake angle ; α_y = End relief angle

r_x = Side rake angle ; α_x = Side relief angle

In DIN system, also, the rake face is specified by the side rake angle and the back rake angle. However, the measurement of these two angles are different from American system.

In this system, the back rake angle is the angle between the rake face and the base plane measured in a plane which is normal to the base plane but parallel to the trace of the end cutting edge angle.

Likewise, the side rake angle is defined as the angle between the rake face and the base plane measured in a plane normal to the trace of the side cutting edge in the base plane.

(d) Orthogonal Rake System (ORS)(See Fig. 7.9 (a) and (b))

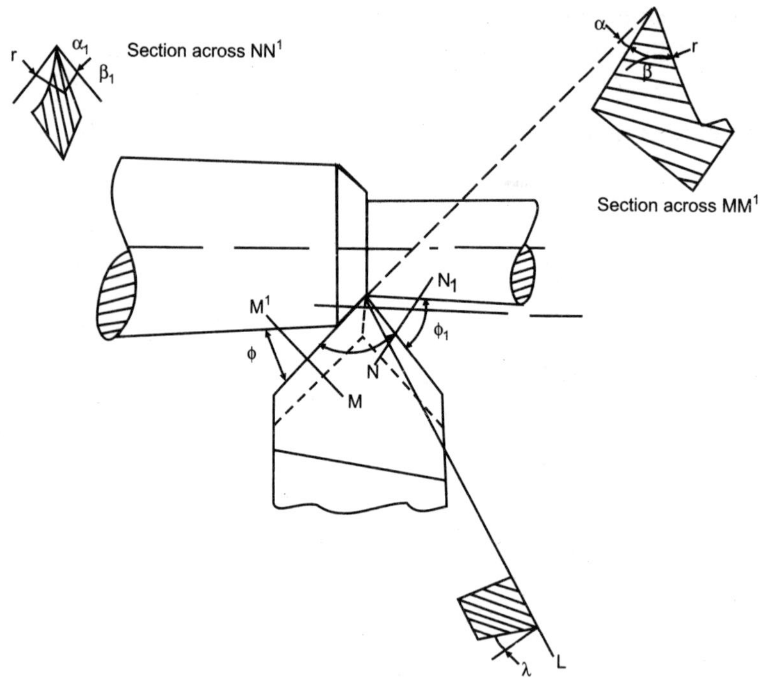

Fig. 7.9 (a) L-M-N-Planes (International System)

α = Side relief angle; β = Wedge angle; r = Side rake angle

$\alpha + \beta + r = 90°$ $\qquad$ δ = Cutting angle $= \alpha + \beta$

$\therefore \delta + r = 90°$ $\qquad$ $\therefore \delta = 90° - r$

ϕ = Plan approach angle; ϕ_1 = End cutting edge angle;

∂ = Nose angle; $\qquad$ λ = Inclination angle;

α_1 = End relief angle; $\qquad$ r_1 = End rake angle;

β_1 = End relief angle;

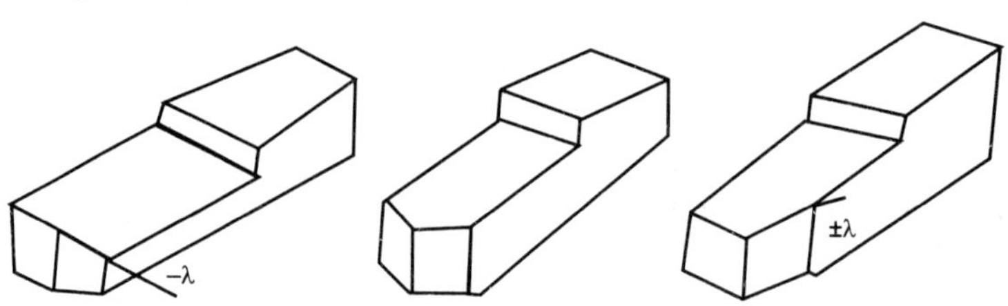

Fig. 7.9 (b) Angle of Inclination of the Side (Main) Cutting Edge

This is also known as International system (ISO). In this system, the side rake angle is defined as the angle between the base plane and the rake face, measured in a plane normal to the side cutting edge. The back rake angle is the angle between the base plane and the rake face measured in a plane normal to the end cutting edge.

7.7 TYPES OF METAL CUTTING PROCESS

In the metal cutting operation, the tool is forced into the work-piece to remove unwanted material from the wor-kpiece. There are two methods of metal cutting, depending upon the movement of the cutting edge with respect to the direction of relative work-tool motion. These are:

(a) Orthogonal cutting process—Two-dimensional cutting (Fig. 7.10 (a))

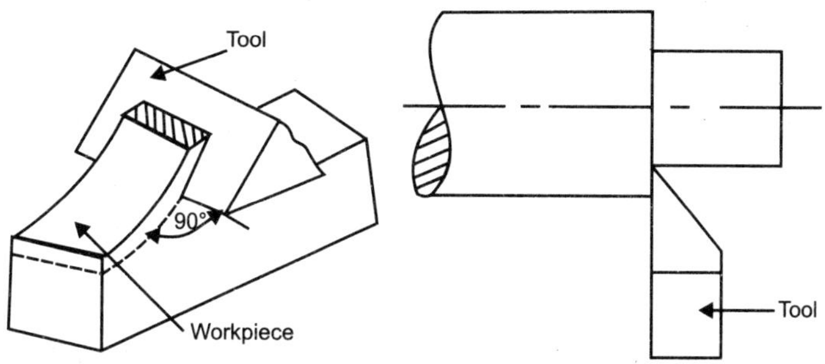

(a) Orthogonal cutting

(b) Oblique cutting process—Three-dimensional cutting (Fig. 7.10 (b))

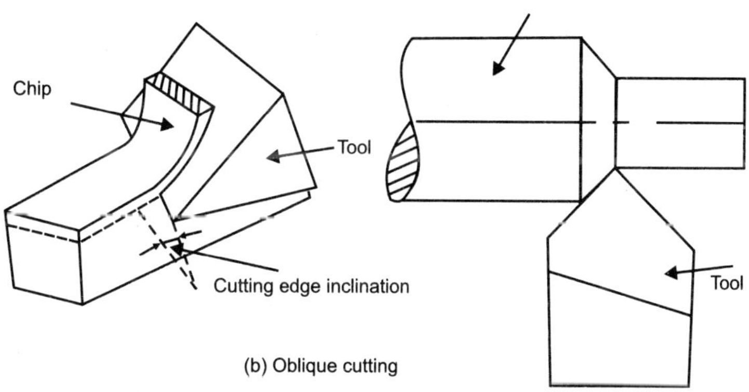

(b) Oblique cutting

Fig. 7.10 Methods of Metal Cutting Processes

7.8 COMPARISON OF ORTHOGONAL AND OBLIQUE CUTTING PROCESSES

Orthogonal Cutting

(a) The cutting edge of the tool remains normal to the direction of tool feed.
(b) The cutting edge clears the width of the work-piece on either sides
(c) The direction of chip flow velocity is normal to the cutting edge of tool

(d) Only two components of cutting forces act on the tool. These two components are perpendicular to each other.

(e) Unsuitable for efficient chip removal

Oblique Cutting

(a) The cutting edge of the tool is inclined at an acute angle to the direction of tool feed.

(b) The cutting edge may or may not clear the width of the work-piece.

(c) The direction of chip blow velocity is at an angle with the normal to the cutting edge of the tool.

(d) Three mutually perpendicular components of cutting forces act at the cutting edge of the tool.

(e) Suitable for efficient metal removal.

7.9 CHIP FORMATION

While machining, the metal in front of the tool rake face is compressed, first elastically and then plastically, finally sheared from the parent metal and comes as in the form of chip. The type of chip produced during metal cutting depends upon the machining conditions and material being cut. The variables which influence the type of chip produced as follows.

(i) Properties of the material
(ii) Depth of cut
(iii) Feed rate
(iv) Effective rake angle of the tool
(v) Cutting speed
(vi) Type and quantity of cutting fluid.

7.10 TYPES OF CHIPS

Three basic types of chips formed are:

(a) Continuous chips
(b) Discontinuous chips
(c) Continuous chip with built up edge.

(a) Continuous Chips (Fig. 7.11)

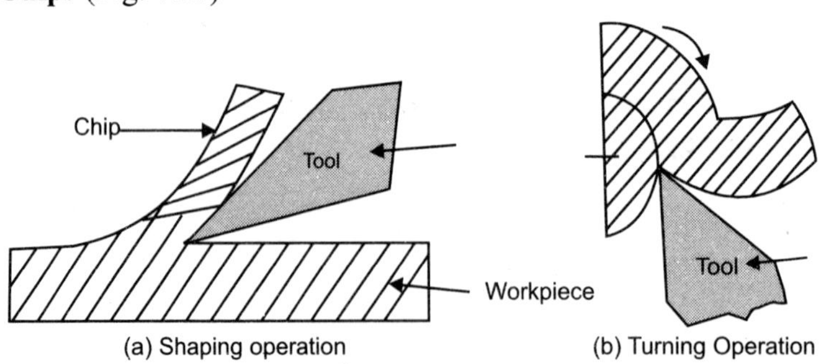

Fig. 7.11 Continuous Chips

These chips are in the form of long coils having same thickness throughout. The chips are produced due to the plastic deformation of metal without rupture. Factors responsible for these chips are:

(i) Ductile material
(ii) High cutting speed
(iii) Large rake angle
(iv) Efficient cutting fluids
(v) Low friction between tool face and chips.

(b) Discontinuous Chips (Fig. 7.12)

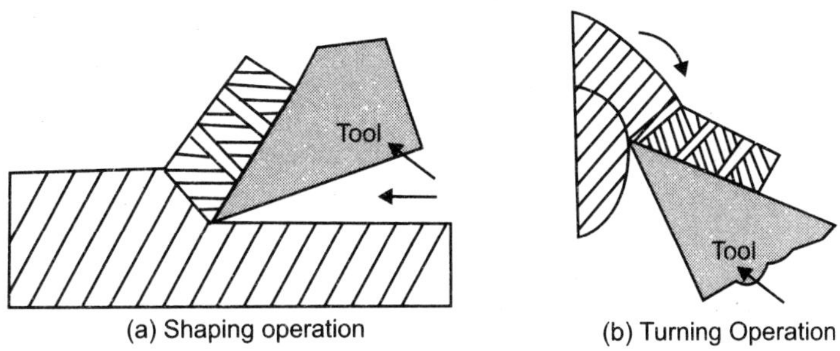

Fig. 7.12 Discontinuous Chips

The chips are small individual segments which may adhere loosely to each other. These segments are formed due to rupture of the metal ahead of the tool.

Factors responsible for these chips are

(i) Low cutting speed
(ii) Brittle materials
(iii) Small rake angle.

(c) Continuous Chip with Built up Edge: (See Fig. 7.13)

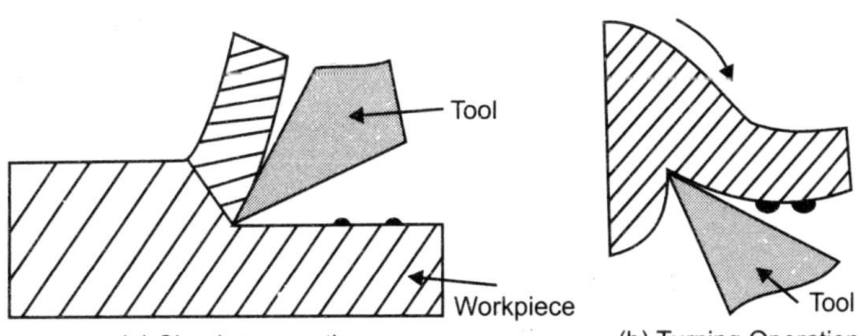

Fig. 7.13 Continuous chip with built up edge

During machining, the temperature and pressure is very high and causes the chip material to weld to the tool face near the nose. This is called build up edge. The accumulated built up of chip material break away, part will adhere to the chip, part will adhere to the work-piece.

Factors are:

(i) Ductile material

(ii) Small rake angle

(iii) Cause feed.

7.11 CHIP CONTROL

Continuous chips are produced while machining ductile materials. The disposal of such long ribbon like chips has never been problem while machining with high speed steels (HSS). But introduction of carbides and ceramics leads to increase in machining speeds and chip production is 3 to 5 times to that of HSS tools. At such high speed of machining, the chips leave the cutting edge at a very high velocity which are dangerous to the operator since the chips are hot and sharp. This leads to the development of a device to break the chip. This is known as chip breaker. The chip breaker is located near the cutting edge where it obstruct the flow of chip and makes them to curl immediately. The curled chips break either because of the tight curling or due to their hitting on the face to the cutting tool.

(a) Reasons for using the Chip Breaker

(i) The chips are dangerous to the operator since they are hot and their edges are sharp.

(ii) Long chips are difficult to dispose and occupy more space.

(iii) Hot and sharp edged chips may damage the instruments and machine tool painting.

(iv) The curled chips small about the work-piece and machine.

(v) The chips interfere with the flow of content and thus the use of coolent becomes less effective.

(vi) The emerging chips spoil the newly generated surface.

(b) Disadvantages

(i) It increases the cost of the cutting tool.

(ii) The cutting force increase and hence power consumption increases.

(c) Types of Chip Breakers

Various types of chip breakers are as follows:

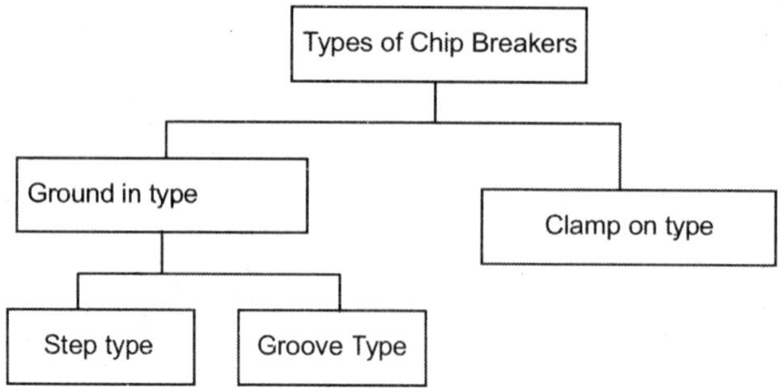

Step type Chip Breaker (Fig. 7.14 (a)

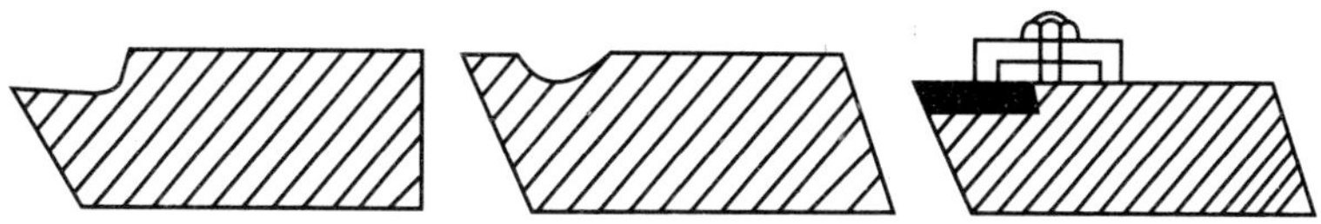

Fig. 7.14 Chip Breakers

In step type breaker, a step is ground on the face of the tool among the cutting edge. This step deflects back the chip into the work-piece or the side of the tool where it breakes.

Groove type chip breakers (See Fig. 7.14 (b))
In this chip breaker, a small groove is cut behind the cutting edge. With the development of inserted carbide tool, it is now possible to provide the groove on the carbide tip during manufacturing itself.

Clamp type chip breaker (See Fig. 7.14 (c))
In clamp type chip breaker, a thin carbide-faced plate or clamp is brazed or screwed on the face of the tool. In screw type, the provision for the adjustment of the plate can be effectively used over a large cutting rangers and materials.

7.12 CHIP THICKNESS RATIO (See Fig. 7.15)

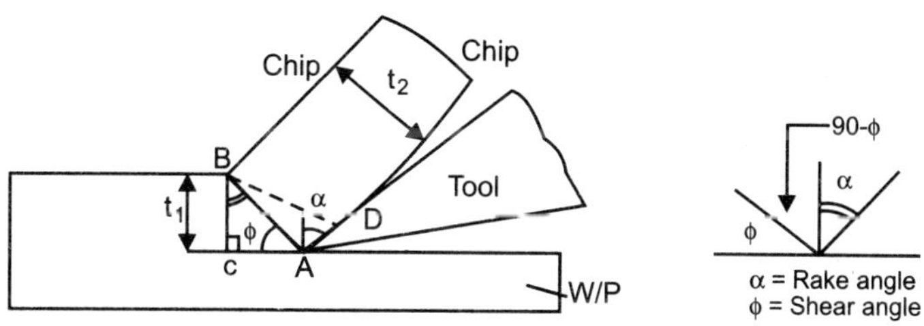

Fig. 7.15 Determination of Chip Thickness Ratio

Let t_1 = thickness of the chip before cutting

t_2 = chip thickness after cutting

chip thickness ratio $r = t_1/t_2$

when metal is cut, there is no change in volume of the metal cut

$$t_1 b_1 l_1 = t_2 b_2 l_2$$

where t_1 = chip thickness before cutting;

b_1 = width of cut

l_1 = length of chip before cutting

t_2 = chip thickness after cutting,

b_2 = width of chip after cutting

l_2 = length of chip after cutting

when there is no side flow of metal, then $b_1 = b_2$

$$\therefore \quad t_1 l_1 = t_2 l_2$$

$$t_1/t_2 = \frac{l_2}{l_1} = r$$

In case side flow is to be considered, the thickness ratio is to be multiplied by λ. Side flow factor is to be obtained when the length ratio

$$\lambda = {}^{b_1}\!/_{b_2}.$$

The chip thickness ratio can also be expressed in different way

From the right angle ΔABC

$$\sin\phi = \frac{BC}{AB}$$

$$\therefore \quad AB = \frac{BC}{\sin\phi} = \frac{t_1}{\sin\phi} \qquad ...(1)$$

From right angle triangle ABD

$$\sin(90 - \phi + \alpha) = \frac{BD}{AB}$$

$$AB = \frac{t_2}{\sin(90 - (\phi - \alpha))} = \frac{t_2}{\cos(\phi - \alpha)} \qquad ...(2)$$

From (1) and (2) equations

$$\frac{t_1}{\sin\phi} = \frac{t_2}{\cos(\phi - \alpha)}$$

$$\frac{t_1}{t_2} = \frac{\sin\phi}{\cos(\phi - \alpha)}$$

$$r = \frac{\sin\phi}{\cos\phi\cos\alpha + \sin\phi\sin\alpha}$$

$$\frac{r\cos\phi\cos\alpha + r\sin\phi\sin\alpha}{\sin\phi} = 1$$

$$r\frac{\cos\alpha}{\tan\phi} + r\sin\alpha = 1$$

$$r\frac{\cos\alpha}{\tan\phi} = (1 - r\sin\alpha)$$

$$\tan\phi = \frac{r\cos\alpha}{1 - r\sin\alpha}$$

$$\tan\phi = \frac{r\cos\alpha}{1 - r\sin\alpha}$$

7.13 FORCES ON THE CHIP (See Fig. 7.16)

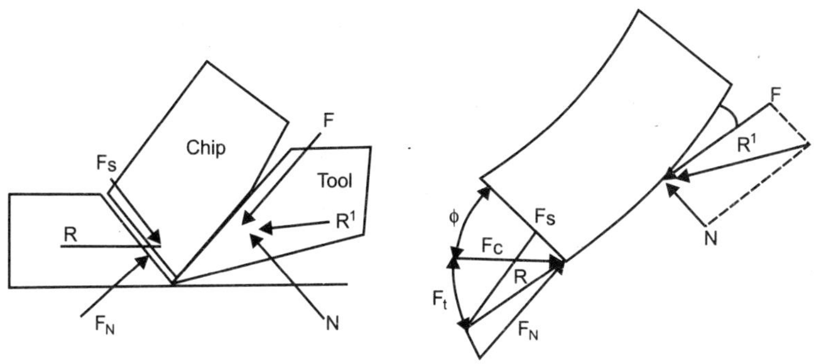

Fig. 7.16 Force Components on the Chip

The relationships amongsts the various forces (Fig. 7.16) has been worked out by Merchant with a number of assumptions.

(a) The tool is very sharp and there is no contact along the clearance face.

(b) The chip does not flow to either side (no side spread).

(c) A continuous chip without built up edge is produced.

(d) The cutting velocity remains constant.

(e) The chip behaves as a free body in stable equilibrium under the action of two equals, opposite and almost collinear resultant forces.

Force F_s acts along the shear plane and is the resistance to shear of the metal informing the chip. F_N is the force normal to shear plane. This is backing up force on the chip provided by the work-piece. R is the resultant of $F_s \propto F_n$.

Force F is the frictional resistance of the tool acting downwards against the motion of the chip as it moves along the tool face. The normal force N is normal to the tool face. The resultant of these two forces is R^1 and is the force exerted by the tool on the work-piece. The forces R^1 and R are equal in magnitude, opposite in direction and collinear

$$\overrightarrow{R^1} = \vec{F} + \vec{N}$$

$$R = \overrightarrow{F_S} + \overrightarrow{F_n} = \overrightarrow{F_C} = \overrightarrow{F_t}$$

Merchant represented the various forces inside a circle. The diameter of the circle is equal to R or R^1 passing through the tool point (Fig. 7.17).

In this diagram, the horizontal component is the cutting force F_c and the vertical component is the thrust force F_t. These two are measured by using a dynamometer t_1, t_2 and ϕ by calculation therefore other components are expressed in terms of known parameters.

As chip slides over the tool face under pressure, the kinematic co-efficient of friction (μ) may be expressed as

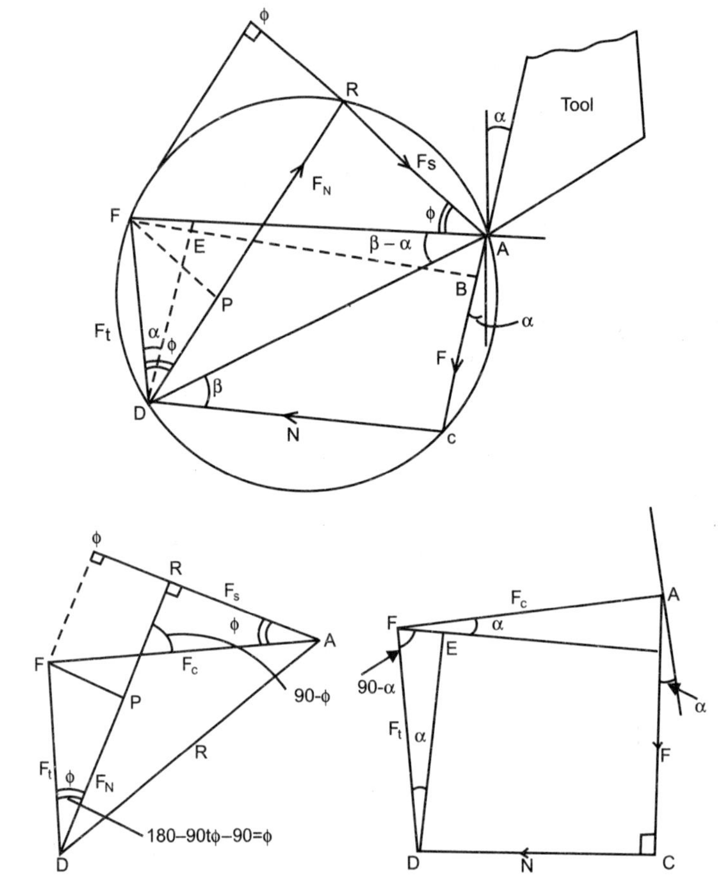

(a) Part of Merchant's force diagram (b) Part of Merchant's force diagram

Fig. 7.17 Merchant Circle Diagram

$$\mu = \frac{F}{N} = \tan\beta$$

Other force relationships:

1. $F = F_t \cos\alpha + F_c \sin\alpha$

$N = F_c \cos\alpha - F_t \sin\alpha$

as $F = AB + BC = AB + DE$

$= F_C \sin\alpha + F_t \cos\alpha$

$N = FB - FE$

$= F_C \cos\alpha - F_t \sin\alpha$

2. $F_s = F_C \cos\alpha - F_t \sin\phi$

$F_N = F_t \cos\phi - F_c \sin\phi$

as $F_s = A\phi - \phi R = A\phi - FP$

$$= F_c \cos\phi - F_t \sin\phi$$

$F_N = DR = DP + PR = DP + F\phi$

$$= F_t \cos\phi + F_c \sin\phi$$

3. $F_c = AD\cos(\beta - \alpha)$

$= R\cos(\beta - \alpha)$

$F_s = R\cos(\phi + \beta - \alpha)$

$$\frac{F_c}{F_s} = \frac{\cos(\beta - \alpha)}{\cos(\phi + \beta - \alpha)}$$

$$F_c = \frac{F_s \cos(\beta - \alpha)}{\cos(\phi + \beta - \alpha)}$$

4. $$\frac{F}{N} = \frac{F_c \sin\alpha + F_t \cos\alpha}{F_c \cos\alpha - F_t \sin\alpha}$$

$$= \frac{F_t + F_c \tan\alpha}{F_c - F_t \tan\alpha}$$

also $$\frac{F}{N} = \tan\beta = \mu$$

and $$\frac{F_t}{F_c} = \tan(\beta - \alpha)$$

7.14 VELOCITY RATIO (See Fig. 7.18)

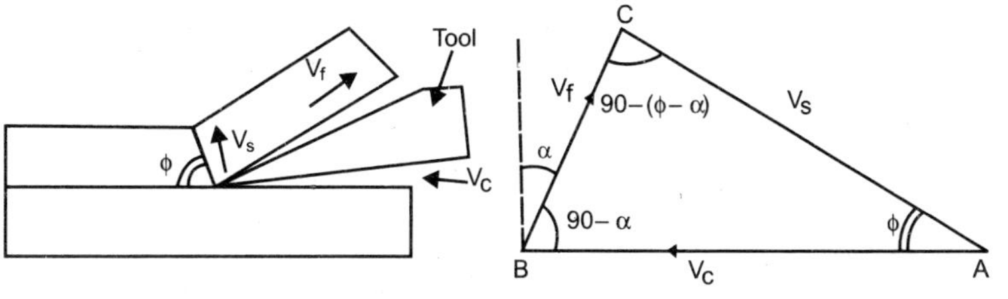

Fig. 7.18 Velocity Relationship

As the tool advances, the metal gets cut and chip is formed. The chip slides over the rake surface of the tool.

Let V_c = Velocity of the tool relative to the work-piece.
It is called cutting velocity

V_f = Velocity of the chip related to tool.
= It is called chip flow velocity.

V_s = Velocity of displacement or formation of the newly cut chip elements relative to the work-piece
It is called velocity of shear

Using trigonometric principles

$$\frac{V_c}{\sin 90-(\phi-\alpha)} = \frac{V_f}{\sin\phi} = \frac{V_s}{\sin(90-\alpha)}$$

$$\frac{V_c}{\cos(\phi-\alpha)} = \frac{V_f}{\sin\phi} = \frac{V_s}{\cos\phi}$$

$$\therefore \qquad V_s = V_c \cdot \frac{\cos\alpha}{\cos(\phi-\alpha)}$$

$$V_f = \frac{V_c \sin\phi}{\cos(\phi-\alpha)}$$

7.15 MACHINABILITY OF METALS

The ease with which a given material may be worked with a cutting tool is Machinability.

7.15.1 Machinability Criteria Depends on many Factors such as

(i) Chemical composition of work-piece material
(ii) Structure of work-piece
(iii) Mechanical properties
(iv) Physical properties (tensile, ductility)
(v) Cutting conditions such as cutting speed, feed etc.

7.15.2 Evaluation of Machinability

(i) Tool life
(ii) Intensity of cutting force
(iii) Quality of surface finish
(iv) Form and size of chip
(v) Temperature of cutting.

7.16 TOOL LIFE

During machining, the cutting edge of the tool gradually wears out and it needs regrinding as the wear increases, the tool looses its efficiency. So the "tool life" is defined as the time elapsed between two- successive grinding of the tool. It is expressed in minutes.

The various ways of expressing tool life are given below:

(a) Time of actual operation
(b) Volume of metal removed
(c) Number of pieces machined
(d) Equivalent cutting speed.

7.16.1 Factor's Affecting Tool Like

(a) Cutting speed
(b) Feed rate
(c) Depth of cut
(d) Hardness of work-piece
(e) Tool material
(f) Tool geometry
(g) Type of cutting fluid
(h) Rigidity of workpiece machine tool system
(i) Nature of cutting.

7.16.2 Tool Life Equation

F.W. Taylor developed the relationship between the tool life and cutting speed.

$$VT^n = C$$

where V = Cutting speed in meters/min

T = Tool life in minutes

n = Tool life index

= 0.1 to 0.15 for HSS tools

= 0.2 to 0.4 for tungsten carbide tools

= 0.4 to 0.6 for ceramic tools

C = Constant

These values When plotted on a log-log graph gives a straight line.

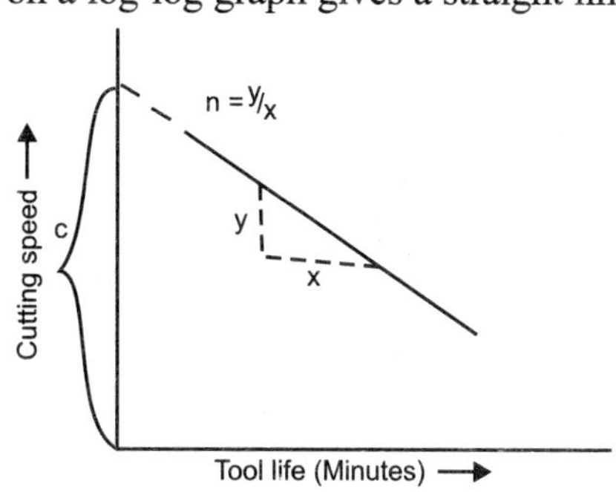

Fig. 7.19 Tool Life Plot

The tool life plot can be used for two things. Firstly, it can be used to determine the value of Exponent (n) and Constant C.

It can be used to predict the values of tool life at other cutting speeds. The slop of the tool life plot is the value of 'n' and the intercept of the y-axis, when the plot is extended backwards to meet the y-axis is the value of C.

The effect of depth cut and feed can be included in the above formula as

$$VT^{n}f^{n_1}d^{n_2} = C.$$

7.17 TOOL WEAR

The tool material on the rake face and the flank will gradually wear out and even fracture. Various wear mechanisms are described below.

(a) Shearing at High Temperature (Plastic Shear)

(See Fig. 7.20)

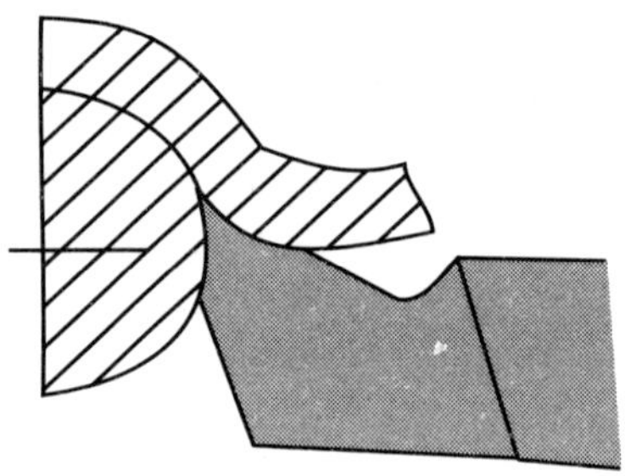

Fig. 7.20 Wear by Plastic Shear

While machining, the tool material softens at high temperature. At the same time, the chip flowing on the rack face gets work hardened so much to exert frictional stress sufficient to cause yielding by shear of the hard tool material.

(b) Diffusion Wear (See Fig. 7.21)

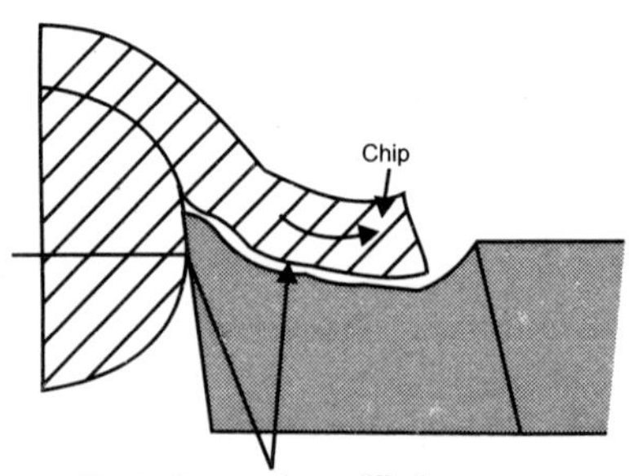

Fig. 7.21 Wear by Diffusion

While machining, the temperature at the interface is so high such that the atoms from hard metal (tool) diffuse into soft material (chip) thereby increasing the latter's hardness and abrasiveness.

On the contrary, atoms from the softer metal (chip) may also diffuse into hard material (tool) weakening the surface layer of the latter to such an extent that the particles on it are dislodged and torn out.

(c) Adhesive Wear (See Fig. 7.22)

On account of friction, high temperature and pressure, particles of the softer matieral (chip) adhere to a few high spots of the harder metal (tool). In due course more particles join up and built up edge is formed. Sooner or later some of these fragments which may have grown up in microscopic size are torn from the surface of the hard metal (tool). It appears as if the surface of hard metal (tool) has been nibbled away and made uneven.

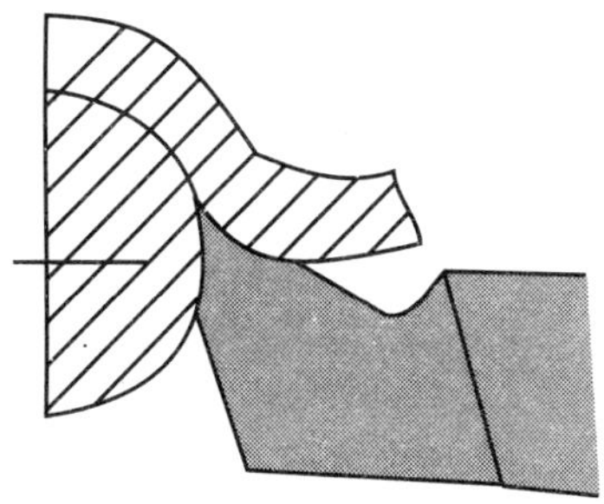

Fig. 7.22 Adhesive Wear

(d) Abrasive Wear (See Fig. 7.23)

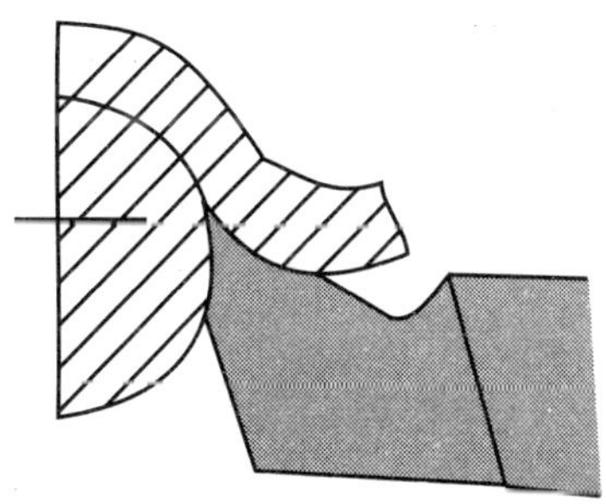

Fig. 7.23 Abrasive Wear

The chip material sliding over the tool may contain appreciable concentration of hard particles. These hard particles act as small cutting edges like those of a grinding wheel on the surface of the tool which in due course, is wornout through abrasion. These particles dragged along or rolled over the surface. While moving, the particles plough grooves into the surface of tool.

(e) Fatigue Wear

When two surfaces slide in contact with each other under pressure, asperities on one surface interlock with those of other. Due to the frictional stress, compressive stress is produced on one side of each interlocking asperity and tensile stress on other side. After a given time, a pair of asperities have moved over, the above stresses are relieved. New pair of asperities are, however soon formed and the stress cycle is repeated. Thus, the material of the hard metal near the surface under goes cyclic stresses. This phenomenon causes surface cracks.

(f) Electrochemical Effect

A thermo electric e.m.f. is setup in the closed circuit due to the formation of a hot junction at the chip tool interface. This current may assist the wear process at the rack face. For example, by aiding the diffusion of carbon ions from the carbide tool to the flowing chip.

(g) Oxidation Effect

Grooves or notches at the rack face and flank are due to reaction with oxygen in the atmosphere to form abrasive oxides while machining.

(h) Chemical Decomposition

Localized chemical reaction may occur that weakens the tool material through formation of weak compounds or dissolution of the bond between the binder and the hard constituents. For example, in carbide tool, the weakened particles are easily torn away by the asperities of the mating surface.

7.18 KINDS OF TOOL DAMAGE (See Fig. 7.24)

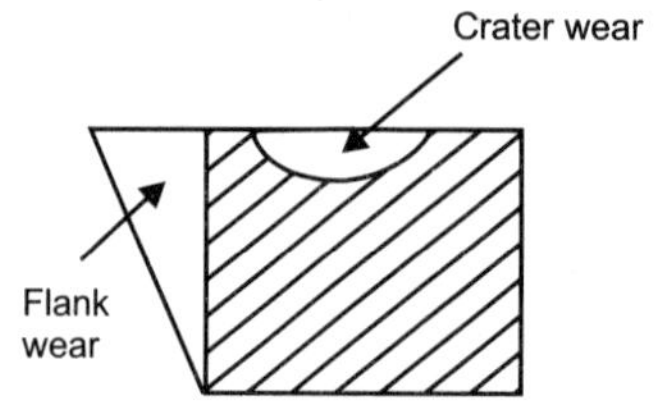

Fig. 7.24 Tool Damage

Generally, tool will fail on account of the following:

(a) Flank wear

(b) Crater wear

(c) Chipping.

(a) **Flank Wear**

This wear produces wear lands on the side and end flanks of the tool on account of the rubbing action of the machined surface.

(b) **Crater Wear**

It occurs on the rake face of the tool in the form of pit called crater. The crater is formed at some distance from the cutting edge. The cratering is a temperature dependent phenomenon caused by diffusion, adhesion etc. The crater significantly reduce the strength of the tool and thereby lead to its failure.

(c) **Chipping**

Chipping refers to the breaking away of small chips from the cutting edge. If a tool on account of impact, excess of plastic deformation, thermal stresses etc.

7.19 CUTTING FLUIDS

With the development of newer tool material like tungsten, carbide, ceramics, and use of high cutting speeds during machining, a large amount of heat is generated. The major part of heat is taken away by the chip and the remaining is shared by the workpiece and tool. This heat reduces the life of tool and influences the dimensional accuracy of the work-piece. The heat generated during machining should be effectively removed using proper type of cutting fluid (coolant). The use of cutting fluid improves tool life, surface finish and reduces the machining forces.

(a) Functions of Cutting Fluids (Coolants)

(i) It cools the cutting tool and work-piece. The heat produced is carried away by the coolant by supplying adequate quantity of coolant.
(ii) It washes away the chips from cutting zone.
(iii) It lubricates the chip-tool interface. Thus reduces the kinetic coefficient of friction and keeps down the cutting force.
(iv) They improve surface finish
(v) The use of coolant results in better surface finish
(vi) Use of coolant, reduces the thermal distortion of the work.

(b) Properties of Cutting Fluids

(i) They should have high heat absorption
(ii) They should possess good lubricating properties
(iii) They should not form easily
(iv) They should be odourless and transparent
(v) They should be stable in use and storage
(vi) They should be non-corrosive to work and machine
(vii) They should passess high flash point
(viii) They should be low viscosity
(ix) They should be low cost
(x) They should not be given the moving parts of the machine tool.

7.20 TYPES OF CUTTING FLUIDS

(a) Gaseous type
(b) Liquid type
(c) Oil based cutting fluids.

(a) Gaseous Type

Air (compressed air), carbon dioxide, argon are some of the examples.

These are applied through a nozzle under pressure. Liquid CO_2 possesses excellent property of extracting the heat from a body due to sublimation of CO_2. However, its use is very limited because of higher cost.

(b) Liquid Type Cutting Fluids

These are two general types, *water base* and *mineral oil base*

(*i*) *Water based cutting fluids*: Early attempts to improve cooling and lubricating properties of water included with the addition of soft soap but there are now oils where the desired effect and form emulsion with water.

(*ii*) *Mineral oil based cutting fluids*: These oils are known as soluble oils. Soda solutions are used on grinding operations as it has good flushing action and cooling effect.

(c) Oil Based Cutting Fluids

These are known as straight oils. These are classified into sub groups.

(i) Mineral oils
(ii) Fatty oils
(iii) Compound oils
(iv) Sulphurised oils
(v) Chlorinated oils.

7.21 METHODS OF APPLICATION OF CUTTING FLUIDS

The method of application of a cutting fluid is very important.

There are three methods commonly used.

(a) Flood application method
(b) Mist application method
(c) High jet method.

(a) *Flood Application Method*: In machine tools, the system consists of a pump mounted on a tank containing the cutting fluid. The outlet of the pump is connected to a nozzle through a flexible pipe. The nozzle can be adjusted to direct the stream of the fluid at the desired point (cutting zone). This is most commonly used method for all machine tools.

(b) *Mist Application Method*: In this method, the fluid is passed through a specially designed nozzle so that it forms very fine droplets of cutting fluid or produces a mist. This mist is directed at the cutting zone at a high velocity. This method of application has an advantage over flood method because during mist formation the temperature of fluids falls due to expansion and so it can absorb more heat.

(c) *High Jet Method*: This is a recent method of applying cutting fluid. In this method, a narrow jet at high velocity is directed at the flank surface of the cutting tool.

7.22 SELECTION OF CUTTING FLUIDS

The selection of cutting fluid for a certain operation depends on a number of parameters.

(a) Work-piece material

(b) Machining operation

(c) Cutting tool material

(d) Cost of the cutting fluid.

Example 7.1: In orthogonal cutting, cutting force = 150 kgs : Feed force = 50 kgs, r = 0.44, Rake angle 10°, cutting speed = 100 m/min : calculate (a) shear angle . (b) F, N, Fs, Ft and μ.

Solution: $\tan\phi = \dfrac{r\cos\alpha}{1 - r\sin\alpha} = \dfrac{0.44\cos 10°}{1 - 0.44\sin 10°} = \dfrac{0.4333}{0.932359} = 0.4691$

$$= 25/3°$$

$$\phi = 25°$$

$$\begin{aligned}
F &= F_t \cos\alpha + F_c \sin\alpha \\
&= 50 \cos 10° + 150 \sin 10° \\
&= 49.240 + 26.047 \\
&= 75.28 \text{ kgf} \\
N &= F_c \cos\alpha - F_t \sin\alpha \\
&= 150 \times 0.985 - 50 \times 0.174 \\
&= 139.3 \text{ kgf} \\
F_s &= F_c \cos\phi - Ft \sin\phi \\
&= 150 \times 0.906 - 50 \times 0.4225 \\
&= 114.9 \text{ kgf} \\
F_n &= F_t \cos\phi + F_c \sin\phi \\
&= 50 \times 0.906 + 150 \times 0.4225 \\
&= 108.9 \text{ kgf}
\end{aligned}$$

Coefficient of friction $= \mu = \dfrac{F}{N} = \dfrac{75.28}{139.3} = 0.54$

Example 7.2: In turning operation, a tool life of 80 minutes is obtained at the cutting speed of 30m/min and 8 minutes at the speed of 60m/min determine.

(a) Tool life equation

(b) Cutting speed for 4 minutes tool life.

$$V_1 = 30;\ T_1 = 80;$$

$$V_2 = 60;\ T_2 = 8$$

$$VT^n = C$$

$$V_1 T_1^n = V_2 T_2^n$$

$$30 \times (80)^n = 60 \times (8)^n$$

$$\frac{30}{60} = \left(\frac{8}{80}\right)^n$$

$$\frac{1}{2} = \left(\frac{1}{10}\right)^n$$

$$n = 0.3$$

$$VT^{0.3} = c$$

$$V_1T_1^{0.3} = V_3T_3^{0.3}$$

$$30 \times (80)^{0.3} = V_3 \times (4)^{0.3}$$

$$V_3 = \frac{30 \times 80^{0.3}}{4^{0.3}}$$

$$= 73.9 \text{ m/min.}$$

QUESTIONS

1. Define various tool angles used in a single point cutting tool with neat sketch?
2. Describe the tool geometry as per ASA and ORS systems?
3. What are the various chips formed in metal cutting. Briefly describe conditions favourable for their formation?
4. Distinguish between orthogonal cutting and oblique cutting?
5. What is tool signature?
6. What are the chip breakers? Sketch and Explain.
7. What are the assumptions made in Merchant circle? Draw the circle and show various cutting forces?
8. What are the essential properties of coolant?
9. Explain various methods of applying cutting fluids?
10. Explain Taylor's tool life equation?
11. Describe various types of tool wear?

8

Lathe

8.1 INTRODUCTION

The lathe is a machine tool which shapes a product by removing the extra material from it in the form of chips by rotating the work against a single point tool. The work is clamped either in a chuck or in between the centres. The most common type of surface produced is cylindrical

8.2 PRINCIPAL PARTS OF LATHE

The principal parts of a lathe are labelled and shown in Fig. 8.1. A brief description of these parts are as follows:

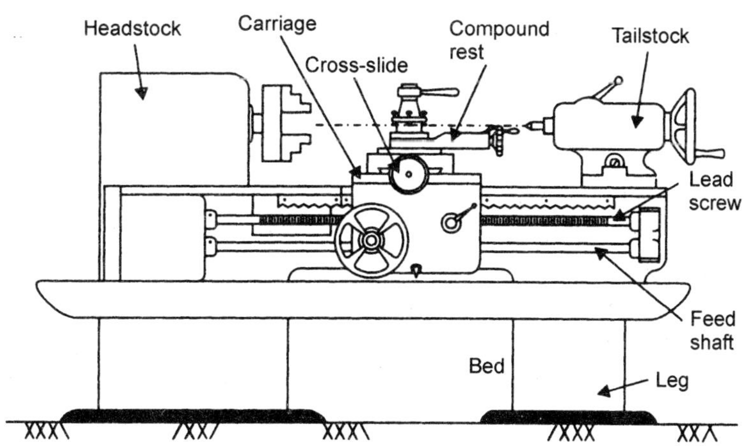

Fig. 8.1 Parts of a Lathe

8.2.1 Bed

The lathe bed forms the base of the machine. It is made of cast-iron or alloysteel. It consists of flat or inverted V shaped inner and outer guideways to guide the carriage, headstock and tailstock. The height of the lathe bed should be appropriate to enable the technician to do his job easily and conformably. It is rigidly by ribs which is shown in Fig. 8.2. Many lathes are made with a gap in the bed. This gap is used to swing extra large diameter pieces.

8.2.2 Headstock

The headstock is fixed at the left -hand side of the lathe bed on the inner guideways. It supports the spindle. The spindle is driven through the gearbox which is housed within the headstock. The function of gearbox is to provide a number of speeds to the spindle. The spindle is always hollow to feed barstock through that hole for continuous production. The nose of the spindle is threaded to mount the chuck or face plate.

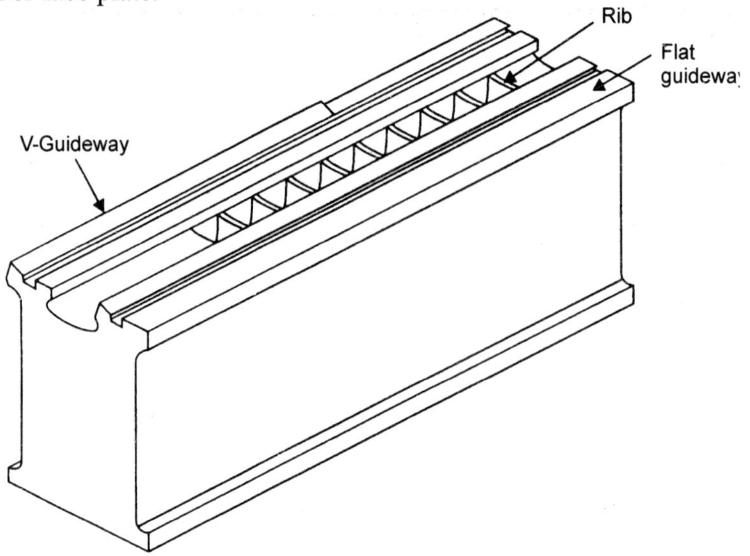

Fig. 8.2 Lathe Bed

8.2.3 Tailstock

It is located on the inner guideways at the right-hand side of the operator. The main purpose of the tailstock is to support the free end of the work piece when it is machined between centre. It is also used to hold tools for operations such as drilling, reaming, tapping etc. To accommodate different lengths of work, the body of the tailstock can be adjusted along the guideways by sliding it to the required position and can be clamped by bolt and plate. The body is bored to act as barrel which carries quill that moves in and out of the barrel. The movement of the quill is achieved by means of a handwheel and a screw which are engaged with a nut fixed in the quill. The hole in the open side of the quill is tapered to mount lathe centres or other tools like twist drills or boring bars. The upper body of the tail-stock can be moved towards or away from the operator by means of the adjustment screws to offset the tailstock for taper turning. A tailstock is illustrated in Fig. 8.3.

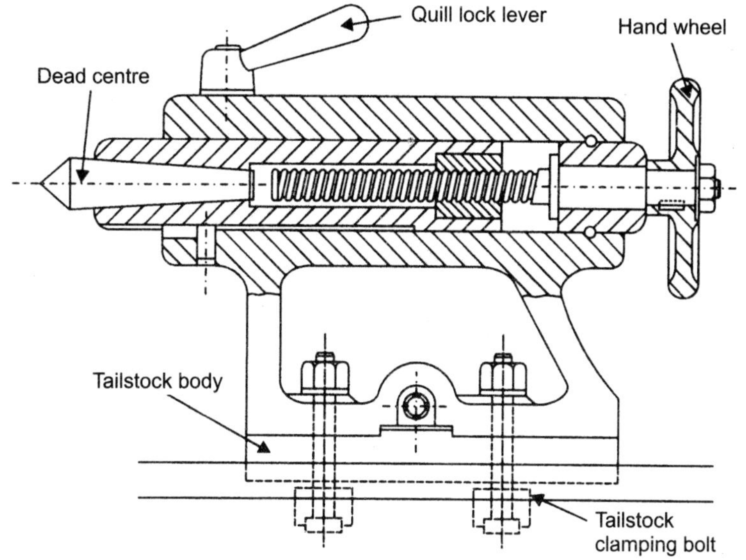

Fig. 8.3 Tailstock

8.2.4 Carriage (Fig. 5.4)

The carriage of a lathe has several parts that support, move and control the cutting tool. It consists of the following parts (a) saddle (b) cross slide (c) compound rest (d) tool post (e) apron.

(a) Saddle

The saddle is an H-shaped casting that fits over the bed and slides along the guideways. It carries the cross lide and tool post.

(b) Cross Slide

The cross slide carries the compound rest and toolpost. It is used to give depth of cut.

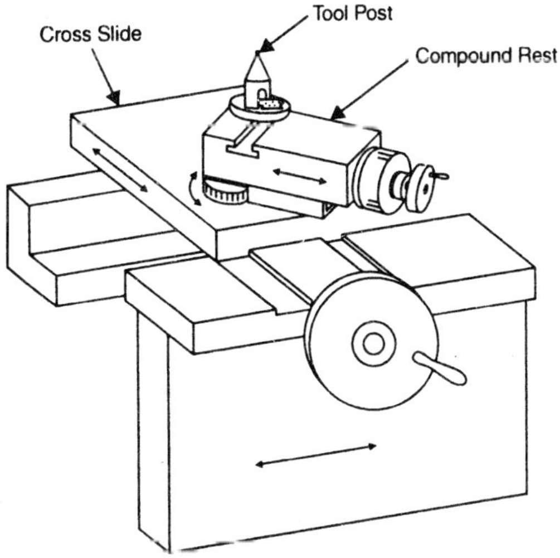

Fig. 8.4 Lathe Carriage

(*c*) *Compound Rest*

It has a circular base with graduations and mounted on cross slide. It is used for turning short tapers and angular cuts.

(*d*) *Tool Post*

This is located on the top of the compound rest to hold the tool and to enable it to be adjusted to a convenient working position. The various types of tool posts used in a lathe are shown in Fig. 8.5.

Single Screw Tool Post: This consists of a round bar with a slotted hole in the centre for the fixing the tool by means of a set screw. The tool post with concave ring and convex rocker slides in aT-slot on the top of the compound rest. The height of the tool point can be adjusted by tilting the rocker and clamping in position by means of set screw.

Four Way Tool Post: In this type of tool post four sides are open to accommodate four tools at a time.

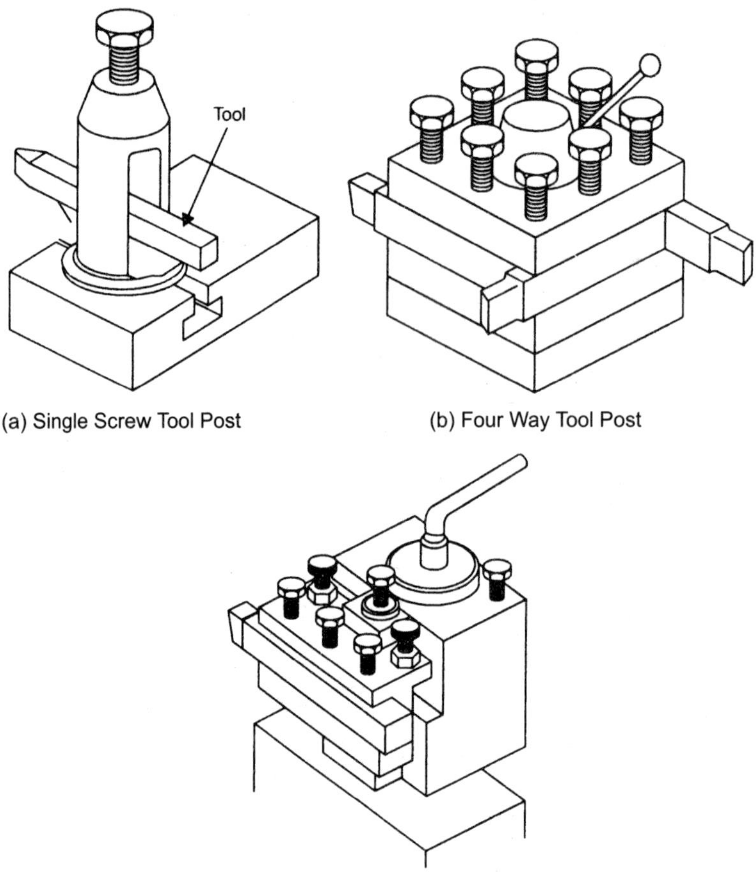

(a) Single Screw Tool Post (b) Four Way Tool Post

(c) Quick Change Tool Post

Fig. 8.5 Tool Posts Used in Turning

Quick Change Tool Post: Modern lathes are provided with this type of tool posts. Instead of changing the tools, the tool holder is changed in which the tool is fixed. This is expensive and requires a number of tool holders. But it has the advantage of ease setting of centre height and rigidity of the tool.

8.2.5 Apron

The apron is fastened to the saddle and hangs over the front of the bed. It contains gears, clutches for transmitting motion from feed rod to the carriage and also contains split nut which engages with lead screw while thread cutting.

8.3 SPECIFICATIONS OF LATHE

A lathe can be completely specified by the following factors :

(i) Height of the centres.

(ii) *The swing over bed*: Largest diameter of work that will rotate without touching the bed. It is generally twice of height of centres.

(iii) *The length between centres*: It is the greatest length of the work that can be held between the headstock and tailstock centres.

(iv) *Type of bed*: It may be straight or gap bed

(v) *Spindle speed range*: Number of speeds

(vi) Width of bed

(vii) Metric thread pitches

(viii) Cross feed and longitudinal feeds

(ix) Cross slide travel

(x) Tailstock sleeve travel

(xi) Horsepower of the main motor and RPM.

(xii) Shipping Dimensions: Length* Height* Width* Weight.

8.4 TYPES OF LATHES

There are various types of lathes differ in size, design, purpose, method of drive etc. as follows:

(i) Speed lathe

(ii) Engine lathe

(iii) Bench lathe

(iv) Tool room lathe

(v) Capstan and turret lathes

(vi) Special purpose lathes.

(i) Speed Lathe

It is the simplest of all lathes. It has only few parts like headstock, tailstock, and tool post mounted on a light bed. It has no gear box, leadscrew and carriage. These lathes are used for wood turning, metal spinning and polishing.

(ii) Engine Lathe

Derives its name from the early lathes that were driven by steam engines. It has compound rest, carriage, leadscrew, feed rod, gear box, in addition to headstock, tail stock and bed. It may be belt driven lathe, or motor driven lathe. The lead screw is used in cutting threads, whereas feed rod is used in automatic feed of the carriage or the tool.

(iii) Bench Lathe

This is a small lathe mounted on a bench. It performs all operations of an engine lathe. It is used mainly for small precision works. Only difference is in size.

(iv) Tool Room Lathe

This is similar in appearance to centre lathe, put it is built more accurately and has a wide range of speeds and is equipped with many extra accessories and attachments. It is mostly used for precision work on tools, jigs, dies etc.

(v) The Capstan and Turret Lathes

These are development of the engine lathe and are used for production work. The main features of this type of lathe is that the tail stock of an engine lathe is replaced by a hexagonal turret. A number of identical parts can be produced in the minimum time.

(vi) Special Purpose Lathes

These are specially designed lathes used for carrying out various operations that can not be done on ordinary lathes. The lathe which is especially desigend for turning crank shafts is known as a crankshaft lathe. The lathe which is powerful, massive and capable of turning axles is known as Axle Lathe or Wheel Lathe. Lathes which are used for duplicating profiles are known as duplicating lathes. The lathe in which various movements to the slides are given by electric motors which are controlled by the codes punched on tape is known as Numerical Control Lathe. This NC Lathe can do all operations which can be done on an ordinary lathe. The cutting tools are preset to the positions specified by the NC programmer.

8.5 LATHE OPERATIONS

The operations that can be performed on a lathe are known as turning operations. These are grouped into the following categories.

Operations performed in a lathe when holding the work-piece between centres (a) Plain turning (b) Taper turning (c) Threading (d) Grooving (e) Knurling.

Operations performed in a lathe when work-piece is held by chuck or face plate (a) Drilling (b) Reaming (c) Borning.

Operations performed by using special attachments (a) Grinding (b) Milling.

Some of the typical operations that can be done in a lathe are illustrated in Fig. 8.6.

8.5.1 Turning

Figure 8.6 (a) and (b) shows the turning operation. It is also called simple or plain turning. In this operation excess material is removed from the work piece to produce a cylindrical surface. It may be rough turning or finish turning according to the depth of cut given.

8.5.2 Facing Operation

Figure 8.6 (c) shows the facing operation in which the work is rotated and the tool is fed in a direction perpendicular to the axis of the work. This is used to cut the work to the required length and to provide flat surface with the axis of the work.

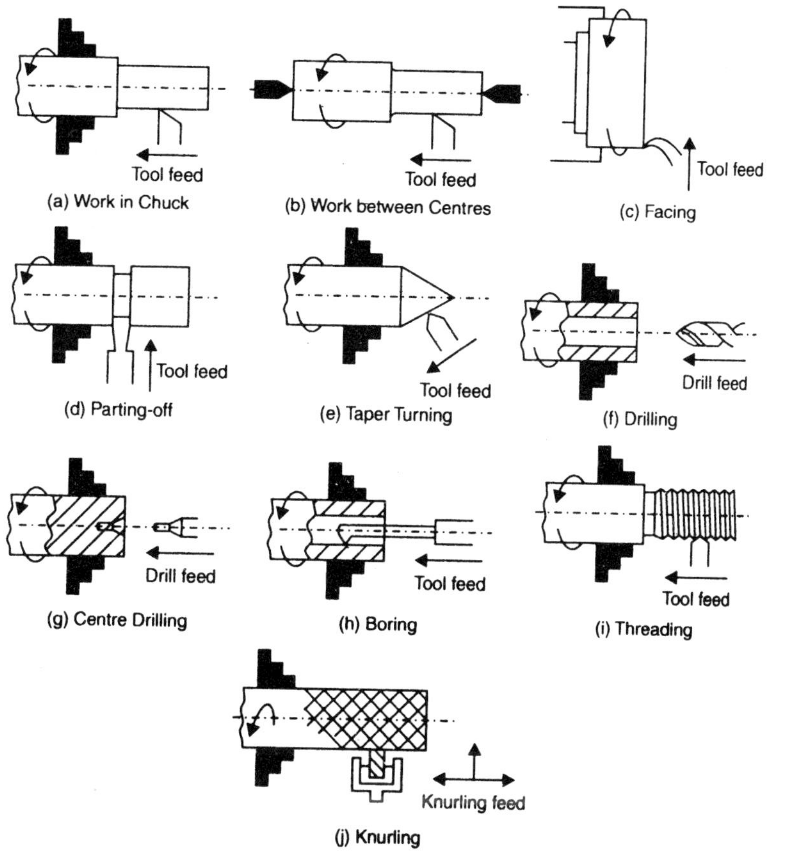

Fig. 8.6 Lathe Operations

8.5.3 Parting Off

Figure 8.6 (d). It is also called cutting off operation. It is used for cutting away the required length from the bar stock. The tool used in this operation is called parting off tool.

8.5.4 Taper Turning

In a lathe, taper turning means to produce a conical surface by gradual reduction in diameter from a cylindrical work-piece. A taper may be turned by any one of the following methods:

(a) Swivelling the compound rest
(b) Tailstock set over method
(c) Using a taper turning attachment
(d) Using a form tool
(e) By combination of longitudinal and cross feed.

(a) Taper Turning by Swivelling of Compound Rest (Fig. 8.7(a) and (b))

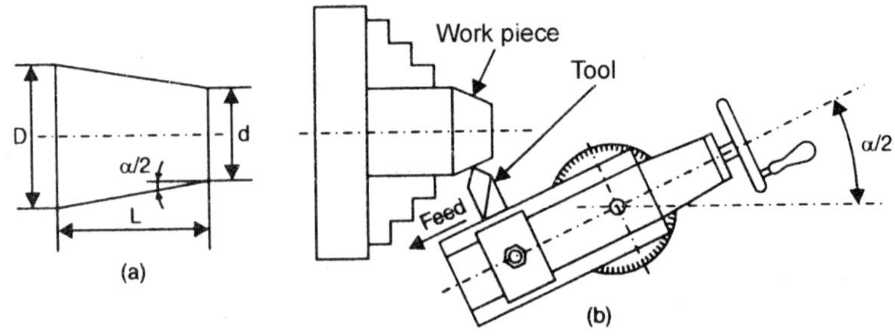

Fig. 8.7 Taper Turning by Swivelling Compound Rest

$$\tan \alpha/2 = (D - d)/2L$$

$$\alpha/2 = \text{Half of the taper angle}$$

In this method, the work is held in the chuck or between centres. The compound rest carrying the tool is swivelled to the desired angle (half of the taper angle) with respect to the work. The tool is fed manually by rotating the handwheel of the compound rest.

Advantages

(i) Easy setting
(ii) Steep tapers can be produced
(iii) Internal and external tapers can be made

Disadvantages

(i) Only hand feed
(ii) Taper length is limited to the movement of the top slide.

Example 8.1: Determine the angle at which the compound rest will be swivelled to turn taper on the given work-piece having the following dimensions (i) Large diameter 45 mm (ii) Small diameter 30 mm (iii) Length of the work-piece 200 mm.

Solution:

$$\tan \alpha/2 = (D - d)/2L$$
$$= (45 - 30)/2 \times 200$$
$$= 0.03$$
$$\alpha/2 = 2°9'$$

Hence, thc compound rest must be swivelled at 2° 9'.

(b) Tail Stockset over Method (Fig. 8.8)

In this method, the body of the tailstock is moved on its base towards or away from the operator by a set over screw by the set over amount.

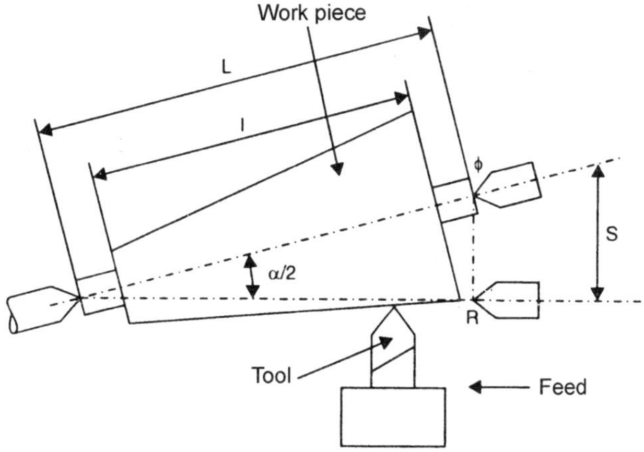

Fig. 8.8 Taper Turning by Tailstock Set over Method

$\phi R = p\phi \sin \alpha/2$, where $\phi 2$ is set over

Set over $= L \sin \alpha/2$

For practical purpose $\sin \alpha/2 = \tan \alpha/2$

Set over $= \text{L} \tan \alpha/2 = L * (D - d)/2\text{I}$

$= (D - d)/2 *$ (total length of the job)/ Taper length

If the taper is turned on entire length, then L = I

Set over $(D - d)/2$

Advantages

(i) Good surface finish can be obtained

(ii) Power feed can be used

Disadvantages

(i) Offset setting is difficult

(ii) Only external tapers can be produced

(c) *Using a Taper Turning Attachment* (Fig. 8.9)

Taper turning attachment is provided on a few modern lathes. The job is held between centres and the tool moves at an angle. The movement of the tool is guided by the attachment.

Taper-turning attachment is attached to the rear side of the lathe by means of bracket. It consists of swivel slide which can be adjusted to the required angle. A guide block sliding in this swivel slide is connected to the rear end of cross slide. To turn a taper, the workpiece is held between centres. The swivel slide is set at half the included angle ($\alpha/2$). The cross slide fixing screw is loosened. As the carriage travels in longitudinal direction, the tool on cross slide will follow straight angular path set by guide block.

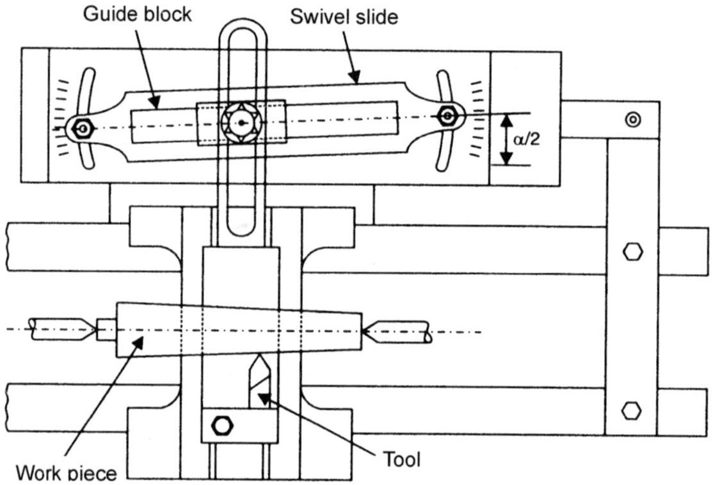

Fig. 8.9 Taper Turning by Taper Turning Attachment

Advantages

(i) Good surface finish can be obtained.

(ii) Lengthy tapers can be produced.

(iii) The alignment of the lathe centres is not disturbed.

Disadvantages

Only limited taper angles (10°) can be turned.

(d) Taper Turning by Form Tool Method (Fig. 8.10)

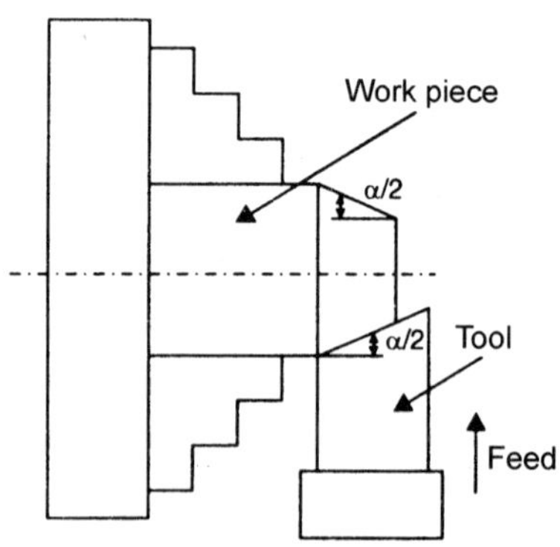

Fig. 8.10 Taper Turning by Form Tool Method

A broad nose tool having straight cutting edge is set on to the work at half taper angel and fed straight into the work to produce tapered surface. This method is limited to short tapers.

(*e*) ***Taper Turning by Combination Feeds*** (Fig. 8.11)

This is a special method of taper turning longitudinal and cross feeds are engaged simultaneously causing the tool to follow diagonal path to produce taper only to the work-piece.

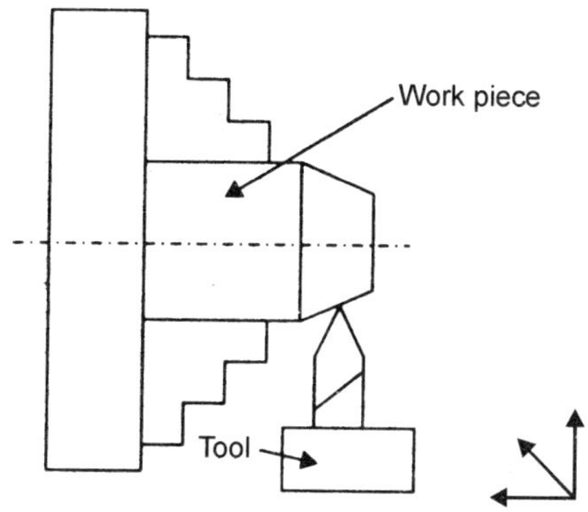

Fig. 8.11 Taper Turning by Combining Feeds

8.5.5 Thread Cutting on a Lathe (Fig. 8.12 (a) & (b))

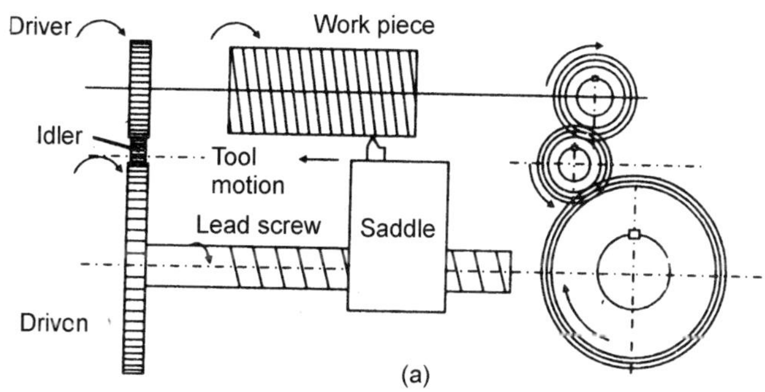

Fig. 8.12(a) Gear Train for Thread Cutting

Thread cutting is the process of producing a helical grove on a cylindrical or conical surface. The necessary condition for cutting screws is that in one revolution of the spindle (work), the tool traverse a distance equal to the pitch of the thread to be cut. This is achieved by a gear train between the lead-screw and the lathe spindle. The gear ratio for cutting screw threads may be given as:

Gear Ratio = Driver/Driven

= T.P.I. of lead screw/T.P.I. to be cut on job

= Pitch of the job/pitch of lead screw

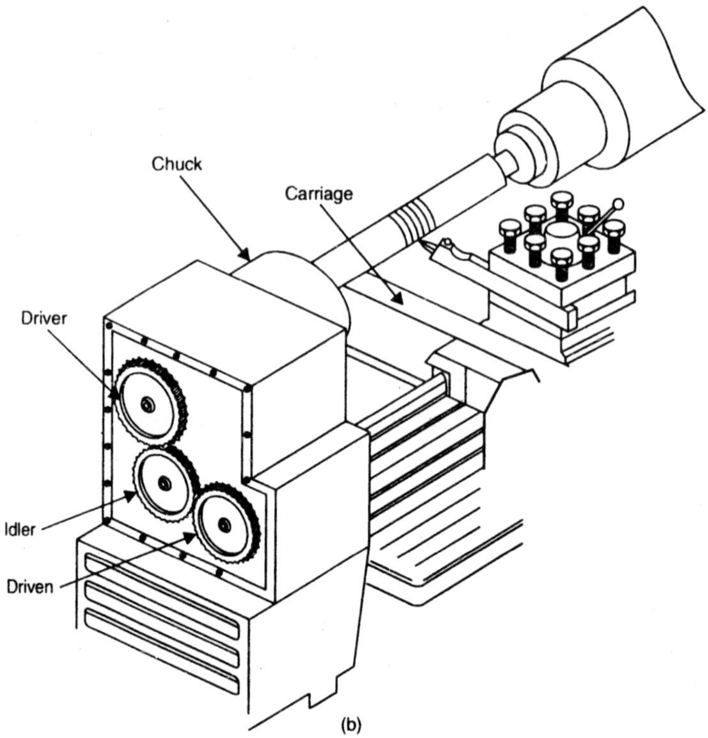

Fig. 8.12(b) Gear Train for Thread Cutting

Example 8.2: Calculate suitable gear train for cutting 8TPI in a lathe with a lead screw having 4TPI (Fig. 8.13).

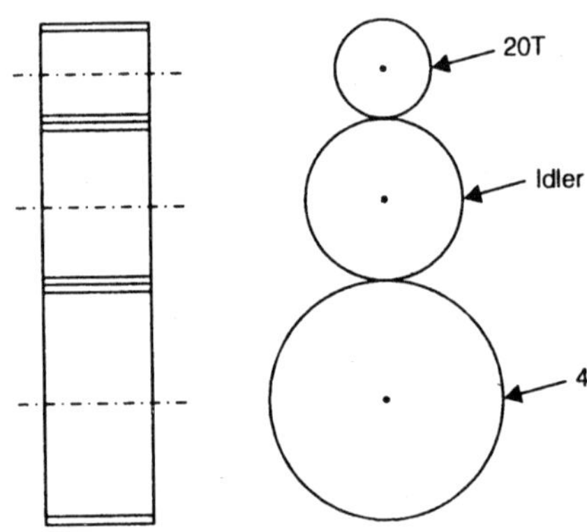

Fig. 8.13 Gear Train

Solution:

$$\text{Ratio} = \frac{\text{Drive teeth}}{\text{Driver teeth}} = \frac{\text{T.P.I of head screw}}{\text{T.P.I. to be cut}}$$

$$= \frac{4}{8} = \frac{4 \times 5}{8 \times 5} = \frac{20}{40}$$

Driver gear on spindle = 20 teeth

Driven gear of lead screw = 40 teeth

Example 8.3: Calculate gear train for 4 mm pitch on 11mm pitch of lead screw.

Solution:

Ratio = Driver teeth/Driven teeth

= Pitch or work / Pitch of lead screw

= 4/11 = 20/55

Driver gear on main spindle = 20 teeth

Driven gear on lead screw = 55 teeth

Gear train for cutting metric threads on a lathe with an English Lead Screw: In order to produce Metric threads with English lead screw, an additional gear with 127 teeth is incorporated in gear train and the formula for gear train is as follows:

Gear Ratio = Pitch to be cut/Pitch of lead screw

= Pc/PL×25.4 = Pc×5/PL×25.4×5 (1 inch = 25.4 mm)

= 5Pc/127 PL

Example 8.4: Calculate gears to cut 6 mm pitch on a lathe having lead screw of 4 TPI (Fig. 8.14)

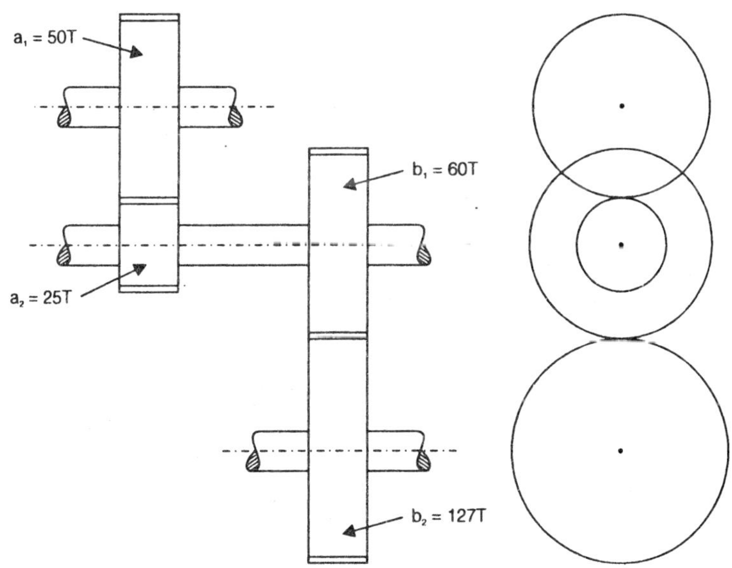

Fig. 8.14 Gear Train

Solution:

Pitch to be cut PC = 6 mm

Number of threads on lead screw = 4 TPI

Therefore, Pitch = 1/4 = 0.25"

Gear ratio = 5Pc/127PL = 5×6/127×25

= 1×30/127×25

= 50×60/25×127 = a1×b1/a2×b2

= Driving gears / Driven gears

8.5.6 Back Gear Mechanism of a Lathe

With direct speed, a three stepped pulley permits three speeds (N1, N2, N3)

When back gear mechanism is engaged, three speeds (N4, N5, N6) are obtained. The back gear mechanism is shown in Fig. 8.15

Pinion is attached to the cone pulley which is mounted freely on lathe spindle and gear D is keyed to the lathe spindle. Gears Band C are compound gears and are meshed with pinion A and gear D respectively, when back gear is engaged. When the back gear is disengaged, gears B and C are moved away and cone pulley is locked to gear D by means of lock pin. When higher speed is required, belt is shifted to smaller step.

If slower spindle speeds are required, the back gear is engaged. It consists of disconnecting the cone pulley from gear D by pulling out the lock pin and bring the back gear into position so that the pinion a meshes with gear B and gear C with gear D. Thus the speed from cone pulley is transmitted to spindle through gear train (A-B-C-D). Generally the back gear is engaged during thread cutting.

8.5.7 Knurling (Fig. 8.6 (j))

Knurling is the process of embossing a diamond or straight shaped pattern on the job.

The knurling may be diamond knurling, straight knurling, cross knurling, concave knurling and convex knurling. The purpose of knurling is to provide

(i) Good grip and make for positive handling.

(ii) Good appearance.

(iii) For raising the diameter to a small range for assembly to get a press fit. for example, the winding knob of a wrist watch has knurl on it

8.5.8 Grooving

This is the process of reducing the diameter of a work-piece over a narrow surface. This is often at the end of a thread or adjacent to a shoulder to leave a small margin.

8.5.9 Parting-off (Fig. 8.6 (d))

Parting-off or cutting-off is the operation of separating the finished work-piece from a bar stock.

8.5.10 Drilling (Fig. 8.6 (f))

Drilling is the process of making hole in a work-piece. Before drilling, the work should be faced and centre drilled. The drill bit is held stationary in the tail stock. The drill is fed into the work which revolving in a chuck.

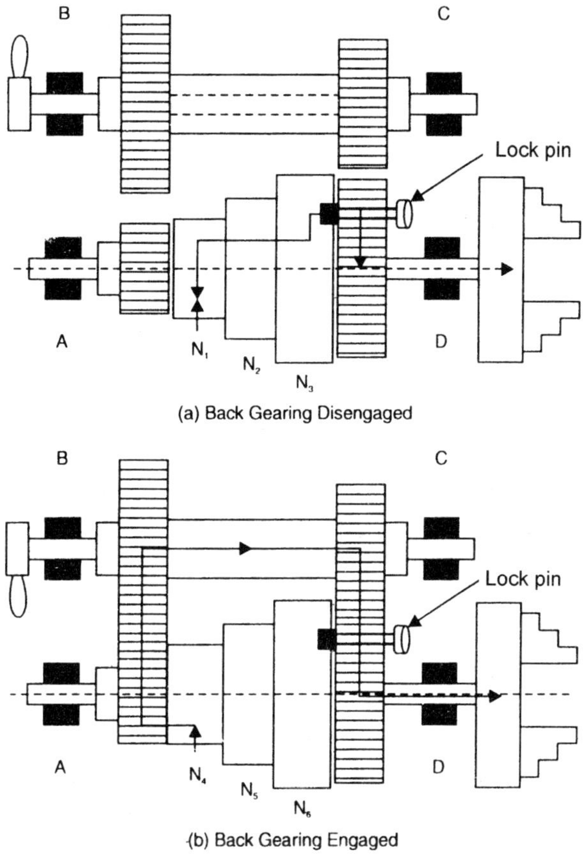

Fig. 8.15 Back Gear Mechanism of a Lathe

8.5.11 Boring (Fig. 8.6 (h))

This is the operation of enlarging a drilled hole. Boring tool is held in the tail stock.

8.5.12 Milling

Milling is the operation of removing metal by feeding the work against the rotating milling cutter. The milling cutter may be held by chuck or by the attachment mounted on carriage.

8.5.13 Grinding

Both internal and external surfaces of the work-piece may be ground by using grinding attachment mounted on the cross slide.

8.6 LATHE ACCESSORIES

The devices used for holding and supporting the work and the tool on the lathe are called accessories. These include chucks, catch plates and carriers, collets, face plate centres, mandrels, jigs and fixtures etc.

8.6.1 Chucks

These are used for holding the work-piece on lathe during the operation. The commonly used chucks are:

(a) Three-jaw universal chuck
(b) Four-jaw independent chuck
(c) Combination chuck
(d) Magnetic chuck
(e) Collet chuck
(f) Air or Hydaulic operated chuck
(g) Drill chuck

(a) Three jaw universal chuck

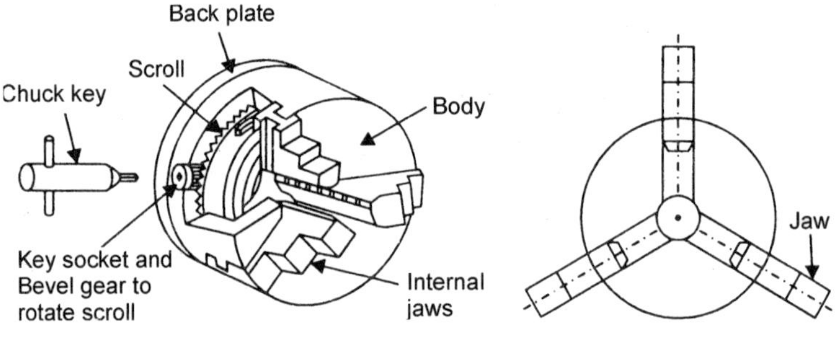

Fig. 8.16 Three Jaw Universal Chuck

In three jaw universal chuck or self centering chuck (Fig. 8.16) all three jaws move together in equal amounts to clamp the work. Therefore the job is automatically centred. The movement is achieved by rotating chuck key in any one of the three pinions which meshes with the teeth cut on the under side of the scroll disc. The scroll disc having a spiral groove cut on the top face meshes with the teeth of jaws. The chuck is used for holding cylindrical or hexagonal shaped work-pieces.

(b) Four jaw independent chuck

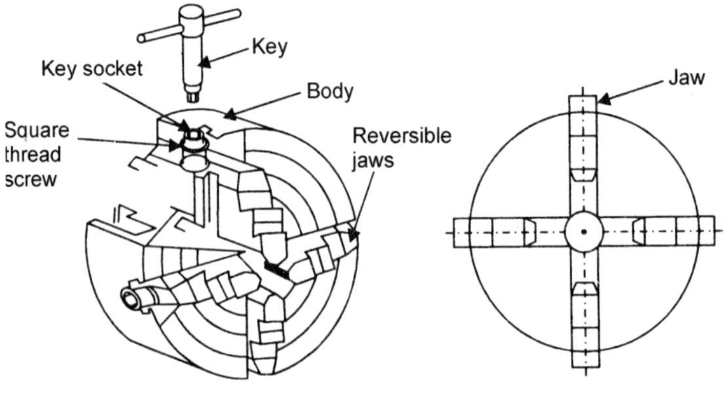

Fig. 8.17 Four Jaw Independent Chuck

In four jaw independent chuck (Fig. 8.17), each jaw is moved independently by rotating the screw which meshes with the teeth cut on the underside of the jaw. These are used for holding square, octagonal or large irregular components.

(*c*) *Combination chuck*

Combination chuck carries the combination of both the above principles. It is provided with four jaws which can be operated either by the scroll disc or individually by separate screws.

(*d*) *Magnetic chuck*

These are used to hold the steel work pieces that are too thin to be held in ordinary chuck. The face of the chuck is magnetized by permanent magnets contained within the chuck.

(*e*) *Collet chuck*

This provides a quick means of holding the bar stock. Draw in type collets are in common use. Their front portion is splitted which provide a spring action and hence the grip.

(*f*) *Air or hydraulic operated chuck*

In these chucks air or hydraulic pressure is used to press the jaws against the job. The pressure is provided by a cylinder and piston mechanism mounted at the back of head stock and controlled by a valve by operator.

(*g*) *Drill chuck*

This is used for holding straight shank drill, reamer or tap for drilling, reaming or tapping operations. This may be held either in head stock or tail stock. This has self centering jaws and is operated by key.

8.6.2 Face Plate

It is a large circular disc having a threaded hole at the centre so that it can be screwed to the nose of the lathe spindle. It contains open slots or T-slots in its face. The work piece is mounted on it with the help of bolts, T-nuts and other means of clamping. It is used for holding work pieces that can not be conveniently held by chucks (See Fig. 8.18).

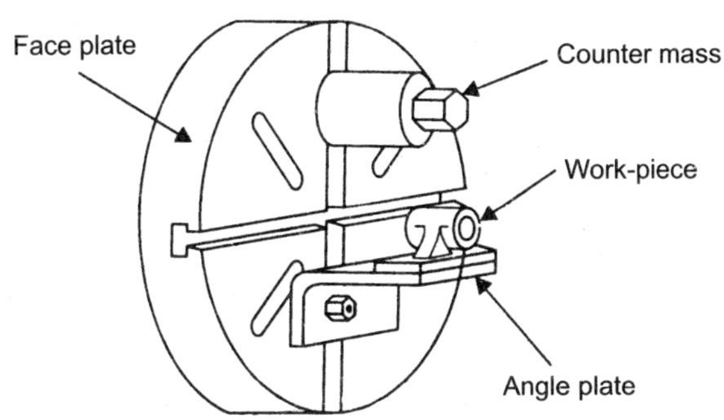

Fig. 8.18 Facc Plate

8.6.3 Angle Plate

It is used for holding work in conjunction with a face plate. When mounting of the work directly on face plate is not possible, the angle plate is used (See Fig. 8.18).

8.6.4 Lathe Centres (Fig. 8.19)

Lathe centres are hardened steel devices used for holding and locating the work to be turned. The centre that is fitted in the head stock spindle is called live centre. The centre that is used in tail stock is called dead centre. The various forms of lathe centres are shown in Fig. 8.19.

(a) Ordinary Centre

This is used for most general works.

(b) Ball Centre (Fig. 8.19(b))

This has ball shape at the end to minimise the wear and strain. This is particularly suitable for taper turning.

(c) Tipped Centre (Fig. 8.19(c))

Hard alloy tip is brazed into steel shank. The hard tip is wear resistant.

(d) Half Centre (Fig. 8.19(d))

The half centre is similar to ordinary centre except that a little less than half of the centre has been ground away. This feature facilitates facing of the bar ends without removal of the centre.

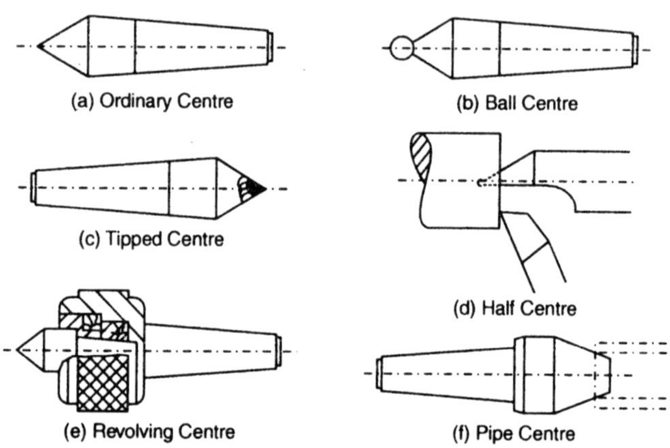

Fig. 8.19 Lathe Centres

(e) Revolving Centre (Fig. 8.19(e))

The ball and roller bearings are fitted into the housing to reduce friction and to take up end thrust. This is used in tail stock for supporting heavy work revolving at a high speed.

(f) Pipe Centre (Fig. 8.19(f))

This is used for supporting pipes, shells and hollow end jobs.

8.7 DRIVE PLATES AND CARRIERS

These are used to drive a work piece when it is held between centres. Drive plate is screwed to the nose of the headstock spindle. Carrier (dog) is attached to the end of the work piece by a set screw Fig. 8.20 shows drive plates, carriers and their assembly.

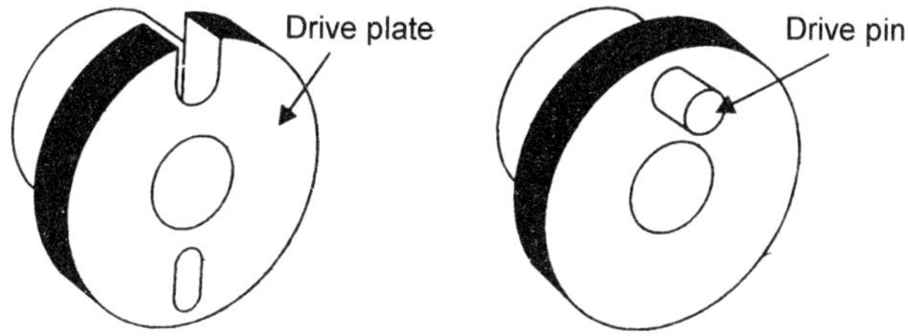

(a) Types of Drive Plates

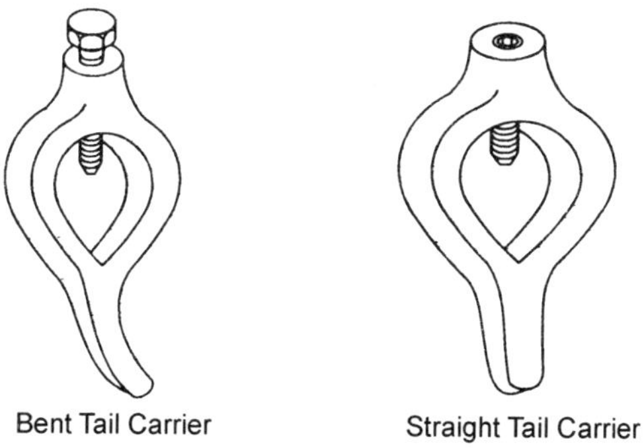

(b) Types of Carriers

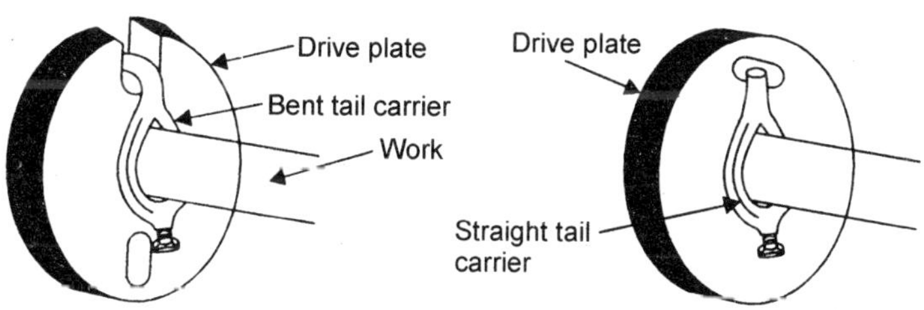

(c) Drive Plate and Carrier Assembly

Fig. 8.20 Drive Plates and Carriers

8.8 MANDRELS

Mandrels are used to hold and rotate hollow works between centres. It is made of high carbon steel and slightly taper. The work is forced to fit on the mandrel. The mandrel is rotated by lathe dog and catch plate and the work is driven by friction. The common types of mandrels are shown in Fig. 8.21.

The plain mandrel body is slightly tapered in order to grip the work piece (See Fig 8.21(a)). This type of mandrel is suitable for only size of bore. For different sizes, different mandrels are used. The stepped mandrel (Fig. 8.21(b)) facilitates the use of the same for various jobs having different sizes of holes. Screwed mandrel (Fig. 8.21 (c)) is threaded at one end with collar. Work pieces having internal threads are screwed to it against the collar for machining. Expansion mandrel (Fig. 8.21(d)) consists of a tapered pin which driven into sleeve which is parallel outside and tapered inside. The sleeve has three longitudinal slots. The sleeve is first placed in the work with the pin removed. The tapered pin is then pressed from one end into the sleeve and the sleeve expands gripping the work piece to be machined.

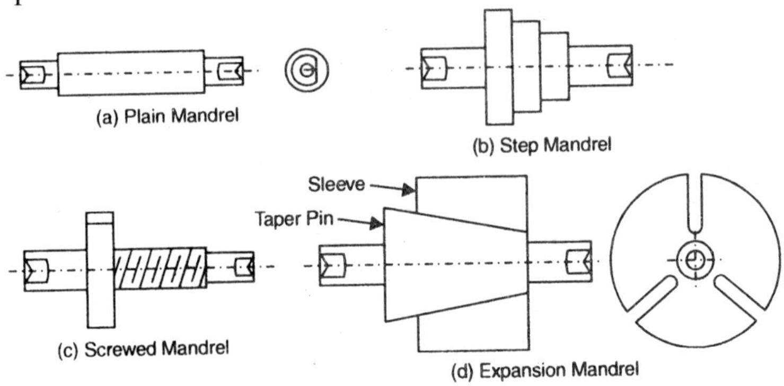

Fig. 8.21 Types of Mandrels

8.9 STEADY REST

The steady rest consists of a cast iron frame made of two parts hinged together on one side. The upper part can be swung back for inserting or removing the job without disturbing the position of the steady rest. It can be clamped at any desired position on lathe bed guideways.

8.10 FOLLOWER REST

The follower rest performs the same function as steady rest, but it is attached to the saddle and moves along the tool. It prevents the job from spring away when cut is given.

8.11 LATHE ATTACHMENTS

There are a number of attachments used on a lathe to increase efficiency and production. The commonly used attachments are:

(i) Stops
(ii) Grinding attachment
(iii) Milling attachment
(iv) Taper turning attachment
(v) Copying attachment
(vi) Relieving attachment

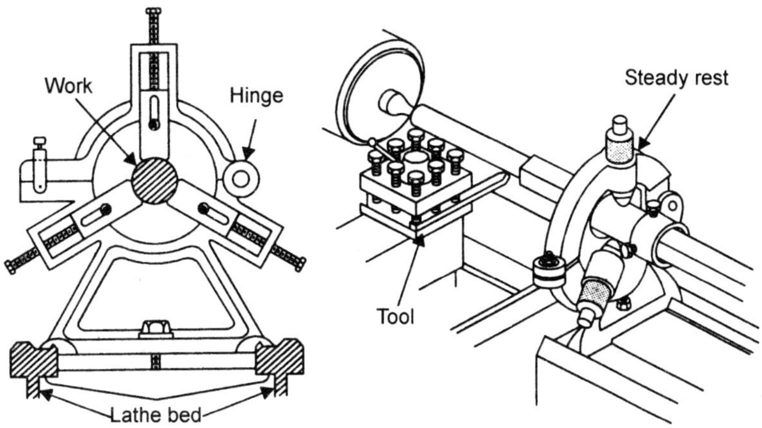

(a) Steady Rest (b) Steady Rest Used to Support Long Job

Fig. 8.22 Steady Rest

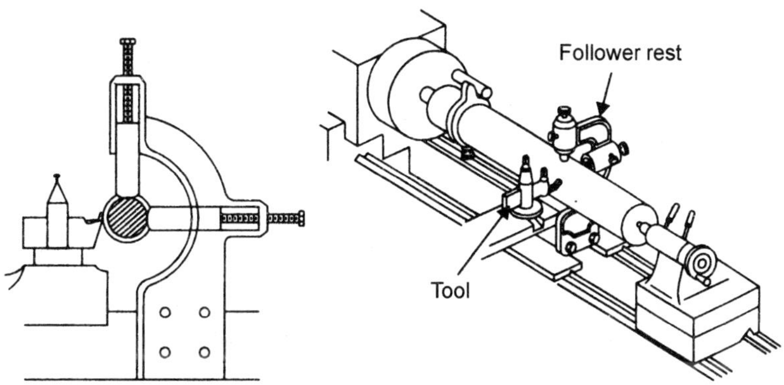

(a) Follower Rest (b) Follower Rest Used to Support Job

Fig. 8.23 Follower Rest

8.11.1 Stops

These are used on the carriage and the cross slide to position them accurately. These are used for repeated works. These stops save set up time and gives more accurate works.

8.11.2 Grinding Attachment

It consists of a grinding wheel driven independently by a small motor which is mounted on the cross slide. The job is held as usual in a chuck or between centres and the rotating grinding wheel is fed against the job instead of the usual cutting tool.

8.11.3 Milling Attachment

Milling attachment is mounted on compound rest in the place of the tool post. It consists of a slide swivel vise. The base of swivel vise is graduated in degrees and can be set at any required angle. This attachment is used for face milling, keyway cutting, T-slot cutting etc.

8.11.4 Taper Turning Attachment

This is used for producing tapers on cylinders. Similarly, various attachments like copying attachment, relieving attachment, etc. can be used on a lathe.

8.12 LATHE TOOLS

Cutting tools used for machining are classified into two groups:

(a) Single point cutting tool
(b) Multi point cutting tool

Single point cutting have only one cutting edge, while multi point cutting tool will have several cutting edges (drill, milling cutter). Single point cutting tools are used on lathes.

Classification

Single point lathes tools are classified in many ways. Main classification is as follows:

(a) According to the direction of feed (Fig. 8.24)

These tools may be right hand and left hand tools. In right hand tool, the cutting edge is on the left hand side of the operator. The right hand tool cuts from right to left i.e., tool is feed from tai stock to head stock. In left hand tool, the cutting edge is on right side of the operator. The left hand tool cuts from left to right.

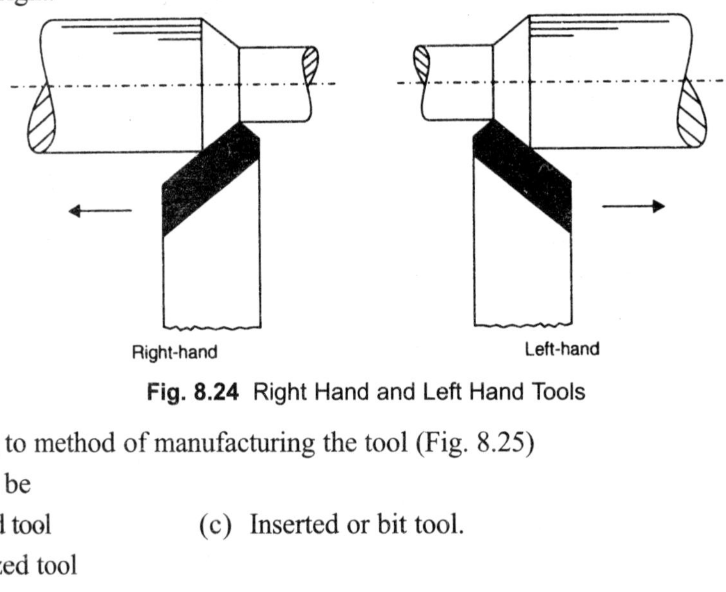

Fig. 8.24 Right Hand and Left Hand Tools

(b) According to method of manufacturing the tool (Fig. 8.25)

These tools may be

(a) Solid tool
(b) Brazed tool
(c) Inserted or bit tool.

Solid tool is made of either high speed steel or carbide bar and the cutting edge is formed by grinding one end of the bar. In braze tool, carbide tip is brazed to a shank of low grade material. In inserted bit tool, the carbide or ceramic bit of a square or rectangular shape is held mechanically in a tool holder.

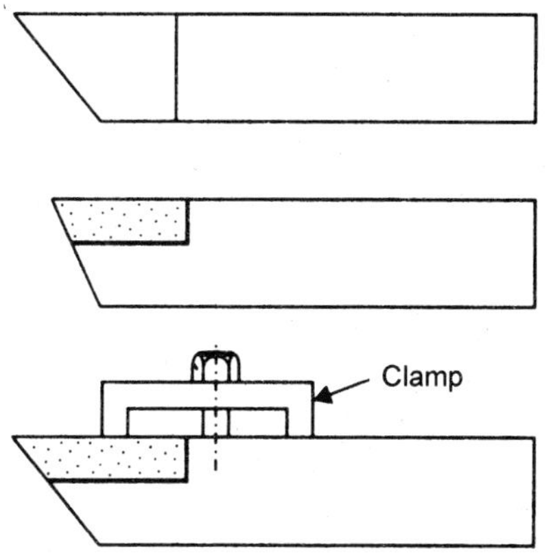

Fig. 8.25 Manufactured Tools

(c) According to the method of using the tool (Fig. 8.26)

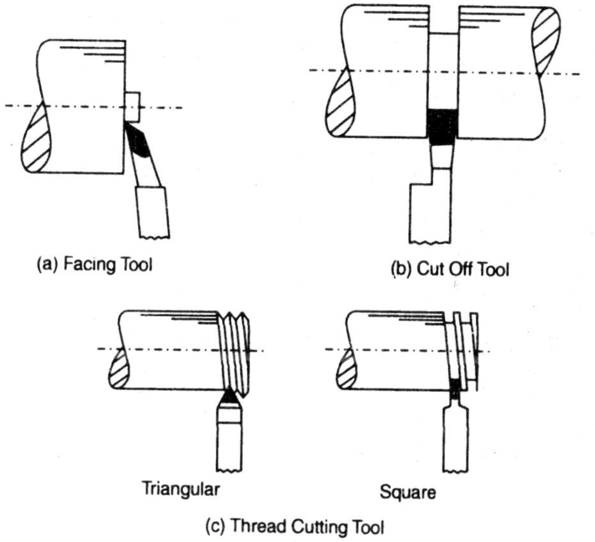

Fig. 8.26 Method of Using Tool

Single point tools are classified as turning, facing, cutting off, boring, grooving etc.

8.13 TOOL NOMENCLATURE

Cutting tool nomenclature comprises the various parts of a tool and various tool angles. The complete nomenclature of a single point cutting tool is shown in Fig. 8.27(a) and (b).

Face: It is the surface over which the chip flows.

Flank: It is the surface below the cutting edge.

Nose: The nose is the junction of the side and end cutting edge.

Side Cutting Edge: It is formed by the intersection of the flank and the side flank. It does the main work in cutting.

End or Auxiliary Cutting Edge: It is the intersection of face end flank.

Tool Angles: In a single point tool, there are various angles, each of them has a definite purpose.

Back Rake Angle: It measures the downward slope of the top surface of the tool from nose to the rear along the longitudinal axis. Its purpose is to guide the direction of chip flow. The size of the angle depends upon the material to be machined.

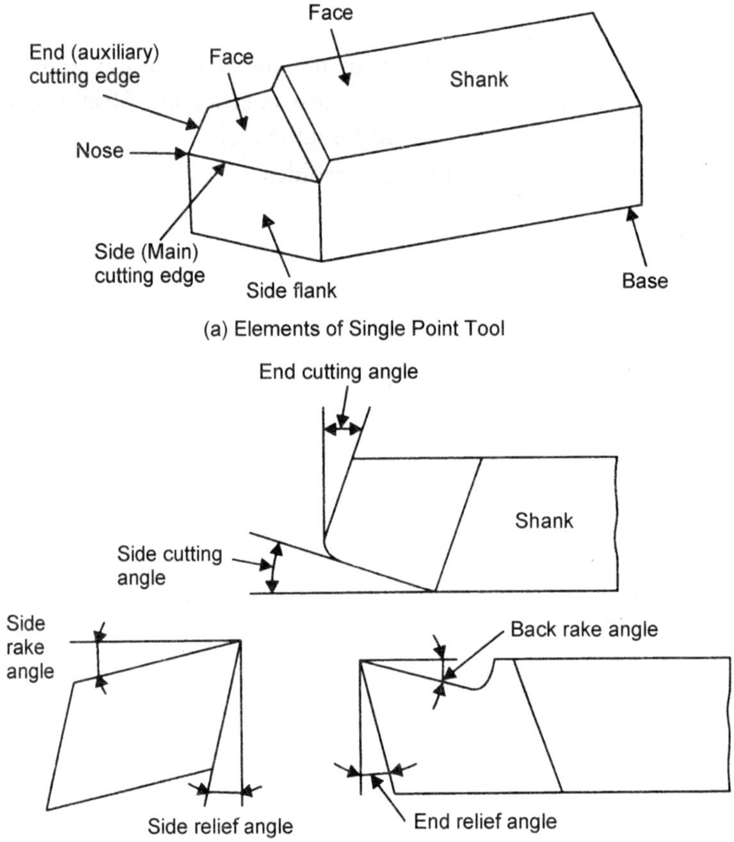

(b) Tool Angles

Fig. 8.27 Single Point Cutting Tool Nomenclature

Back rake angle may be positive, neutral or negative. The angle is positive if the face slopes downwards from the tip towards the shank. It is used to cut low tensile strength and non-ferrous materials. The angle is negative if the face slopes for high tensile strength materials, heavy feed and interrupted cuts.

Side Rake Angle: It measures the slope of the top surface of the tool to the side in a direction perpendicular to the longitudinal axis. It also guides the direction of the chip away from the job.

Side Relief Angle: It is the angle made by the flank of the tool and a plane perpendicular to the base just under the side cutting edge. This angle permits the tool to be fed side ways into the job, so that it can cut without rubbing.

End Relief Angle: It is the secondary relief angle between a plane perpendicular to the base and the end flank.

Side Cutting Edge Angle: It is the angle between the side cutting edge and the longitudinal axis of the tool.

Nose Radius: It is the curve formed by joining the side cutting and end-cutting edges. The angle so formed is called nose angle and the radius of the curve is called nose radius.

8.14 CUTTING TOOL MATERIALS

The material used for tools must be harder than the metal to be cut and must possess wear resistance, hot hardness, high thermal conductivity, low coefficient of friction, machinability. The following tool materials are most commonly used for lathe tools.

1. High carbon steel
2. High speed steel
3. Stellite(Cast alloys)
4. Cemented carbides
5. Ceramics
6. Diamond

8.14.1 High Carbon Steel

This material was used for making tools upto 1870. The carbon content in this type of material is low and varies from 0.8 to 1.5 per cent. The carbon steel tools are easy to manufacture and their cutting edges can be sharpened easily. They lose their hardness rapidly at temperature greater than 200°C. They are used for cutting softer material at low speed. They are particularly used in the manufacture of hand tools like taps, files, hacksaw blades, wood working tools, knives etc.

8.14.2 High Speed Steel

High speed steel is a carbon steel to which alloy elements like tungsten, chromium, vanadium, molybdenum and cobalt are added in order to increase its hot hardness and wear resistance. These are basically two types of H.S.S, namely T(Tungsten) type in which tungsten is the major alloying element(12-18%) and M(Molybdenum) type having molybdenum as the chief alloying element(8-12%). T-type HS was developed earlier. However, because of the relative scarcity of tungsten, M-type was invented. It is cheaper and therefore more widely used. Drills, milling cutters, gear hobs are made of high speed steels.

8.14.3 Stellite (Cast Alloys)

Stellite is the trade name of a non-ferrous cast alloy composed of cobalt, chromium and tungsten. It consists of cobalt(40-55%), chromium (25-35%), tungsten (1.5-3%) and carbon(0-0.5%). Cobalt acts as a solvent or matrix with chromium as the major alloying element. The material is made by melting the elements together and then cast in moulds. Hence these are known as cast alloys. The cast alloys retain their hardness upto 750°C. Because this tool material is castable, it is specially used for making form tools.

8.14.4 Cemented Carbides

Cemented Carbides are made by powder metallurgy technique. These withstand temperature upto 1000°C. These are classified into two main types namely straight tungsten carbides and alloyed tungsten carbides. Straight tungsten carbides consists of tungsten carbide (85-95%) and cobalt (5-15%). Alloyed tungsten carbides have additions of titanium and niobium etc. Cemented carbides used in the form of small tips.

Coated Carbides: Coating of aluminium and zirconium oxides deposited on the tool surface at high temperature retard the diffusion wear of the tool. Similarly crater wear can be reduced by coating a thin layer of titanium carbide or hafnium nitride.

8.14.5 Ceramics

Ceramics are made by compacting aluminium oxide powder in a mould and sintered. The main constituent is aluminium oxide upto 10% additions usually of oxides of magnesium, titanium and chromium are often made to obtain superior properties. These withstand temperatures upto 1400°C. The hardness of ceramics is greater than the cemented carbides, but it is more brittle. The ceramic tips are supplied in 'throw away' form.

8.14.6 Diamond

The diamond tool is harder than any other material. It is also chemically inert. Diamond tool is either a single crystal or polycrystalline. Single crystal diamond is used from machining non-ferrous metals like aluminium, brass, copper and bronze. It is also used for machining non-metallic materials like plastic, epoxy resins, hard rubber and also precious metals like gold, silver and platinum. Polycrystalline diamond is used for machining glass, reinforced plastics, eutectic and hyper-eutectic alloys etc.

8.15 CUTTING PARAMETERS

8.15.1 Cutting Speed (*V*)

Surface speed at which the work piece passes the cutter. It is expressed in meters per minute (m/min). In turning, cutting speed is given by the relation.

Cutting Speed $V = \pi DN/1000$

Where,

D = Diameter of job in mm

N = R.P.M. of the job

8.15.2 Feed

It refers to the amount of tool advancement per revolution of the job parallel to the surface of the job to be machined. It is expressed in millimeters per revolution (mm/rev).

8.15.3 Depth of Cut

It refers to the advancement of the tool in the job in a direction perpendicular to the surface being machined. It is expressed as follows:

Depth of cut = $(d_1 - d_2)/2$

Where,

d_1 = Diameter of the uncut surface

d_2 = Diameter of the machined surface

8.15.4 Machining Time

Machining time on lathe depends upon the speed and feed and length of the job. Machining time is given by equation:

$T = L/fN$

Where, T = Machining time (min)

f = feed – mm/rev)

N = R.P.M.

L = Length of the job to be turned (mm)

Example 8.5: Estimate the machining time to turn a MS bar of 40 mm diameter to 35 mm diameter to a length of 300 mm in a single cut. Assume cutting speed 35m/min and feed 0.4 mm/rev.

Solution:

$V = \pi$ DN/10000

$N = 1000 * \text{V}/\pi\text{D} \quad = 1000 \times 35/\pi \times 40$

$= 278.5\text{RPM}$

Machining time $T = L/fN \quad = 300/0.4 \times 278.5 = 2.69$ min

Example 8.6: Estimate the machining time required to turn a stepped shaft from a shaft of 40 mm diameter (see Fig. 8.28). Neglect facing and setting times. The depth of cut should not exceed 2.5 mm. Assume the cutting speed is to be 20 m/min and feed to be 0.3 mm/rev for each cut.

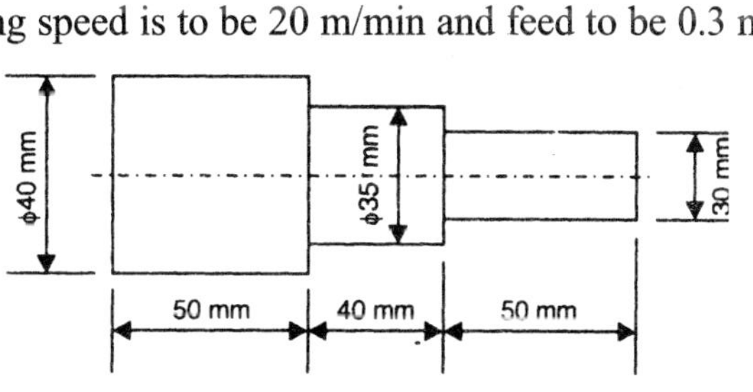

Fig. 8.28 Step Turning

Solution:

Step I: Reduce 40 mm dia to 35 mm dia for a length of 90 mm

$V = \pi\text{DN}/1000$

$N_1 = 20 \times 1000/\pi \times 40 = 159.55$ RPM

T_1 = L/Fnl

$= 90/0.3 \times 159.55 = 1.885$ min

Step II: Reduce 35 mm dia to 30 mm dia to length of 50 mm

$$N_2 = 20 \times 1000/\pi \times 35 = 181.89\text{RPM}$$

$$T_2 = 50/0.3 \times 181.89 = 0.9163 \text{ min}$$

Total machining time $T = T_1 + T_2 = 1.885 + 0.9163 = 2.8$ min.

QUESTIONS

1. Explain with the help of neat sketch the parts of a lathe.
2. How lathes are classified? Describe in brief the different types of lathes.
3. How do you specify a lathe?
4. What are the various methods of Taper turning in a lathe? Explain in detail any one of them.
5. With the help of neat sketch explain thread cutting on a lathe.
6. What are the various operations done on a lathe?
7. Explain the various types of mandrels used in a lathe.
8. Explain the use of the following in a lathe:
   (a) Fa ce Plate (b) Steady Rest (c) Follower Rest.
9. What are the accessories used on a lathe?

9

Capstan and Turret Lathes

9.1 INTRODUCTION

Centre lathe is a very versatile machine tool, but not suitable for the economic manufacturing of identical parts. Hence capstan and turret lathes are developed to turn parts rapidly using pre-set tooling. Once the tools are set, it is possible to operate the machines with semi-skilled operators. These are developed from engine lathe.

9.2 DIFFERENCE BETWEEN CENTRE LATHE AND TURRET LATHE

Although the turret lathe is development of an engine lathe, there are certain differences in their construction, operations and use. They are:

1. The headstock of a turret lathe is heavier than the headstock of engine lathe and is provided with a wider range of speed. When an engine lathe requires a motor of 5hp, the turret lathe will have a motor of 15hp for high rate of production.
2. In turret lathe the tailstock of an engine lathe is replaced by turret. The turret is a six-sided block, each side carries one or more tools.
3. The tool post mounted on the cross slide of a turret lathe is a four-way tool post. In addition to this, there is a rear tool post mounted on the carriage.
4. The feed movement of each tool of hexagonal turret may be regulated by stops or feed trips.
5. In turret lathes two or more cutting tools are used simultaneously whereas in engine lathe only one tool is used at once.
6. A semi skilled operator can operate the turret lathe, once a skilled machinist sets it.
7. Turrret lathes are not usually fitted with lead screw and using die heads and taps normally cuts the threads.
8. The time required to finish a component on a turret lathe is less than on an engine lathe.
9. The chucking and rechucking of jobs is time consuming in an engine lathe while in turret lathe, works are clamped automatically.

10. Turret lathes are suitable for mass production whereas engine lathes are not economical for mass production of identical parts.

9.3 TYPES OF TURRET LATHES

Depending on the position of the turret lathes are classified into two types:

1. Horizontal turret lathe
2. Vertical turret lathe

9.3.1 The commonly used Horizontal Turret Lathes are further classified as

(a) The ram type (capstan) lathe

(b) The saddle type (turret) lathe

(a) The Capstan Lathe

The capstan lathe is a ram type turret lathe. It consists of hexagonal turret mounted on a ram slide. The saddle, which supports the ram, is clamped to the bed at a desired position. This type of machine is light in construction and suitble for machining bars of small diameter. The turret stroke depends on the length of the ram. The principal parts of a capstan lathe are shown in Fig. 9.1.

1. *Bed*: The bed is a long casting provided with accurate guideways on which the headstock, the carriage and turret saddles are mounted.
2. *Headstock*: The headstock is located at the left-hand end of the bed. All geared headstock is most commonly used which provides a wide range of speeds.
3. *Turret*: It is hexagonal shaped tool holder mounted on the ram. The tools used in turret are drills, reamers, boring bars and cutting tools. For each tools, there is a stop screw to control the tool movement.
4. *Turret Saddle*: This supports the ram (auxiliary slide) on which hexagonal turret is mounted. The saddle can be moved over the bed and clamped in any required position.
5. *Cross Slide and Carriage*: The cross-slide is mounted on the carriage. It is equipped with four-way tools post at the front, one rear tool post at the back of the cross-slide.

Carriage moves parallel to the bed. For plane turning, the carriage is moved parallel to the bed.

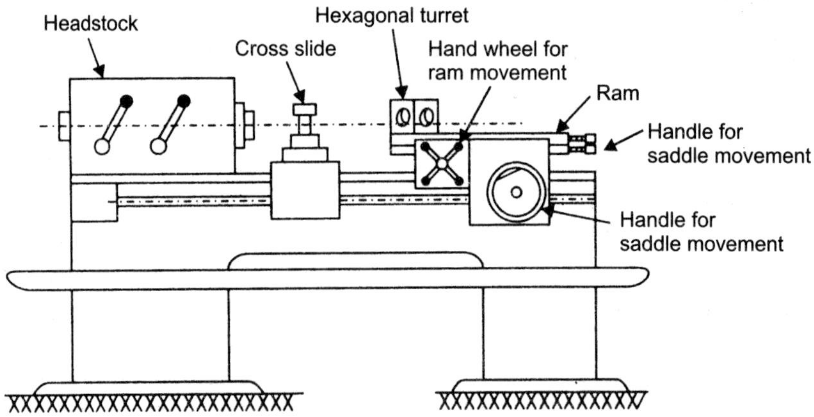

Fig. 9.1 Principal Parts of Capstan Lathe

(*b*) *The Turret Lathe*

Turret lathe is a saddle type lathe in which turret is mounted directly on the saddle. The saddle moves entire length of the bed. It is heavy in construction and more rigid in design. Hence is used for heavy works. The principal parts of a turret lathe are shown in Fig. 9.2.

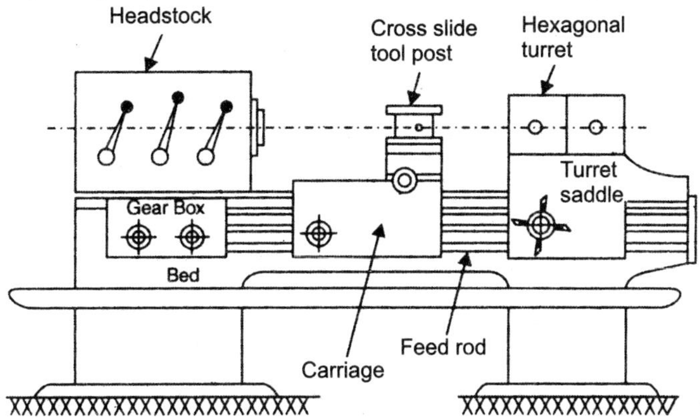

Fig. 9.2 Principal Parts of Turret Lathe

9.4 DIFFERENCE BETWEEN CAPSTAN AND TURRET LATHES (See Fig. 9.3 (a) and (b))

1. In a capstan lathe the turret is mounted on a short slide (ram) which slides on the saddle, whereas in turret lathe the turret is mounted on the saddle which directly slides on the bed guide ways.
2. In a capstan lathe the movement of turret is very short, whereas turret of turret lathe can be moved on the entire length of the bed.

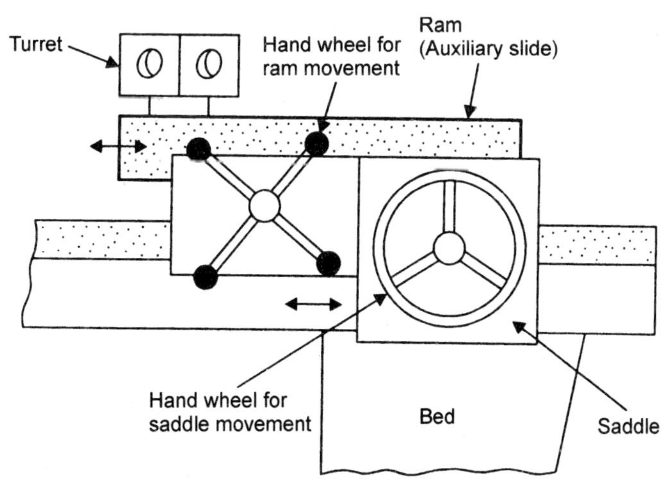

Fig. 9.3(a) Capstan Lathe Slide Arrangement

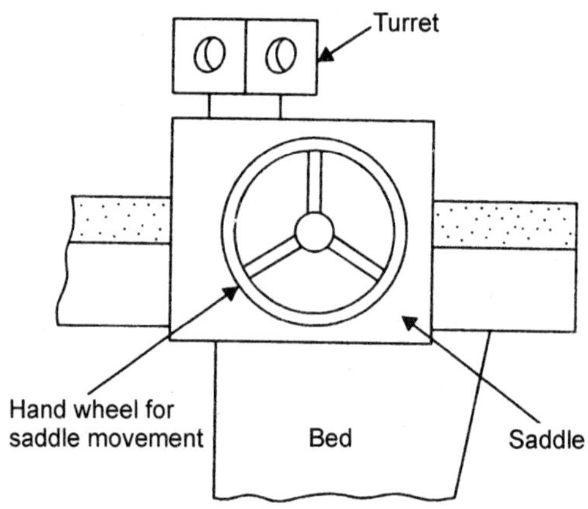

Fig. 9.3(b) Turret Lathe Slide Arrangement

3. In the case of capstan lathe, the ram is fed into the work. The overhanging of the ram from saddle presents non-rigid construction, which is subjected to bending, deflection or vibration. In a turret lathe, turret is mounted on the saddle. This type of construction provides utmost rigidity to the tool.
4. Turret lathe is widely used for chucking operations for forging and castings, whereas capstan lathe is suitable for bar work.
5. Turret lathe is heavier than capstan lathe. Hence turret lathe is suitable for medium to heavy work, whereas capstan lathe is suitable for small to medium work.
6. On a capstan lathe, less fatigue to the operator due to the lightness of the ram, whereas in the case of turret lathe the hand feeding is a laborious process due to the movement of the entire saddle.

9.5 TURRET INDEXING MECHANISM FOR CAPSTAN AND TURRET LATHE (See Fig. 9.4)

Figure 9.4 shows an inverted view of the turret assembly. The plunger locks the index plate by spring pressure and prevents rotatory movement when turret is moving forward for doing the operation. Indexing of turret is done during backward motion of turret. As the turret head moves backward, the actuating cam lifts the plunger out of groove and unlocks the index plate. This is due to riding of the pin on the bevel surface of the cam. When the turret head is still going back, the spring-loaded pawl engages with the groove of the ratchet plate causes the ratchet to rotate (index plate is rotated by one-sixth revolution). Now when the turret head is moving forward, the plunger drops out of the cam and locks the index plate in the next groove. Now the next tool of the turret head is ready for the operation.

As the turret rotates, the stop rod shaft holding six adjustable stop rods will rotate by engaging bevel pinion with bevel gear attached to the turret. The setting of stop rods (longitudinal travel of the tool) is done by unscrewing the lock nuts and rotating the stop rod on the plate against the stop.

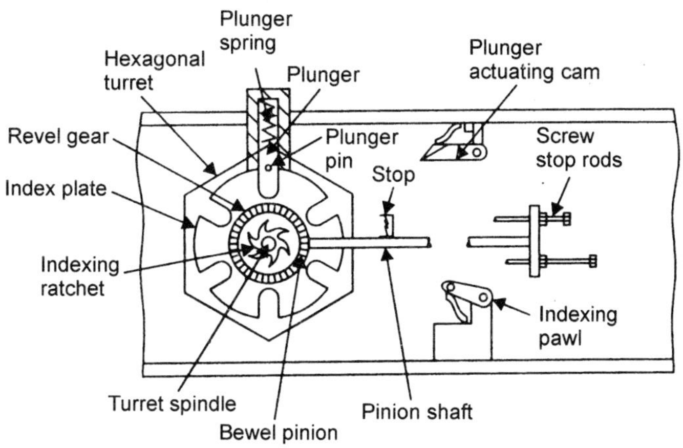

Fig. 9.4 Turret Indexing Mechanism

9.6 BAR FEEDING MECHANISM IN CAPSTAN AND TURRET LATHES

Various methods of bar feeding mechanisms are designed which push the bar forward after each finished component is cut off. Fig. 9.5 shows a simple bar feed mechanism.

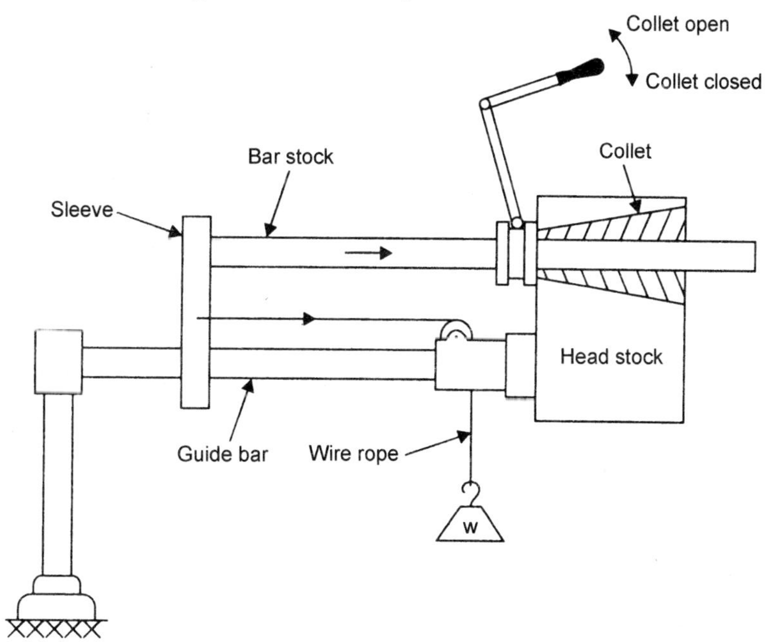

Fig. 9.5

In this one end of guide bar is fixed to the rear side of the headstock and a pedestal supports the other. A rotating sleeve is mounted on the guide bar to carry the rear end of the barstock and the other end of barstock is passed through the spindle to project outside the collet. One end of the rope is connected to sleeve and other end passes over the pulley and carries weight (W). The weight constantly exerts end thrust on the bar chuck and forces the barstock through the spindle, to strike against the bar stop, the moment the collet is opened.

9.7 WORK HOLDING EQUIPMENT

The work holding equipment used on capstan or turret lathes differs slightly from those used on a central lathe.

The commonly used devices are:

1. Collets 2. Chucks 3. Fixtures

9.7.1 Collets

Collets are used to grip the bars passing through the headstock spindle. They are suitable for mass production. Collet grips the work by the spring action of its split jaws (Fig. 9.6(a)). They may be operated manually or by power. The collets are classified by the methods used to close the jaws on the work.

(a) Draw Collet (b) Pushout Collet (c) Dead Length Collet

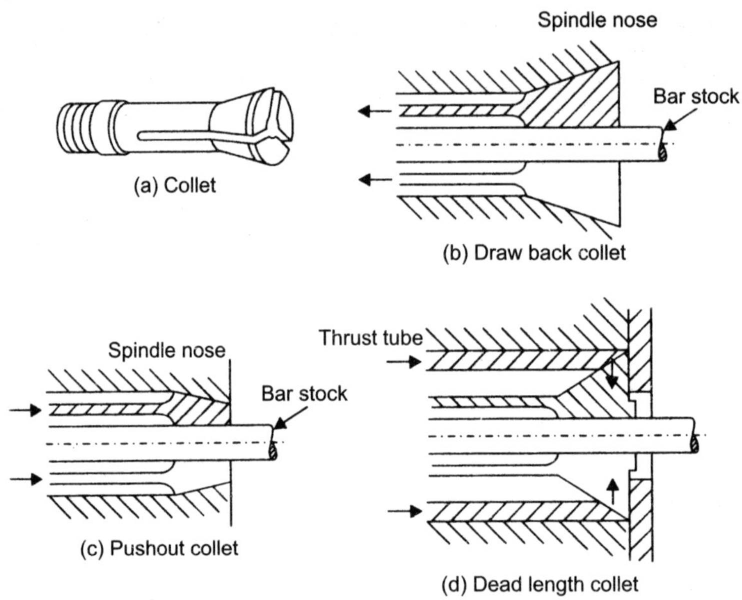

Fig. 9.6 Types of Collets

(a) Draw Collet (*Fig. 9.6(b)*): Collet is pulled by the thrust or collet tube to the left into the taper bore of the spindle nose. This action puts pressure on the tapered section of the collet, forcing them inward and tightly clamping the barstock.

(b) Pushout Collet (*Fig. 9.6(c)*): The thrust tube pushes the spring collet to the right into the tapered seat in the spindle nose.

(c) Dead Length Collet (*Fig. 9.6(d)*): In this design, the spring collet has no axial movement during the operation. This chuck is closed when the thrust tube pushes a sleeve with an internal taper into the taper of the collet forcing it inward to clamp the work.

9.7.2 Chucks

Chucks are used for holding large-sized components, which can not be introduced through head-stock. Both three-jaw and four-jaw chucks are used.

9.7.3 Fixtures

A fixture is specially design chuck for the purpose of holding, locating and machining a large number of identical pieces which can not be held by conventional gripping devices.

9.8 TOOL HOLDING DEVICES

The wide variety of work performed in a capstan or turret lathes in mass production necessitated designing of many different types of tool holders for holding tools for typical operations. The following tool holders are widely used on capstan or turret lathes:

1. Straight cutter holder
2. Adjustable or plain angle cutter holder
3. Multiple cutter holder
4. Offset cutter holder
5. Slide tool holder
6. Knee tool holder
7. Knurling tool holder

9.8.1 Straight Cutter Holder (Fig. 9.7)

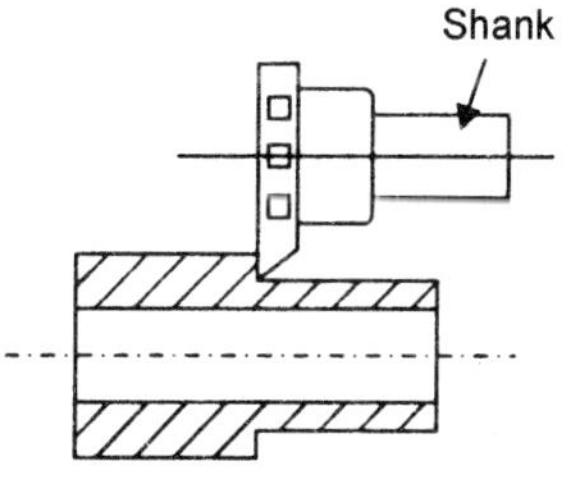

Fig. 9.7 Straight Cutter Holder

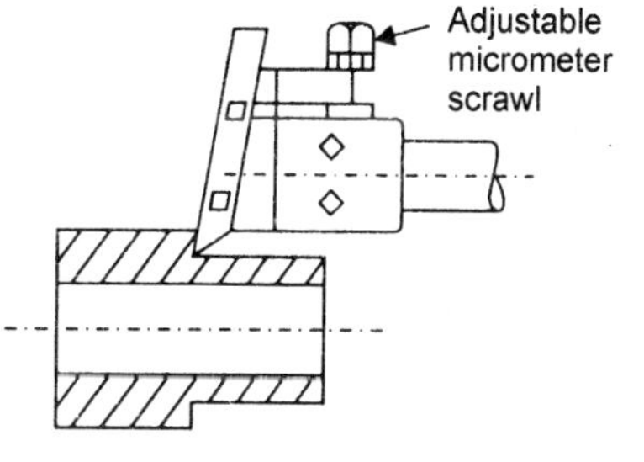

Fig. 9.8 Adjustable or Plain Angle Cutter

This is a simple tool holder in which the tool is held perpendicular to the holder shank axis by three set screws. The shank can be mounted directly into the hole of the turret face.

9.8.2 Adjustable or Plain Angle Cutter Holder (Fig. 9.8)

The cutter has an inclined slot cut into the body into which the tool is fitted. This type of setting of the tool permits the tool to maintain a clearance with the work.

9.8.3 Multiple Cutter Holder (Fig. 9.9)

It can accommodate double tools in the body. This features enables turning of two different diameters simultaneously.

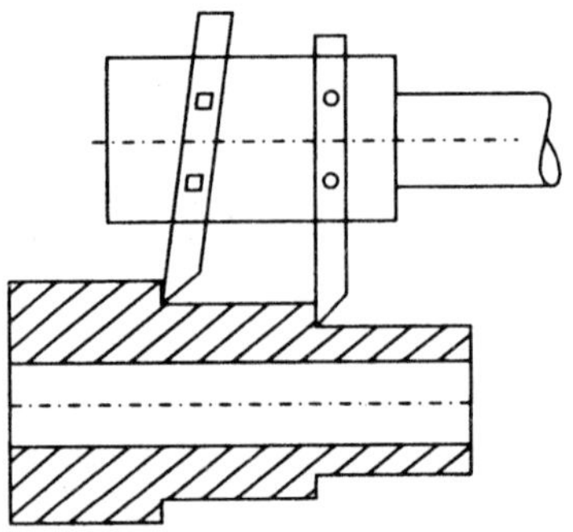

Fig. 9.9 Multiple Cutter Holder

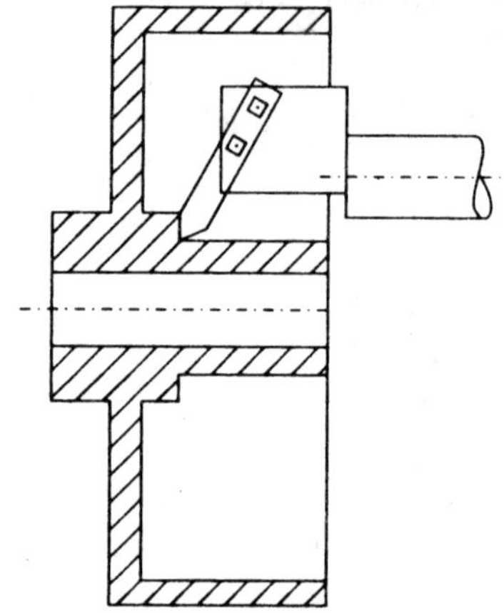

Fig. 9.10 Offset Cutter Holder

9.8.4 Offset Cutter Holder (Fig. 9.10)

In this type of holder, the holder body is made offset with the shank axis. Larger diameter of work may be turned or bored by this type of holder.

9.8.5 Slide Tool Holder (Fig. 9.11)

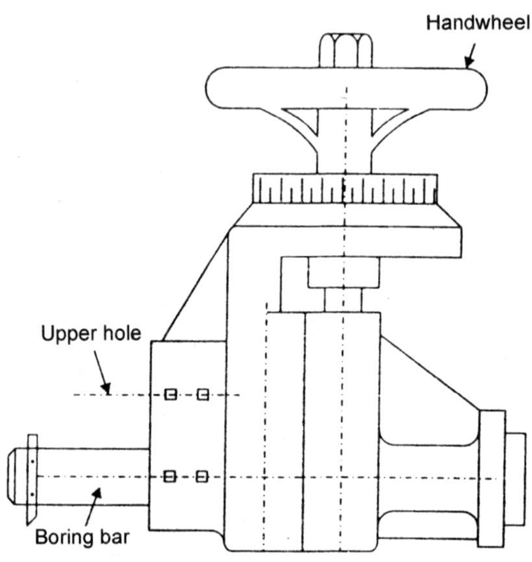

Fig. 9.11 Slide Tool Holder

This holder is very much useful for rough and finish boring, recessing, grooving etc. The slide may be adjusted up or down by rotating hand wheel. Two holes are provided on the sliding unit for holding tools.

9.8.6 Knee Tool Holder (Fig. 9.12)

The Knee tool holders are useful for simultaneous turning and boring and drilling operations. The knee holder is bolted directly on the turret face.

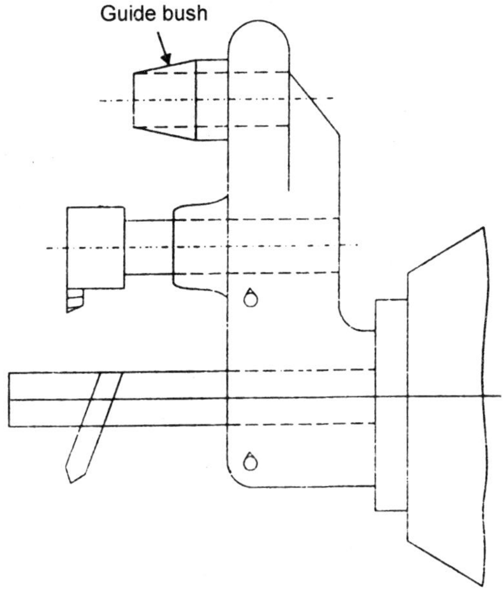

Fig. 9.12 Knee Tool Holder

9.8.7 Knurling Tool Holder (Fig. 9.13)

Knurling tool holder is fitted to the turret face. The knurls can be adjusted to knurl on different diameter work.

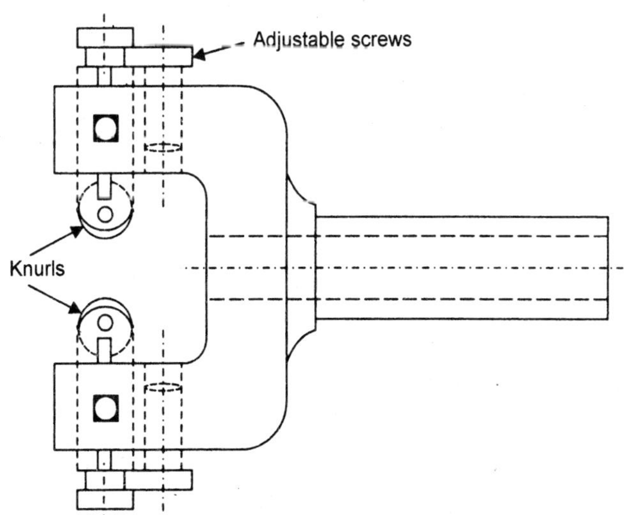

Fig. 9.13 Knurling Tool Holder

9.9 TOOLING LAYOUT

In order to perform any work in a capstan and turret lathe, proper planning for systematic operations should be done in advance before setting the work on it.

Planning Procedure

1. The capacity chart of the machine should be examined.
2. For tooling layout, finished part drawing is required.
3. The proper tool section for different operations should be made.
4. Proper spindle speed, feeds and depth of cut should be calculated.
5. Finally the work and tools are set on the machine according to the above.

Example 1: Prepare a Tool layout for the production of hexagonal bolt.

Solution: The sequence of operations and tooling arrangements are given below. (See Fig. 9.14).

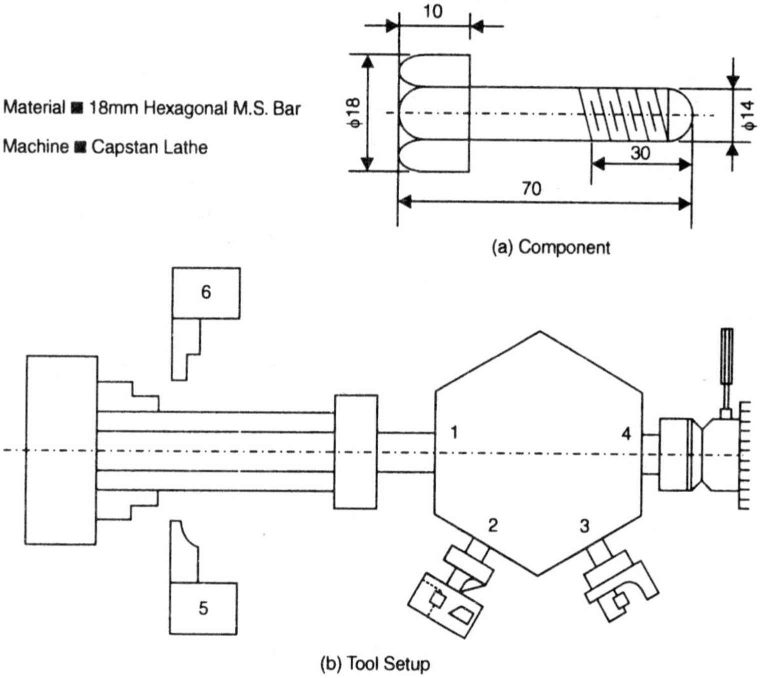

Fig. 9.14 Tooling Layout for Hexagonal Bolt

1. Feed the bar stock to bar stop (turret 1) to a distance of 80 mm. An extra distance 10mm than the bolt length is allowed, 4 mm for the parting off and 6mm for clearance from the collet face so that the tool will not interference with collet.
2. Turn 14mm diameter with the roller steady box turning tool for 60mm length by adjusting the stop (turret 2).
3. Round the end with roller steady ending tool i.e. chamfering (turret 3).
4. Cut threads with die head for 30mm length by adjusting the stop (turret 4).

5. Chamfer the head with chamfer tool using front square tool post (tool 5).
6. Parting-off with parting tool in the rear tool post completes the machining of the component (tool 6).

Process Sheet

Operation No.	*Description of operation*	*Tool position*	*Tools*
1.	Feed bar to stop	1st turret position	Bar stop
2.	Turn 16 mm dia.	2nd turret position	Box turning tool
3.	Chamfer the end of the bolt	3rd turret position	Box ending tool
4.	Thread cutting	4th turret position	Die head
5.	Head changer	Front tool post	Chamfering tool
6.	Parting-off	Rear tool post	Parting-off tool

QUESTIONS

1. Differentiate between engine lathe and turret lathes?
2. Differentiate between capstan and turret lathes?
3. Explain various tool holding devices.
4. Draw a tool layout for producing hexagonal botton capstan lathe.

10

Drilling Machines

10.1 INTRODUCTION

Drilling machine is one of the most important machine tool used in a workshop to produce holes in solid objects. The process of making a hole is called drilling. In drilling operation, work is clamped to the table and the rotating cutting tool, called drill is fed into it.

10.2 TYPES OF DRILLS

Drills are manufactured in a wide variety of types and sizes. The following types of drills are most widely used.

1. Flat drill
2. Straight-Fluted drill
3. Twist drill
4. Oil hole drill
5. Centre drill

10.2.1 Flat Drill

Flat drill is shown in Fig. 10.1. It is a simple drill used to produce holes in softer materials like wood and plastic. It is made of tool steel. It has two cutting edges with cutting angles varies from 90° to 120° and relief angle at the cutting edge is 3° to 8°. The disadvantages of this type of drill is that its diameter is reduced as a result of sharpening the edges. The chips will not come out of the hole automatically.

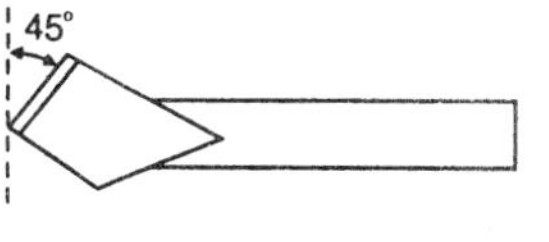

Fig. 10.1 Flat Drill

10.2.2 Straight-Fluted Drill

It is shown in Fig. 10.2. It has grooves or flutes running parallel to the drill axis. Chips will come out from the hole automatically. It is used for drilling brass and non-ferrous materials.

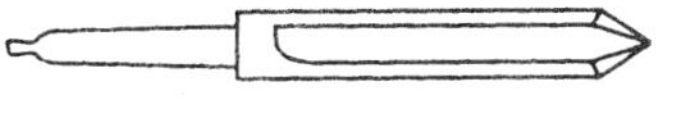
Fig. 10.2 Straight Fluted Drill

10.2.3 Twist Drill

Twist drill is the most widely used tool in modern drilling practice. It consists of a cylindrical body carrying spiral flutes cut on it that run length wise around the body of a drill. Twist drills are usually made of high speed steel. The twist drill consists of two main parts shank and body. The shank is the part held in drilling machine for driving (rotating) the drill. The body is the cutting unit with flutes, cutting edges and drill point. The twist drill bits are classified into two types:

A. Parallel shank twist drill
B. Taper shank twist drill

A. Parallel Shank Twist Drill

Parallel shank drills are held in drill chuck. Depending upon the length of the drill

These are subdivided into three series:

1. Short series (jobber) twist drill
2. Stub series twist drill
3. Long series twist drill

In jobber drill, the diameter ranges from 0.2mm to 16mm. The long series with diameter ranging from 1.5mm to 26mm and stub series diameter ranging from 0.5mm to 40mm are used for special jobs (See Fig. 10.3).

Fig. 10.3 Parallel Shank Drill Bit

Fig. 10.4 Taper Shank Twist Drill

B. Taper Shank Twist Drill

These drills fit into the internal taper of the drilling machine spindle. The shank for these drills conform to Morse Tapers (See Fig. 10.4).

10.2.4 Oil Hole Drill

These are used for drilling deep holes. It has holes through the body of the drill from shank to the point to carry oil directly to the cutting edges. Cutting fluids or compressed air is forced through the holes to the cutting point of the drill to remove the chips, cool the cutting edge and lubricate the machined surface (See Fig. 10.5).

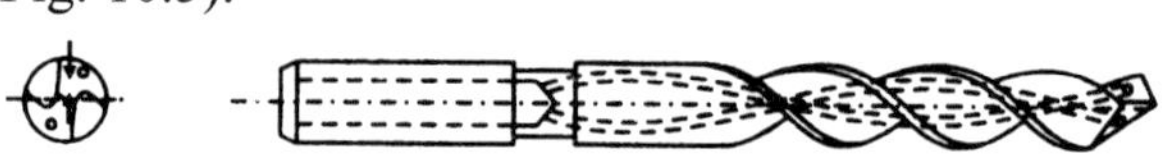

Fig. 10.5 Oil Drill

10.2.5 Centre Drill

It is a two-fluted twist drill with straight shank. It is used to make the centre holes on the end of the shaft (See Fig. 10.6).

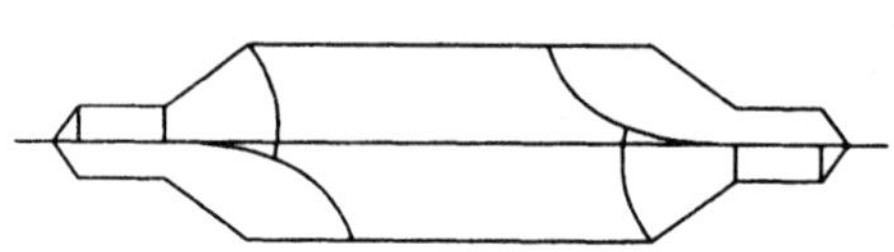

Fig. 10.6 Centre Drill

10.3 TWIST DRILL NOMENCLATURE

Drill nomenclature consists of various parts and geometric parameters. They are shown in Fig. 10.7.

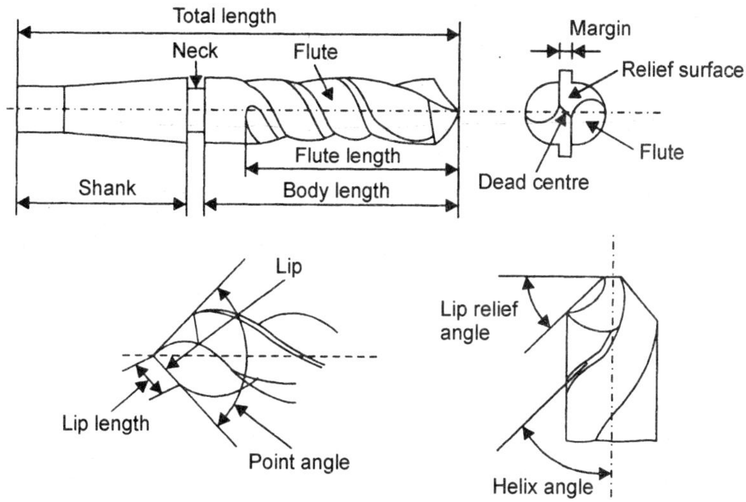

Fig. 10.7 Twist Drill Nomenclature

Body: The fluted portion of a drill.

Shank: It is the part held in the holding device.

Flutets: The helical grooves cut or formed in the body of the drill to provide cutting edges and permit removal of chips and allow the cutting fluid to reach the cutting edges.

Dead Centre: The dead centre or chisel edge of the drill is the sharp edge at the extreme tip end of the drill.

Helix Angle: The angle between the drill axis and the leading edge of the land.

Rake Angle: The angle between the face and line parallel to the drill axis and is equal to the helix angle at the periphery.

Lip Relief Angle: Surface of the point that is relieved just back of the cutting edge.

Point Angle: The included angle of a cone formed by lips.

Recommended values of various angles of a drill are shown in Table 10.1

Table 10.1: Drill Angles

Material	*Point Angle*	*Helix Angle*	*Lip Relief Angle*
Aluminium	90 –140	24 – 48	8 –12
Brass	111	0 – 27	8 – 15
Copper	100 – 118	28 – 40	8 – 15
Cast Iron	118	24 – 40	8 –12
Steel	125 – 135	24 – 32	10 – 12

10.3.1 Drill Size

The size of a standard twist drill is specified by four ways.

(a) Fractional sizes (c) Letter sizes
(b) Number sizes (d) The Metric sizes

(*a*) *Fractional Sizes*: These are also called inch drills. Drill size ranges from 1/64" to 5" diameter. Each successive drill is 1/64", longer up to 13/4", then the diameter of successive drill gradually increases.

(*b*) *Number Sizes*: The drill sizes range from No.1 to No. 80. Number 80 is the smallest having diameter equal to 0.0135 inch and the number 1 is the largest having diameter equal to 0.228 inch.

(*c*) *Letter Sizes*: This size of drills is designed by letters from A to Z. A represents the smallest size and Z the largest.

(*d*) *The Metric Sizes*: The drills are available from 0.20 mm to 100 mm in steps of 0.02 mm upto 1mm, 0.05 mm steps upto 3 mm and afterwards in gradually rising steps.

10.4 TYPES OF DRILLING MACHINES

Drilling machines are manufactured in types and sizes to suit the different type of work. The different types of drilling machines are:

(a) Hand drill
(b) Portable drilling machine
(c) Sensitive drilling machine
(d) Pillar drilling machine
(e) Radial drilling machine
(f) Gang drilling machine
(g) Multi-spindle drilling machine
(h) Numerically controlled drilling machine

(*a*) *Hand Drill* (Fig. 10.8)

Hand drill is used for drilling small holes. The handle of the hand drill is held in the left hand while the right hand turns the crank, which in turn causes the drill to rotate.

(*b*) *Portable Drilling Machine* (See Fig. 10.9)

This is small and compact. It is used for drilling holes in any position, which is not possible with convertical drilling machines. These machines are used for drilling holes up to 18 mm diameter.

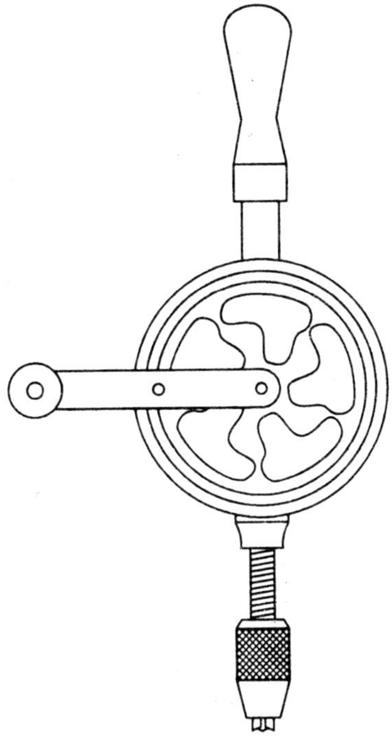

Fig. 10.8 Hand Drill

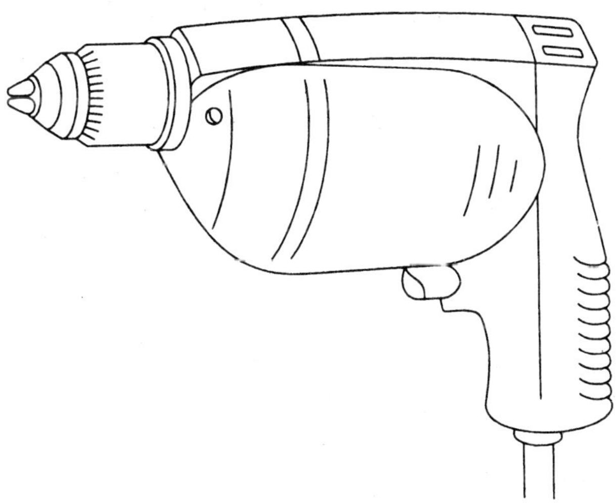

Fig. 10.9 Portable Drilling Machine

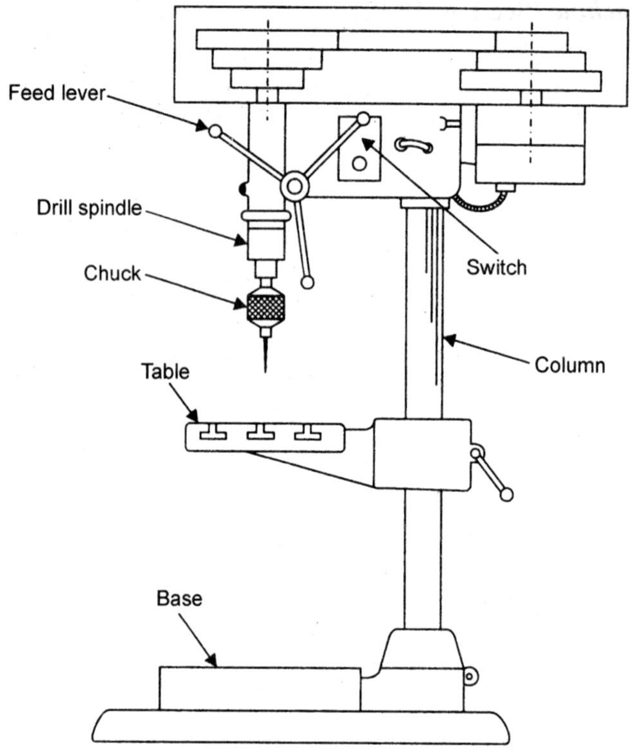

Fig. 10.10 Sensitive Drilling Machine

(c) ***Sensitive Drilling Machine*** (See Fig. 10.10)

It is a small drilling machine designed for drilling small holes. The base of the machine is mounted on a bench. The drive mechanism of sensitive drilling machine consists of V-belt drive from motor shaft to drill spindle. Three or four stepped cone pulley is provided to give a required speed range. No gears are used in the drive. The handle through a rack and pinion arrangement gives vertical movement to the spindle (see Fig. 10.11).

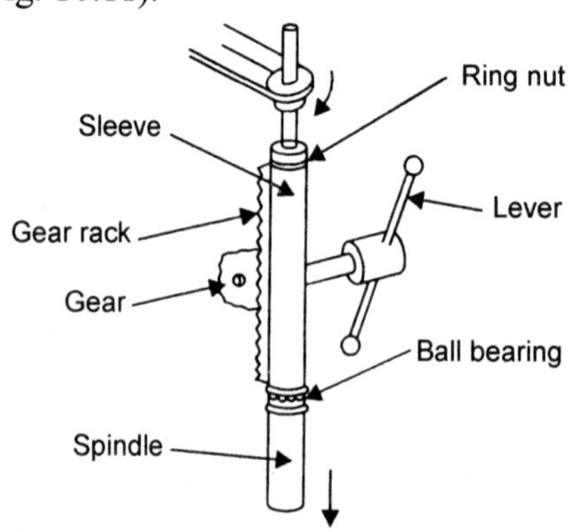

Fig. 10.11 Spindle Feed Arrangement

(d) Pillar Drilling Machine (See Fig. 10.12)

This is also known as upright drilling machine. It is similar to sensitive drilling machine. But it is larger and heavier than a sensitive drilling machine and is provided with power feed arrangement. The vertical column of this machine can be round or box type.

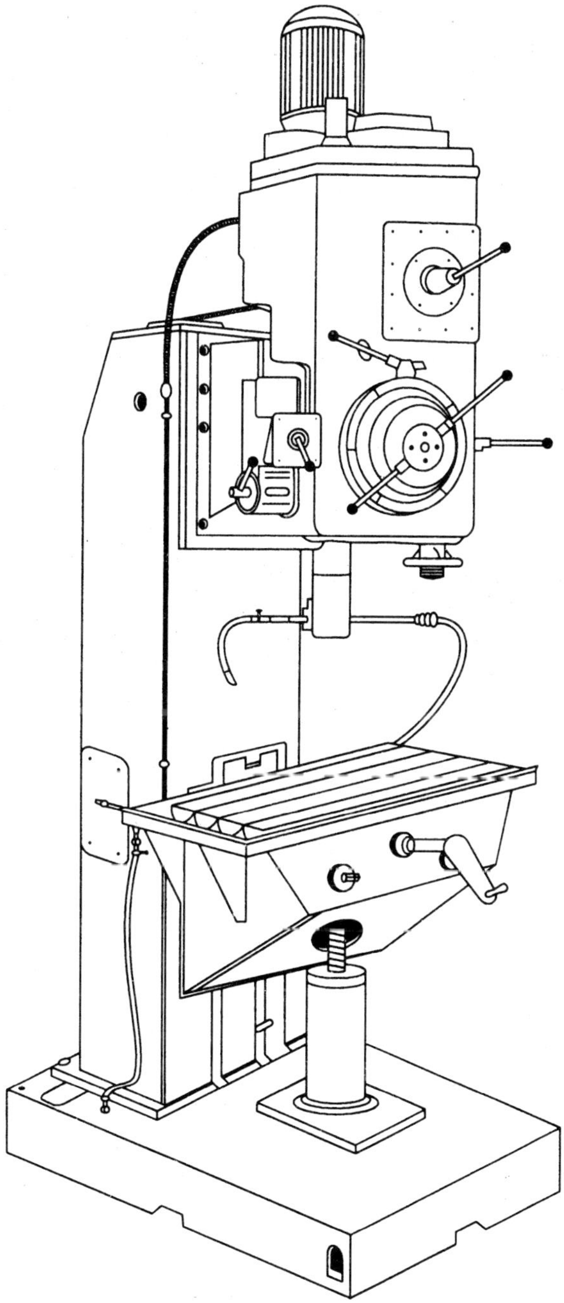

Fig. 10.12 Pillar Drilling Machine

(e) Radial Drilling Machine (See Fig. 10.13)

The radial drilling machine is used for drilling heavy work and specially for the jobs where high degree of accuracy is required. The principal parts of the radial drilling machine are:

(i) Base (ii) Column (iii) Radial arm
(iv) Drill head (v) Spindle sped and feed mechanism

(*i*) *Base*: It is a large rectangular casting and is made rigid to support column and the table. T-slots are provided for clamping the work piece directly so that it will serve the function of the table.

(*ii*) *Column*: It is a cylindrical casting mounted vertically to the base. It supports the radial arm which can be raised or lowered vertically to accommodate work pieces of different heights. A separate motor is used for raising and lowering the arm.

(*iii*) *Radial Arm*: The radial arm is mounted horizontally on the column. It may be swung around to any position over the work table.

(*iv*) *Drill Head*: It is mounted on the radial arm. It can be slide along the arm to locate the spindle with respect to the work.

(*v*) *Spindle Speed and Feed Mechanism*: A motor fitted directly over the drill head drives the drill spindle. Through gear box, different spindle speeds and feeds can be obtained.

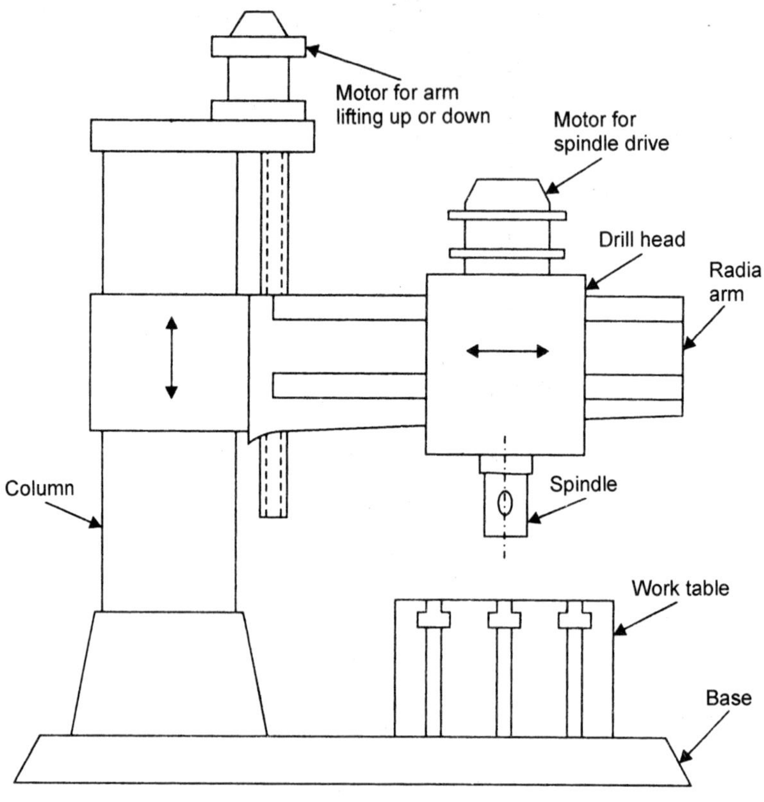

Fig. 10.13 Radial Drilling Machine

(*f*) ***Gang Drilling Machine*** (Fig. 10.14)

When a member of single spindle drilling machine columns are placed side by side on a common base and have a common worktable, the machine is known as gang drilling machine. The gang drilling machine is compact and saves the transfer time of the work-piece from one table to other, thus increases the production rate.

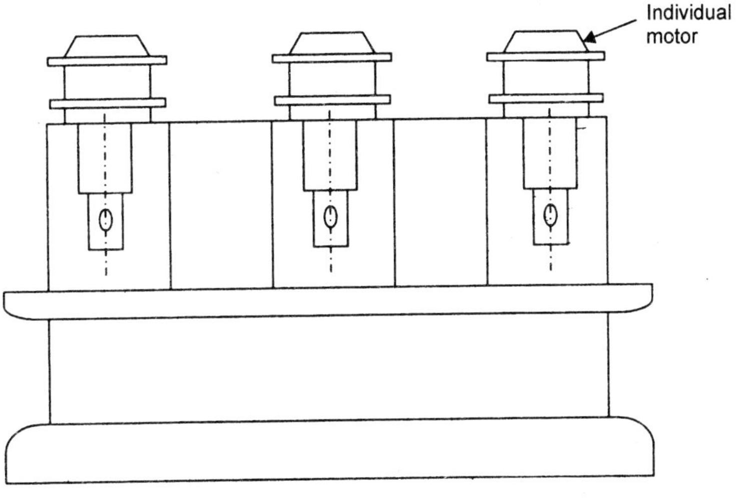

Fig. 10.14 Gang Drilling Machine

(*g*) ***Multi-spindle Drilling Machine*** (See Fig. 10.15)

As the name indicates, this machine has multispindle on a single head. It is a vertical type. The function of a multispindle drilling machine is to drill a number of holes in a piece of a work simultaneously and to reproduce the same pattern of holes in a number of identical pieces in a mass production. The head assembly has the number of spindles driven from pinions surrounding a central gear. Feed motion is obtained by rising the work table. But the feed motion can also be obtained by lowering the drill head.

(*h*) ***Numerically Controlled Drilling Machine***

This is the latest type of drilling machine. In this machine, the table is positioned with the help of numerical controls so as to locate the work accurately under the drill. Programmed tape is used which can be used repeatedly.

10.5 SPECIFICATION OF DRILLING MACHINES

The specification of a drilling machine depends on the type of machine. Small portable drilling machine is specified by the maximum diameter of the drill that it can be hold, whereas the sensitive and upright drilling machine are specified by largest diameter of work piece that can be centred under spindle.

The length of the arm and diameter of the column specify the radial drilling machine.

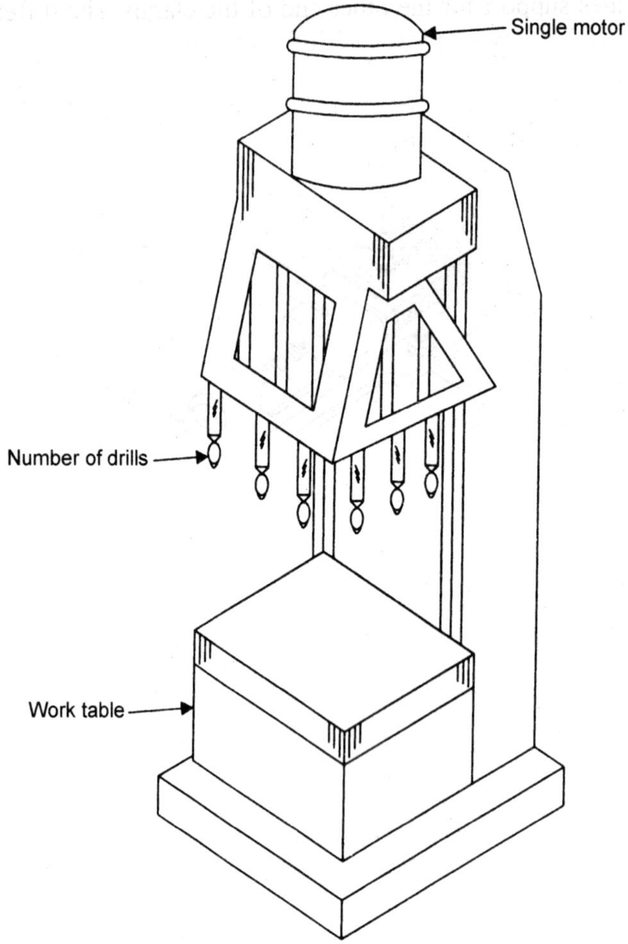

Fig. 10.15 Multi-spindle Drilling Machine

10.6 WORK HOLDING DEVICES

Components to be drilled should be secured firmly on the drilling machine table. The following devices are used for holding the work:

(a) Drill vice	(d) V-Block
(b) Parallel bars	(e) Clamps and T-bolts
(c) Step block	(f) Drill jig

(*a*) *Drill vice*: It is a work holding device in which the work is clamped between a fixed jaw and a movable jaw. The vice is fastened to the table by means of T-bolts. (See Fig. 10.16).

(*b*) *Parallel bars*: Parallel bars are placed below the work so that the drilling is carried out without damaging the vice.

(*c*) *Step block*: Step block is used along with clamps and bolts for holding the work directly on the table. It provides support for the other end of the clamp. The different steps of the step block are used for holding work pieces of different heights. Fig. 10.17 illustrates the used of step block.

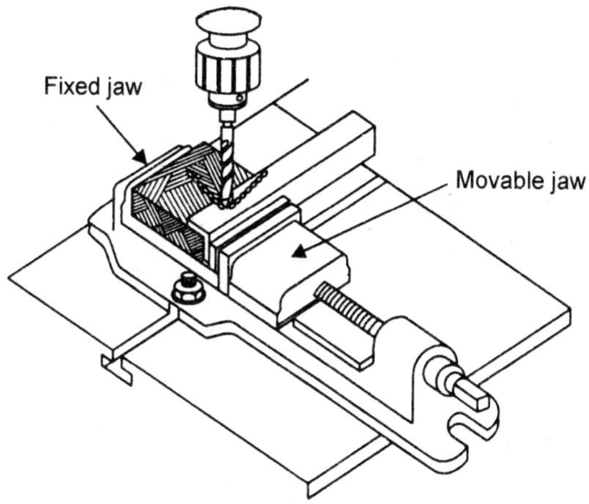

Fig. 10.16 Drill Vice

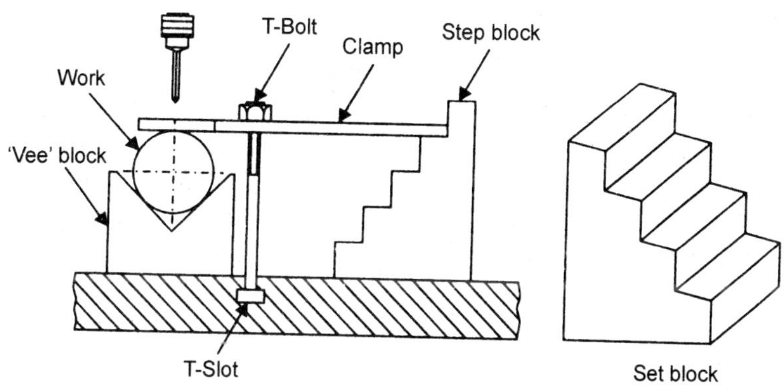

Fig. 10.17 Use of Set Block, T-Bolt and Clamps

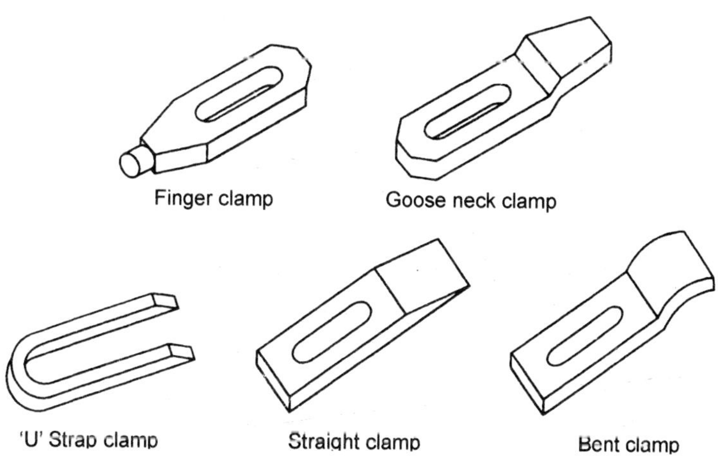

Fig. 10.18 Types of Clamps

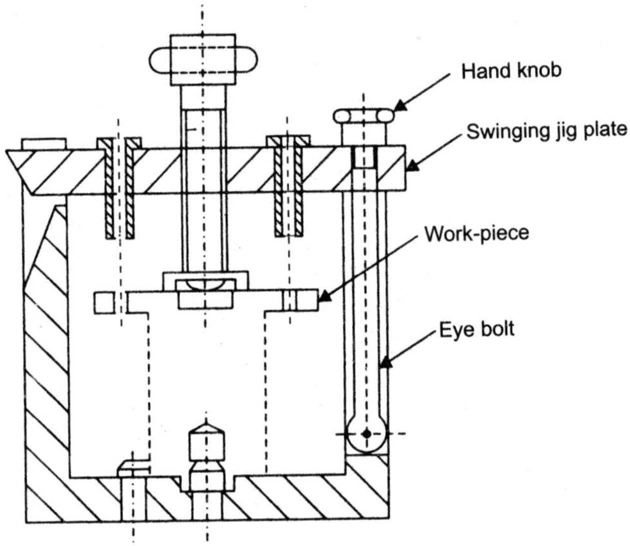

Fig. 10.19 Drill Jig

(*d*) *V-block*: The V-Block is used for holding round work-pieces.

(*e*) *Clamps and T-bolts*: Clamps and T-bolts are used for clamping the work. Various types of clamps used are shown in Fig. 10.18.

(*f*) *Drill Jig*: A drill jig locates the work-piece in proper position and holds it. It also guides the drill so that holes are drilled in the same exact location on all the parts (See Fig. 10.19).

10.7 TOOL HOLDING DEVICES

The following devices are used for holding the drills.

(a) Drill chuck

(b) Drilling machine spindle

(c) Sleeve

(d) Socket

(*a*) *Drill chuck*: It is designed to hold straight shank drills of different sizes. The jaws of the chuck are tightened around the drill by means of chuck key (see Fig. 10.20).

(*b*) *Drilling machine spindles*: All general purpose drilling machines have the spindle bored to a Morse standard taper which is approximately 1:20. The drill may be removed by driving the drift as shown in Fig. 10.21.

(*c*) *Sleeve*: It is used to hold taper shank drills whose taper is less than hole of the spindle. The outer taper of the sleeve matches with spindle hole taper (see Fig. 10.22).

(*d*) *Socket*: It is used for the drills whose taper is larger than sleeve. (See Fig. 10.23).

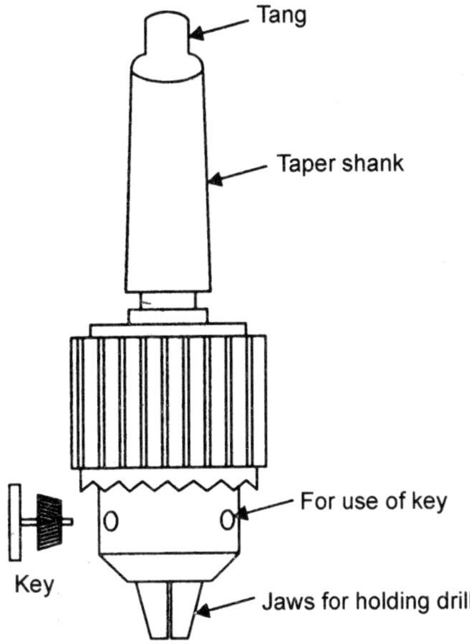

Fig. 10.20 Drill Chuck

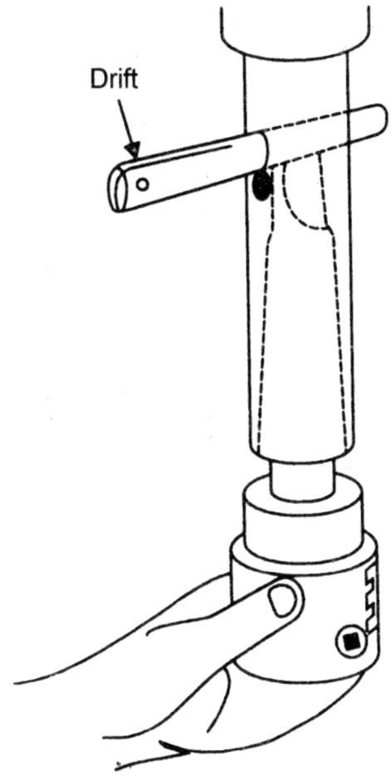

Fig. 10.21 Drilling Spindle

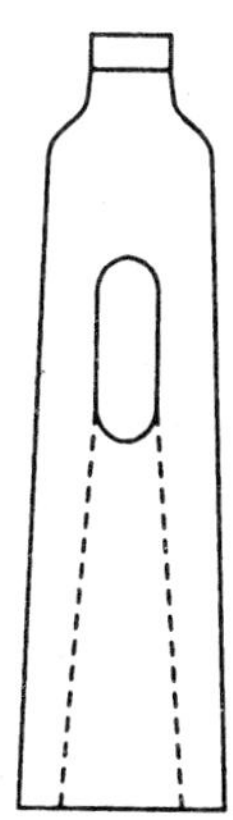

Fig. 10.22 Drill Sleeve

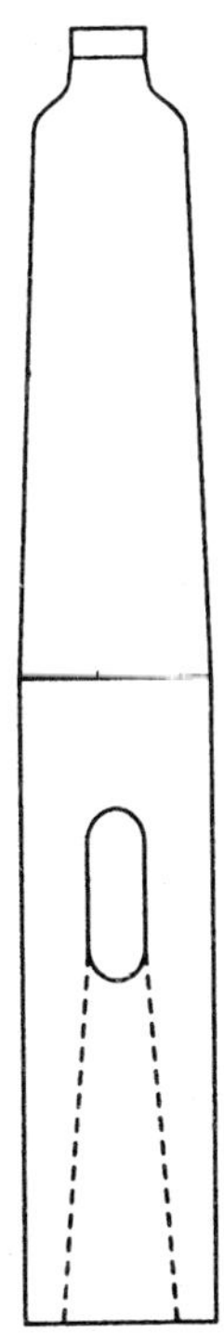

Fig. 10.23 Drill Socket

10.8 DRILLING MACHINE OPERATIONS

A drilling machine is capable of performing various operations by using suitable tools. The following operations are generally performed on a drilling machine (See Fig. 10.24).

1. Drilling
2. Reaming
3. Boring
4. Counter boring
5. Counter sinking
6. Spot facing
7. Tapping
8. Trepaning or Circle cutting

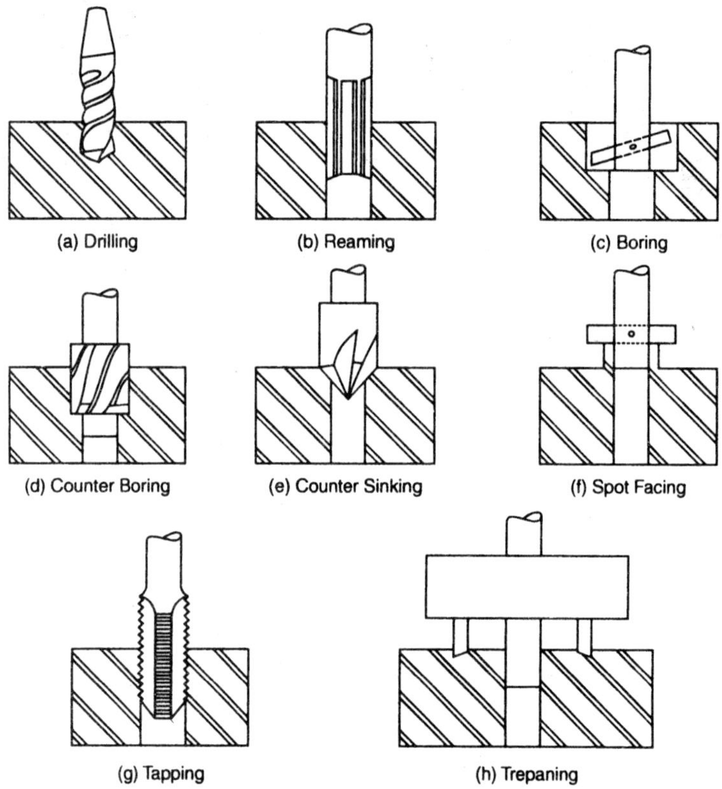

Fig. 10.24 Drilling Machine Operations

1. *Drilling*: Drilling is the process of making hole by rotating of cutting tool called drill. (Fig. 10.24(a)). For accurate location of the hole before drilling, should be marked out and centre punched. For mass production work a drill jig is used which eliminates the marking operation.
2. *Reaming* (Fig. 10.24(b)): It is an accurate way of sizing and finishing a hole, which has been already drilled. Material allowance left in the hole for hand reaming is usually 0.05 to 0.1mm and for machine reaming, it ranges from 0.13 to 0.65mm.
3. *Boring* (Fig. 10.24(c)): It is generally adopted for producing non-standard holes for which drills are not commercially available (enlarging a hole).

4. *Counter boring* (Fig. 10.24(d)): It is the operation of enlarging the hole for part of its depth. The enlarged hole forms a square shoulder with the original hole. This is necessary to accommodate the heads of bolts, studs and pins.
5. *Counter sinking* (Fig. 10.24(e)): It is the operation of making a conical shaped enlargement at the top of the hole. This is done to provide seat for a flat head of screw or counter sink rivet fitted into the hole.
6. *Spot facing* (Fig. 10.24(f)): It is the operation of smoothing and squaring the surface around a hole for the seat for a nut or the head of the screw.
7. *Tapping* (Fig. 10.24(g)): It is the operation of cutting internal threads by using the tool called tap.
8. *Trepaning or circle cutting* (Fig. 10.24(h)): It is the process in which a large hole in a sheet or plate is made with the panning tool. In this process a central hole is drilled. It receives the pilot of the tool, thus prevents the lateral displacement of the tool. On rotation of the tool, a circle is cut from the plate. The advantage of this process is that the central portion of the plate is removed as a solid mass, whereas in drilling, the central portion is removed in the form of chips.

10.9 SPEED, FEED AND MACHINE TIME

Speed: The cutting speed of a drill is defined as the peripheral speed of a drill surface, which is in contact with the work. It is expressed in meters/ min. It depends on number of factors such as work material, tool material, depth of hole use of coolant etc.

Cutting speed $V = \pi DN/100$ m/min

Where D = Work diameter in mm

N = Rotational speed of drill in RPM

The cutting speeds with high speed drills are shown in Table 10.2

Table 10.2: Cutting Speeds in Drilling

Material	*Speed Range m/mm*
Mild Steel	25 – 35
Steel	20 25
Soft Cast Iron	30 – 45
Medium Cast Iron	21 – 30
Aluminium and Alloys	60 – 90
Brass and Bronze	60 – 90
Copper	18 – 30

Feed: It is the distance the drill enters the work for each revolution of the drill spindle. It is expressed in millimeters per revolution.

The Feed per min may be calculated as :

$$f_M = f_r \times N$$

Where

f_M = Feed per minute

f_r = Feed per revolution
N = RPM of the drill

Machining Time in Drilling

$$\text{Time for drilling} = \frac{L}{f_r \times N}$$

Where N = rpm of drill
f = Feed per rev in mm
$L = l + a$

Where l = Thickness of work piece
a = Approach of drill = 0.3d
d = Diameter of drill

Example 10.1: Find the time required to drill 6 holes of 16 mm diameter each on a flange. Assume flange thickness = 30 mm, cutting speed = 20 m/min, feed = 0.2 mm/rev.

Solution:

$$V = \frac{IIDN}{1000}$$

$$N = \frac{1000 \times 20}{3.14 \times 16}$$

$$= 398.08 \text{ rpm}$$

$$\text{Time for drilling one hole} = \frac{L}{f_r \times N}$$

$$L = 1 + a = 30 + 0.3 \times 16$$

$$= 30 + 4.8 = 34.8$$

$$T = \frac{34.8}{0.2 \times 398} = 0.437 \text{ min}$$

Time for 6 holes = 6 × 0.437 min = **2.623 min.**

Example 10.2: A drilling machine has to be designed for 6 speeds ranging from 34 rpm to 353 rpm:

(a) Draw the ray diagram and gear arrangement.

(b) Also deduce the standard series used. [GATE-problem]

Solution:

(a) For Ray diagram see Fig. 10.25

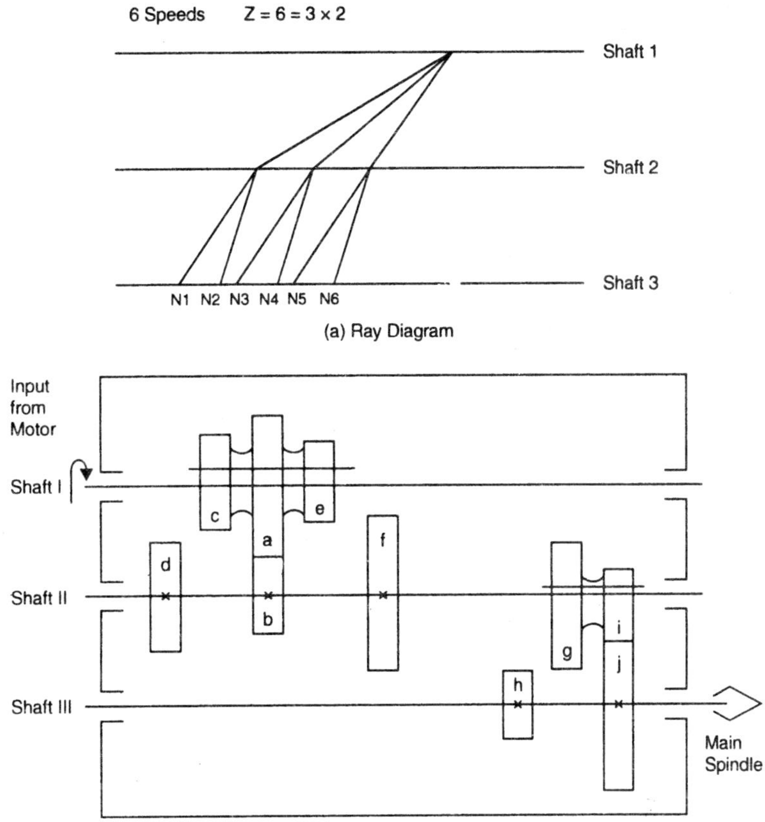

(a) Ray Diagram

(b) Arrangement of Gear Box

Fig. 10.25 Six Speeds Drilling Machine

(b) $\phi = Z - 1 \sqrt{\frac{N_{max}}{N_{min}}} = \sqrt[5]{\frac{353}{34}}$

$= 1.596 = 1.6$

for $\phi = 1.6$

speed are:

$N_1 = 34$ rpm

$N_2 = N_1\phi = 34 \times 1.6 = 54.4$ rpm

$N_3 = N_2\phi = 87$ rpm

$N_4 = 139$ rpm

$N_5 = 222$ rpm

Standard $\phi = 1.58$

Hence, $N_1 = 34$ rpm

$N_2 = N_1\phi = 54$ rpm

$N_3 = N_2\phi = 85$ rpm

$N_4 = 134$ rpm

$N_5 = 212$ rpm

QUESTIONS

1. Draw a neat sketch of twist drill and show various parts.
2. What are the various types of drilling machines? Explain their usage in the workshop.
3. Draw a neat sketch of Radial Drilling machine and label the parts.
4. How a drilling machine is specified?
5. Enumerate different operations that can be done on a drilling machine.
6. List various work holding and tool holding devices used in a drilling machine.

11

Milling Machine

11.1 INTRODUCTION

Milling is the process of removing metal by feeding the work against a rotating multi-point cutter. In milling operation the rate of metal removal is rapid as the cutter rotates at a high speed and has many cutting edges.

11.2 TYPES OF MILLING MACHINES

There are many types of milling machines from simple hand milling machine to the complex tape-controlled machines. Each has a particular field, in which it performs best. However, all these machines can be classified into the following categories.

1. *Column and knee types*
   (a) Plain milling machine
   (b) Vertical milling machine
   (c) Universal milling machine
   (d) Ram-type milling machine
2. *Bed type milling machine*
   (a) Simplex milling machine
   (b) Duplex milling machine
   (c) Triplex milling machine
3. *Plano type milling machine*
4. *Special purpose milling machine*
   (a) Rotary table milling machine
   (b) Drum milling machine
   (c) Profile milling machine

11.2.1 Column and Knee Types

(a) Plain Milling Machine

It is also known as horizontal milling machine. Its principal parts are shown in Fig. 11.1.

Base: It is the foundation of the machine. All other parts are mounted on it. It also serves as reservoir for cutting fluids.

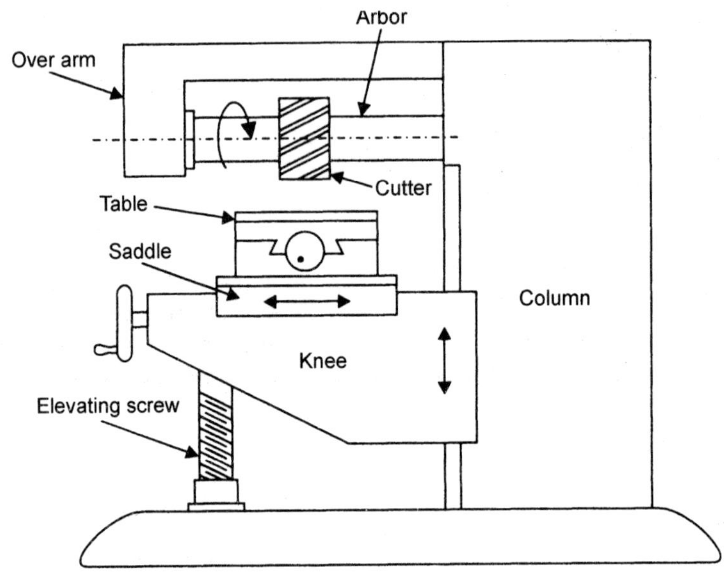

Fig. 11.1 (a) Horizontal Milling Machine

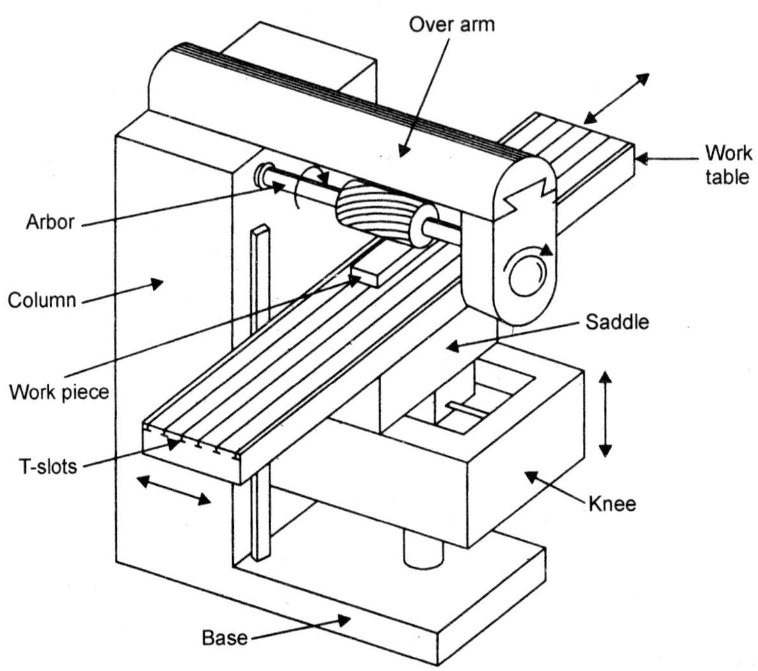

Fig. 11.1 (b) 3-D View of Horizontal Milling Machine

Column: It is the main support of the machine. The motor and other driving mechanisms are housed in it. It supports and guides the kee in its vertical travel.

Knee: The knee projects from the column and slides up and down. It supports saddle and the table. Elevating screw provides its vertical movement (up and down).

Saddle: The saddle supports and carries the table and provides traversed movement.

Table: The table rests on the ways on the saddle and travels longitudinally in a horizontal plane. It supports the work-piece, fixtures etc.

Over arm: It is mounted on and guided by the top of the column. The over arm is used to hold the outer end of the arbor to prevent it from bending.

Arbor : Arbor is an accurately machined shaft. Cutters are mounted on the arbor, which is rigidly supported by the over arm, spindle and end braces. It is tapered at one end to fit the spindle nose and has two slots to fit the nose keys for locating and driving it.

(*b*) *Vertical Milling Machine*

A vertical milling machine can be distinguished from a horizontal milling machine by the position of its spindle which is vertical or perpendicular to the work table. The spindle head, which is clamped, to the vertical column may be swivelled at an angle, permitting to work on angular surfaces. The machine is used for machining grooves, slots and flat surfaces. Generally vertical milling machine is used to perform end milling and face milling operations. The Fig. 11.2 illustrates the vertical milling machine.

(*c*) *Universal Milling Machine*

It is the most versatile of all the milling machine. It is similar to plain milling machine and differs only in respect of the table movement. The table can be swivelled about a vertical axis up to 45°. The capacity of a universal milling machine is increased by the use of special attachments such as dividing head, vertical milling attachment, rotary attachment, slotting attachment etc. The machine can produce spur, spiral, bevel gears, reamers, milling cutters etc.

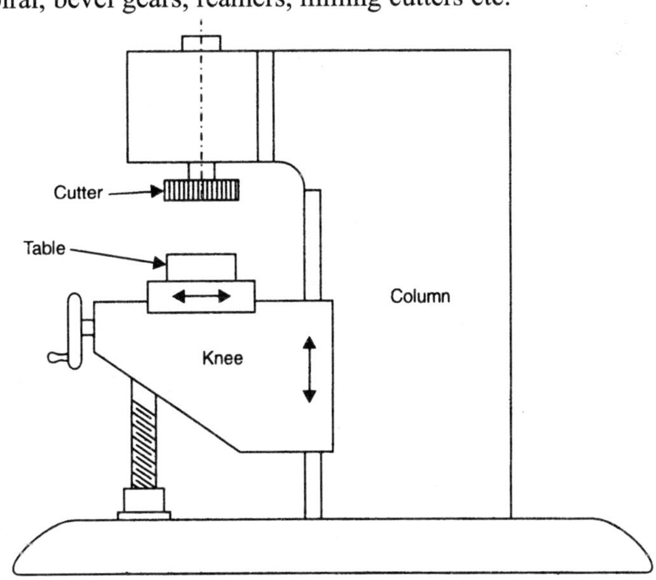

Fig. 11.2(a) Vertical Milling Machine

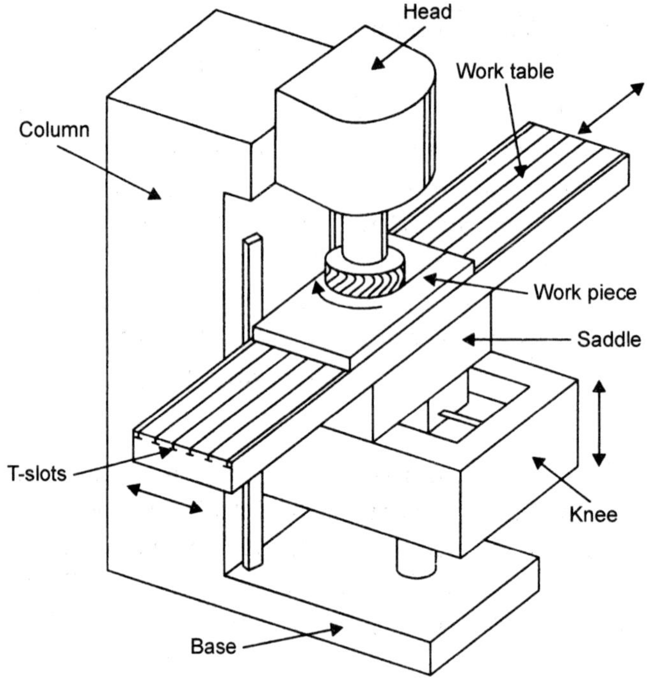

Fig. 11.2(b) 3-D View of Vertical Milling Machine

Comparison between Plain and Universal milling machine:

1. In plain milling machine the table is provided with three movements: longitudinal, cross and vertical. In universal milling machine in addition to these three movements, there is a fourth movement to the table. The table can be swivelled horizontally and can be fed at angle to the milling machine spindle.

2. The universal milling machine is provided with auxiliaries such as dividing head, vertical milling attachment, rotary table etc. Hence it is possible to make spiral, bevel gears, twist drills, reamers etc. on universal milling machine.

3. The plain milling machine is more rigid and heavier in construction than a universal milling machine.

4. The plain milling machine is used for manufacturing operations, whereas universal milling machine is used for tool room work and for special machining operations. Hence generally universal milling machine is used in tool room.

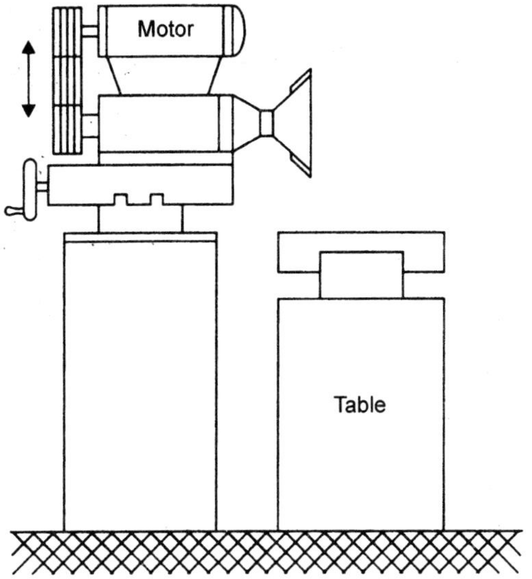

(a) Horizontal Bed Type Milling Machine — SIMPLEX

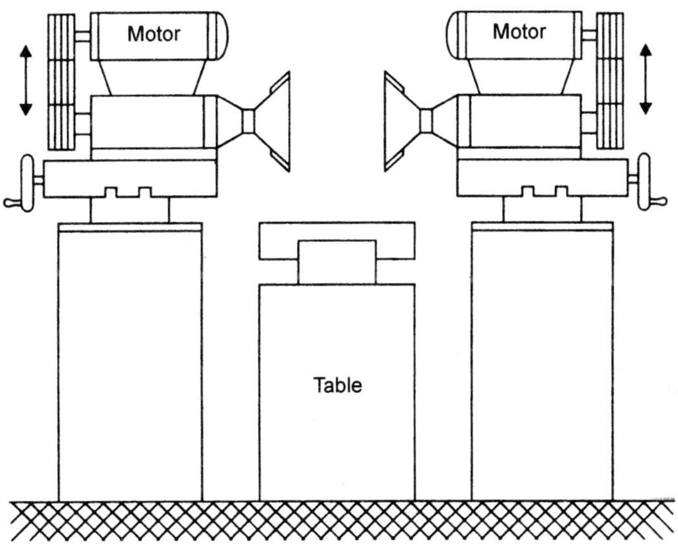

(a) Horizontal Bed Type Milling Machine — DUPLEX

Fig. 11.3 Fixed Bed Type Milling Machine

11.2.2 Bed Type Milling Machines

These are comparatively large, heavy and rigid in construction. The vertical motion is imported to the spindle head instead of the table. Depending upon the number of spindle, these are classified as simplex and duplex machines.

11.2.3 Plano Type Milling Machine (see Fig. 11.4)

It resembles a planer. The essential difference between a planer and plano-miller lies in the table movement. In a planer, the table moves to give the cutting speed, but in a plano-miller, the table movement gives the feed. Hence the table movement in plano-miller is much slower than that of a planing machine.

11.2.4 Special Purpose Milling Machine

These machines are designed to perform a specific type of operation only.

(*a*) *Rotary Table Milling Machine* (see Fig. 11.5)

This is modification to a vertical milling machine. It consists of two vertical spindles mounted with face milling cutters. A number of work pieces are clamped on a circular table which rotates about a vertical axis. The cutters may be set at different heights relative from work so that when one of the cutter is for roughing operation, the other is for finishing operation.

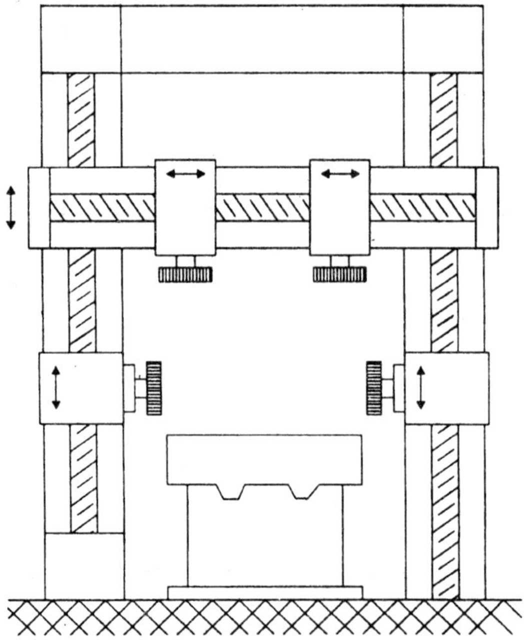

Fig. 11.4 Plano Type Milling Machine

(*b*) *Drum Milling Machine* (see Fig. 11.6)

These machines are used to face the two ends of a work piece simultaneously and in continuous machining cycle. The work pieces are clamped on a central drum, which rotates on a horizontal axis so that both end faces are machine simultaneously, first by roughing and then by finishing cutters.

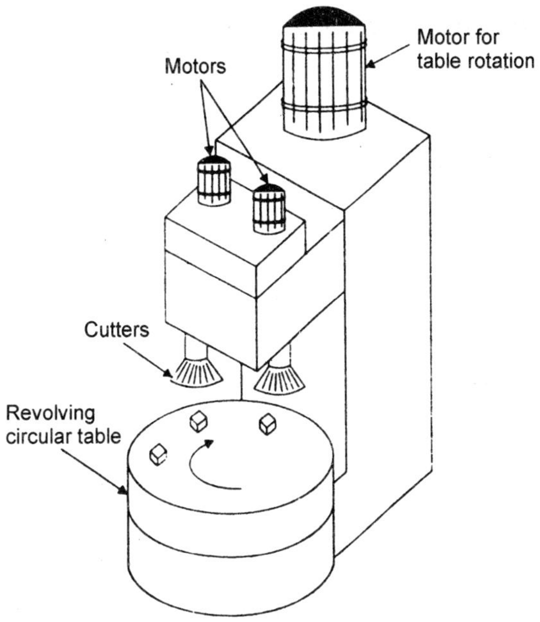

Fig.11.5 Rotary Table Milling Machine

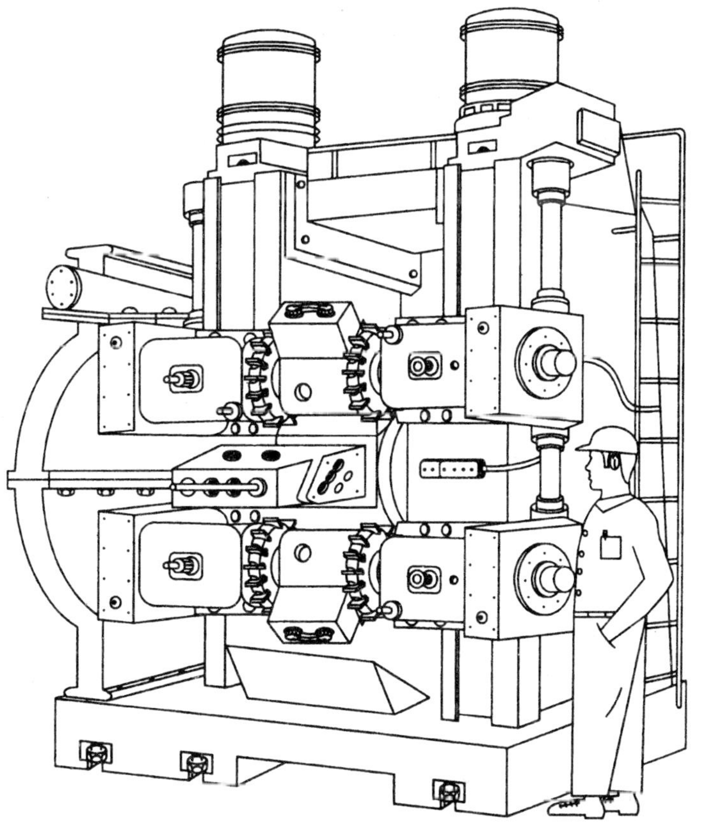

Fig. 11.6 Drum Type Milling Machine

(c) *Profile Milling Machine*

It has the cutter guided by means of a guide pin which is held against and follows the outline or the profile on a guide block (template). It is largely used for making sewing machine parts.

11.3 SIZE OF MILLING MACHINE

The size of the milling machine is denoted by the dimensions of the working surface of the table and its maximum length of longitudinal, cross and vertical travel of the table.

$$\text{Table length} \times \text{Width} = 1100 \text{ mm} \times 310 \text{ mm}$$

$$\text{Power traverse} = \text{Longitudinal} \times \text{Cross} \times \text{Vertical}$$

$$= 650 \text{ mm} \times 230 \text{ mm} \times 400 \text{ mm}$$

In addition to the above dimensions, number of spindle speeds, number of feeds, power available, net weight and floor space required.

11.4 MILLING MACHINE ATTACHMENTS

The range of work that a milling machine can do is greatly increased by the use of attachments. The following are the different attachment used on a milling machine.

(a) *Vertical Milling Attachments* (see Fig. 11.7)

With the use of this attachment, the horizontal and universal milling machines can be made to act as vertical milling machine.

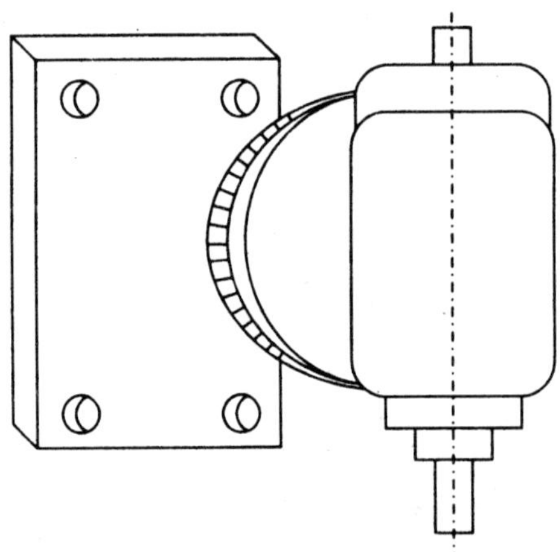

Fig. 11.7 Vertical Milling Attachment

(b) *High Speed Milling Attachment*

This attachment consists of gearing arrangement to increase the spindle speeds by four to six times. This is for operating small diameter cutters efficiently and at the proper cutting speed.

(c) *Slotting Attachment*

This attachment converts the rotary motion of the spindle into reciprocating motion by means of an eccentric or crank arrangement. This is used for key way cutting, grooving and internal gear cutting.

(d) *Dividing Head Attachment*

This is a work holding device, which is mounted, on the machine table. It is used for dividing the periphery of the work-piece into a required number of equal parts.

11.5 MILLING CUTTERS

The milling cutter is a multi-point revolving tool. The teeth of milling cutter may be parallel to the axis of the rotation or at an angle known as helix angle. The helix angle may be right or left hand. Further a milling cutter may be made of single piece (solid cutter) or having removable cutting teeth inserted in a solid body (inserted teeth cutter). Following are the common type of these cutters (see Fig. 11.8).

1. *Plain milling cutter*: These cutters have straight or helical teeth cut on the periphery of a cylindrical surface in Fig. 11.8(a). These are used to machine flat surfaces. These are mounted on horizontal milling machines.

2. *Face milling cutters* (Fig. 11.8(b)): A face-milling cutter is also used for machining flat surfaces. It is mounted on a vertical-milling machine.

3. *Plain slitting saw* (Fig. 11.8(c)): It resembles a plain milling cutter or a side-milling cutter in appearance, but it is of very small width. It is used for cutting-off and slotting operations and somewhat similar to the circular saw blade.

4. *Side milling cutter* (Fig. 11.8(d)): The side milling cutter have teeth on its periphery and also on one of its side. It may have plain, helical or staggered teeth. These cutters are used for side milling and slot cutting.

5. *Angle milling cutter* (Fig. 11.8(e)): This cutter will have their cutting teeth at an angle to the axis of rotation. Angular cutters are classified as single angular cutters or double angular cutters. Single angular cutter is used in cutting dovetail grooves.

6. *T-Slot cutter* (Fig. 11.8(f)): The T -slot milling cutters are special forms of end mills for producing T -slots.

7. *End Mill Cutters* (Fig. 11.8(g)): The end mills have cutting teeth on the end as well as on the periphery of the cutter. The peripheral teeth may be straight or helical. These are used for light milling operations like cutting slots, machining accurate holes and for profile milling operations.

8. *Form milling cutters* (Fig. 11.8(h)): These cutters are designed to cut definite shapes are known as form milling cutters. These cutters can be classified according to their shape as convex or concave cutters, gear cutters etc.

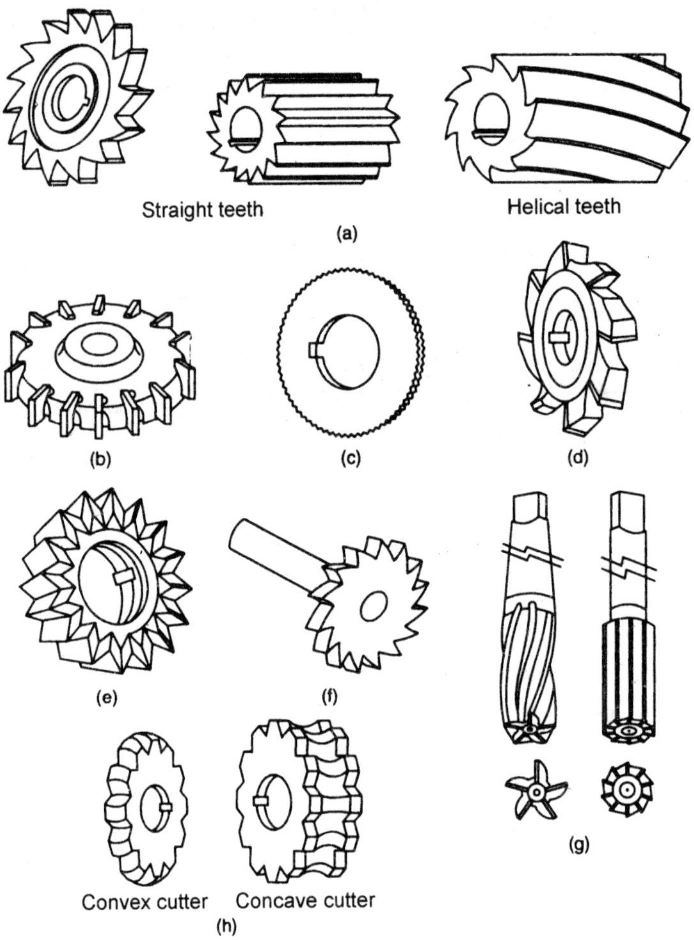

Fig. 11.8 Types of Milling Cutters—(a) Plain milling cutter; (b) Face milling cutter with inserted teeth; (c) Plain metal slitting saw; (d) Side milling cutter; (e) Angle milling cutter; (f) T-slot cutter; (g) End mill cutter; (h) Form cutter

11.6 CUTTER MATERIALS

The milling cutter is made of (a) High speed steels (b) Non-ferrous cast alloys or cemented carbide tips.

In normal work, high-speed steel cutters are more commonly used in production shops. Carbide tipped tools is used for mass production as they last long and yield high production. The tips of carbide are brazed or insert into the high carbon steel body of the cutter. Tips made of ceramics are also used in milling cutters.

11.7 ELEMENTS OF A PLAIN MILLING CUTTER

Fig. 11.9 shows the principal parts and angles of a plain milling cutter.

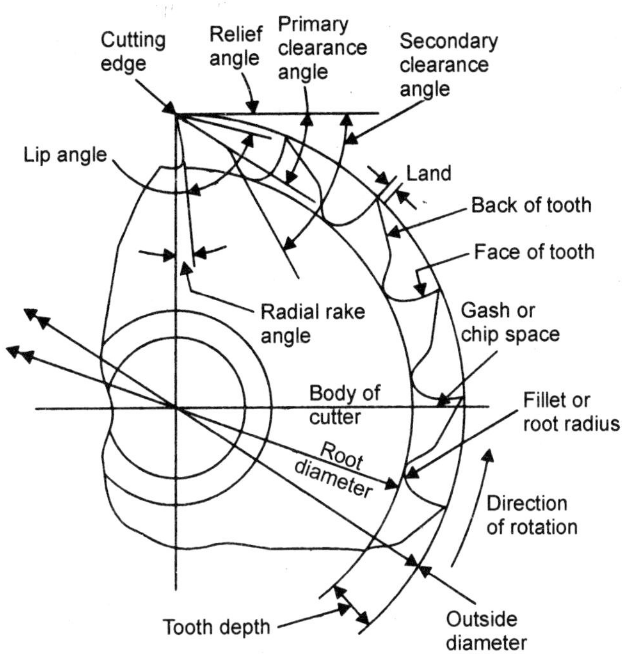

Fig. 11.9 Elements of Plain Milling Cutter

Body of cutter: The main frame of the cutter on which the teeth rest to form an integral at part is known as body of the cutter.

Cutting edge: The edge formed by the intersection of the teeth and the circular land of the surface left by the provision of primary clearance.

Face: The portion of the gash adjacent to the cutting edge on which the chip impinges as it is cut from the work.

Fillet: The curved surface at the bottom of gash which joins the face of one teeth to the back of the tooth immediately ahead.

Gash: The chip space between the back of one tooth and the face of the next tooth

Land: The part of the back of tooth adjacent to the cutting edge which is relieved to avoid interference between the surface being machined and the cutter.

Outside diameter: The diameter of the circle passing through the peripheral cutting edge.

Root diameter: The diameter of the circle passing through the bottom of the fillet.

Cutter angles: Like a single point cutting tool, the milling cutter is also provided with rake, clearance and other angles to remove metal efficiently.

Relief angle: The angle between the land of a tooth and the tangent to the outside diameter at the cutting edge.

Clearance angle: There are two types of clearance angels on a milling cutter–primary clearance and secondary clearance angle.

Primary clearance angle: This is the angle between the surface of land and a tangent to the periphery at the cutting edge.

Secondary clearance angle: This is the angle formed by the secondary clearance surface of the tooth and the tangent to the periphery at the cutting edge.

Lip angle: The included angle between the land and the face of the tooth.

Rake angle (*radial*): The angle measured in the diametrical plane between the face of the tooth and a radial line passing through the cutting edge. It may be positive; negative or zero.

11.8 MLLLING METHODS

(a) Peripheral milling
(c) End milling
(b) Face milling

***(a) Peripheral milling*:** It is the operation performed by a milling cutter to produce a machined surface parallel to the axis of rotation of the cutter. The contact between the cutter and the work surface is over the whole width of the cutter. The cutting force is large. According to the relative movement between the tool and the work, the peripheral milling is classified under two headings:

(*i*) *Up milling*: It is also called as conventional milling. In this method of milling the cutter rotates in a direction opposite to that in which the work is fed as shown in the Fig. 11.10(a). In this method of milling the thickness of the chip is minimum at the beginning of the cut and it reaches to the maximum when the cut terminates. The cutting force in up milling increases from zero to the maximum value per tooth movement of the cutter. The cutter force is directed upwards and this tends to lift the work from the fixture. Pouring the coolant on the cutting edge is not possible. The surface milled by up milling is not smooth.

(*ii*) *Down milling*: The down milling is also called as climb milling. In this method of milling the cutter rotates in the same direction of travel of the work-piece as shown in Fig. 11.10(b). The thickness of the chip is maximum when the tooth begins it's cut and reduces to minimum when the cut terminates. In down milling, the fixture design becomes easier as the direction of the cutting force is such that it tends to seat the work firmly in the work holding devices. The chips are also disposed off easily and do not interfere with the cutting. The coolant can be poured directly at the cutting zone. This result in improved surface finish and diminishes the heat generated.

***(b) Face milling*:** The face milling is the operation performed by a milling cutter to produce flat-machined surfaces perpendicular to the axis of rotation. The peripheral cutting edges of the cutter do the actual cutting, whereas the face cutting edges finish up the work surface by removing a very small amount of material.

***(c) End milling*:** The end milling may be considered as the combination of peripheral and face milling operation. The cutter has teeth both on the end face and on the periphery.

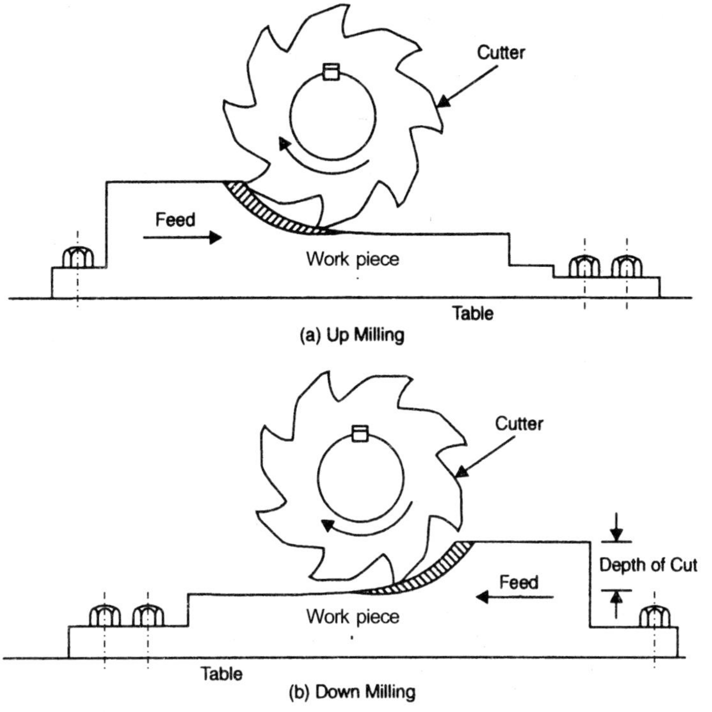

Fig. 11.10

11.9 MILLING OPERATIONS

1. Plain milling
2. Face milling
3. Straddle milling
4. Angular milling
5. Gang milling
6. Form milling
7. Profile milling
8. End milling
9. Helical milling
10. Gear cutting

1. *Plain milling*

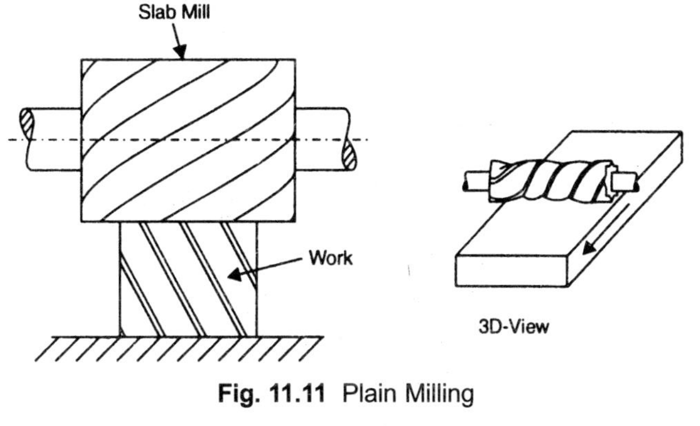

Fig. 11.11 Plain Milling

It is the operation of production of plain, horizontal surfaces parallel to the axis of rotation of a plain milling cutter. The operation is also called slab milling. The plain milling operation is illustrated in Fig. 11.11.

2. *Face milling*

It is the process of milling flat surfaces using a milling cutter such that the surface generated is at right angle to the axis of the cutter as shown in Fig. 11.12.

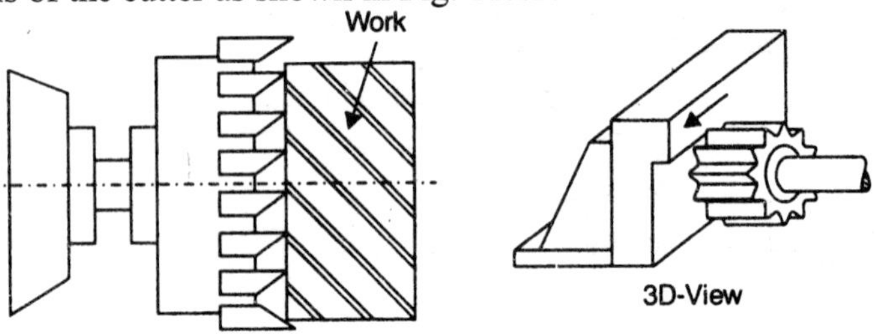

Fig. 11.12 Face Milling

3. *Straddle milling*

It is a milling operation in which a pair of side milling cutters is used for machining for two parallel vertical surfaces of a workpiece simultaneously as shown in Fig. 11.13.

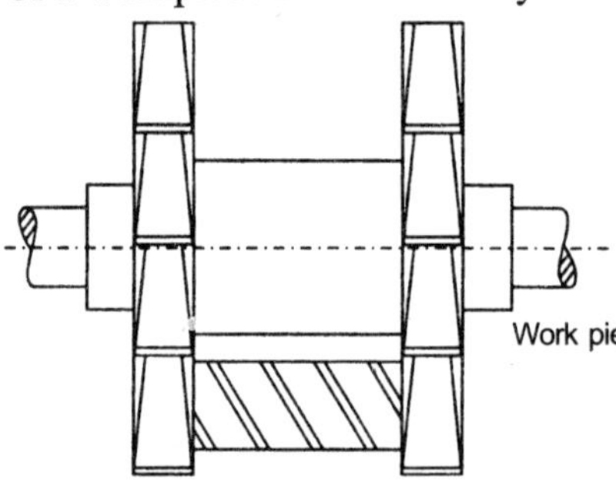

Fig. 11.13 Straddle Milling

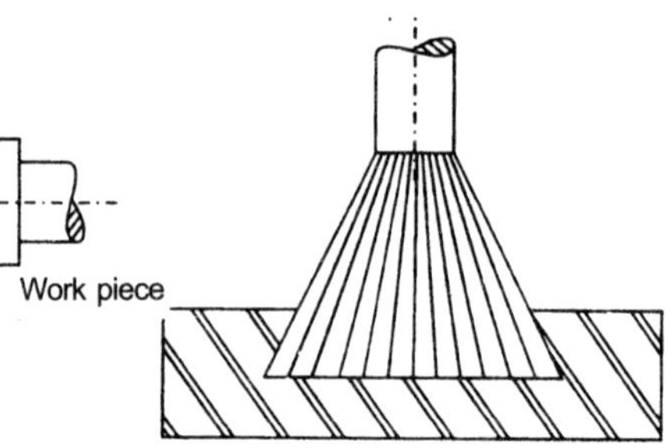

Fig. 11.14 Angular Milling

4. *Angular milling*

It is milling process, which is used for machining a flat surface at an angle, other than a right angle to the axis of the revolving cutter. Fig. 11.14 shows forming a dovetail using angular cutter.

5. *Gang milling*

It is the operation of machining several surfaces of a work-piece simultaneously by feeding the work table against a number of cutters having same or different diameters mounted on the arbor of the machine. This method saves much of machining time. The operation is illustrated in Fig. 11.15.

6. *Form milling*

It is the operation of production of irregular contours by using form cutters. The irregular contour may be convex, concave or any other shape. The form milling operation is illustrated in Fig. 11.16.

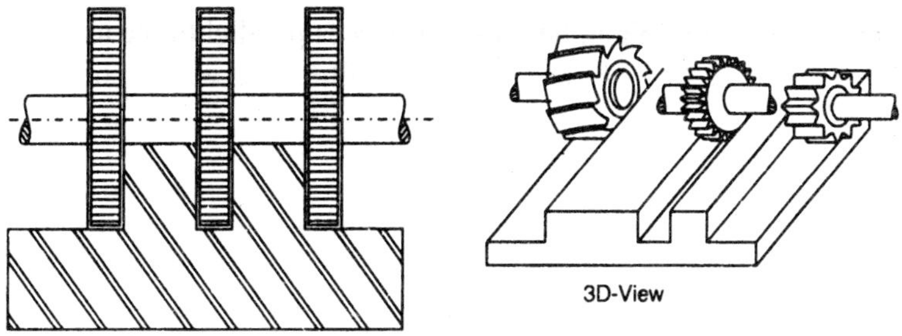

Fig. 11.15 Gang Milling

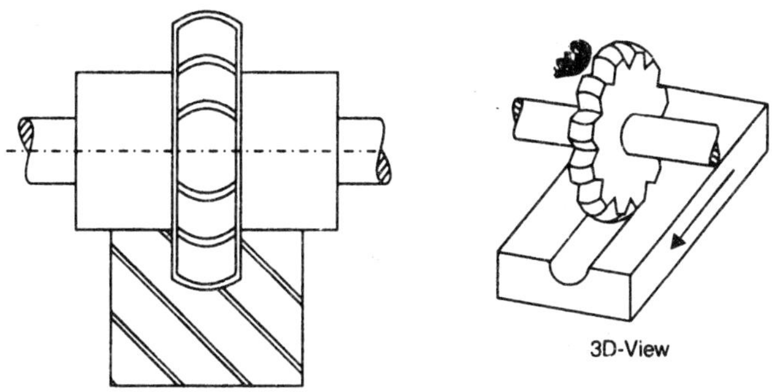

Fig. 11.16 Form Milling

7. *Profile milling*

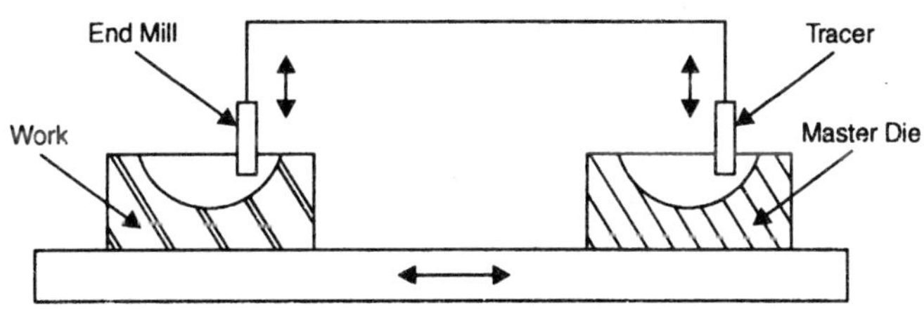

Fig. 11.17 Profile Milling

The profile milling is the operation of reproduction of an outline of template or complex shape of master die on a work-piece. Different cutters may be used for profile milling. An end mill is one of the most widely used milling cutters in profile milling work as shown in Fig. 11.17.

8. *End milling*

It is the operation of production of a flat surface which may be vertical horizontal or at an angle in reference to the table surface. The cutter used is an end mill. The end mill cutters are also used for production of slots, grooves or keyways. The operation is shown in Fig. 11.18.

9. *Helical milling*

It is the operation of production of helical flutes or grooves around the periphery of a cylindrical or conical work piece. The operation is performed by swivelling the table to the required helix angle and then by rotating and feeding the work against rotary cutting edges of a milling cutter. The production of helical gears, cutting helical grooves or flutes on a drill shank or a reamer is the examples of helical milling.

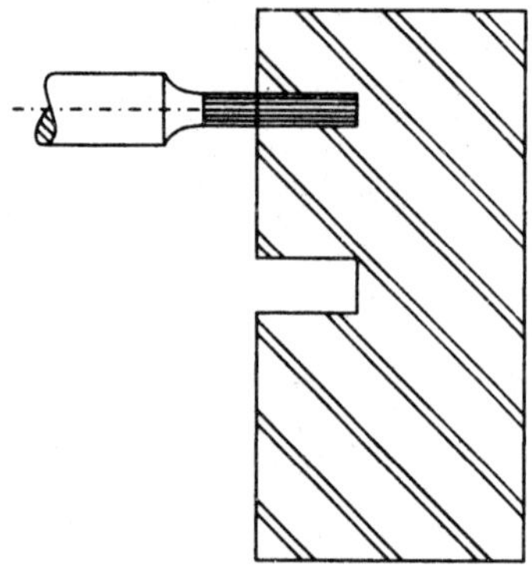

Fig. 11.18 End Milling

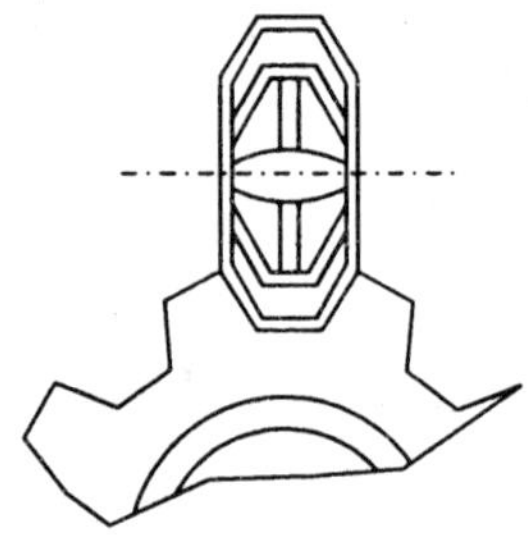

Fig. 11.19 Gear Cutting

10. *Gear cutting*

The gear cutting operation is performed in a milling machine by using a form relieved cutter. The cutter may be cylindrical type or end mill type. The cutter profile corresponds exactly with the tooth space of the gear. Equally spaced gear teeth are cut in a gear blank by holding the work on a universal-dividing head (UDH) and then index it. The gear cutting operation is shown in Fig. 11.19.

11.10 INDEXING AND DIVIDING HEAD

Indexing is the operation of dividing the periphery of work-piece into any number of equal parts. A wide range of indexing operations are used in milling like producing hexagonal and square headed bolts, cutting splines on the shaft, gear cutting. For doing this indexing, an attachment known as dividing head is used with the milling machine. The dividing heads are of three types.

(a) Plain or simple dividing head (c) Optical dividing head

(b) Universal dividing head

(a) Plain or Simple Dividing Head (see Fig.11.20)

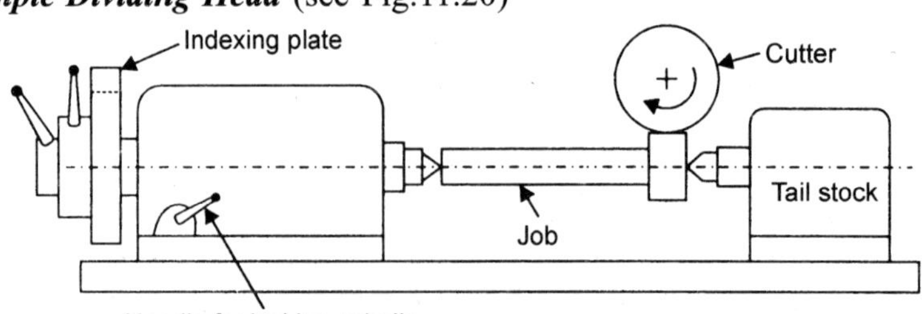

Fig. 11.20 Plain Indexing Head for Direct Indexing

The plain dividing head consists of a cylindrical spindle housed in a frame and a base bolted to the machine table. The indexing crank is connected to the tail end of the spindle directly, and the crank and spindle rotate as one unit. The index plate is mounted on the spindle and rotate with it. The spindle can be rotated through the desired angle and then clamped by inserting the clamping lever pin into anyone of the slots of the index plate. The job is held between two centers, one on the dividing head spindle and the other on the tailstock as shown in Fig. 11.20. In this dividing head, there is no worm and worm wheel.

(*b*) *Universal Dividing Head*

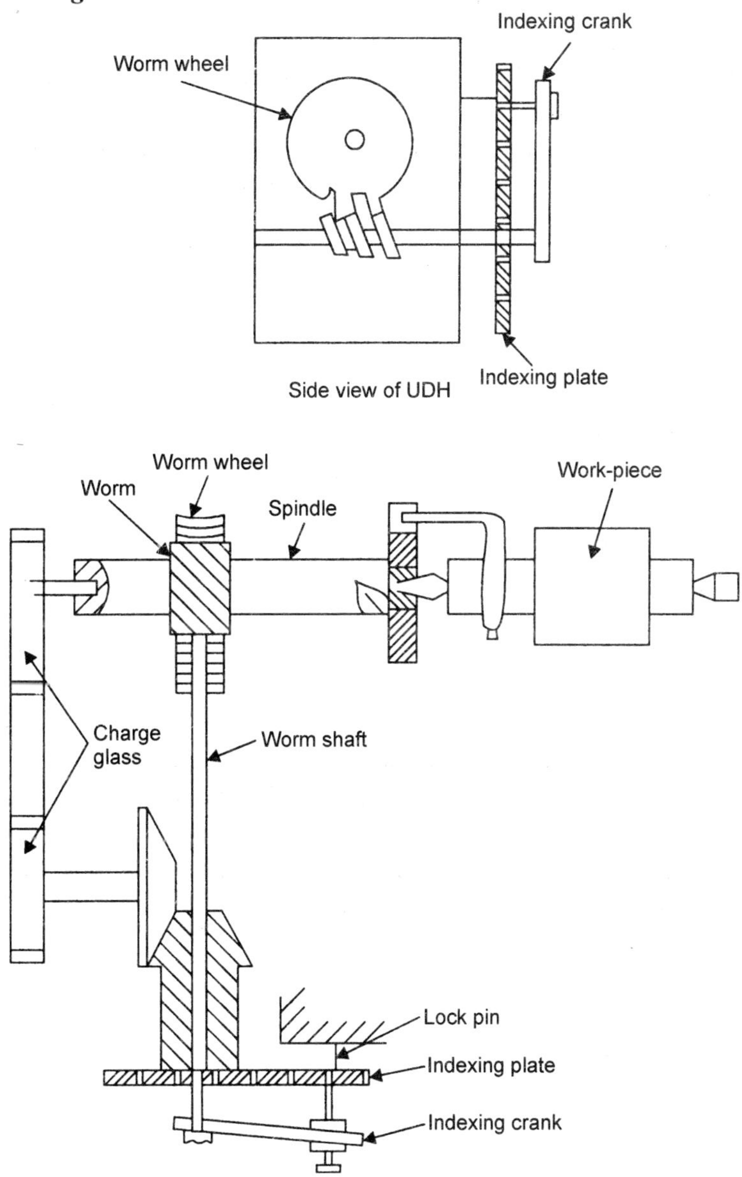

Fig. 11.21 Working Mechanism of a Universal Dividing Head

This dividing head is very useful device for the purpose of indexing work. The working mechanism of UDH is shown in Fig. 11.21. The spindle carrying the worm wheel meshes with the worm, which carries a crank at its outer end. The worm wheel has 40 teeth and the worm is single threaded. Thus 40 turns of crank will rotate the spindle for one complete revolution or one turn of the crank will cause the spindle to be rotated by 1/40 of a revolution. In order to turn the crank a fraction of a revolution, an index plate is used.

Index plate is a circular disc having a different member of equally spaced holes is arranged in concentric circles. The index plate is screwed on a sleeve, which is loosely mounted on the worm shaft. Normally the index plate remains stationary by a lock pin. The index pin works inside the spring loaded plunger. This plunger can slide, radially along a desired hole circle on the index plate.

The dividing head spindle may be connected with the table feed screw through a train of gears to impart a continuous rotary motion to the work piece for helical milling.

(*c*) *Optical Dividing Head*

The optical dividing heads are used for precise angular indexing during machining and for checking the accuracy of various angular surfaces.

11.11 INDEXING METHODS

The common methods of indexing are:

(a) Direct or rapid indexing
(b) Plain or simple indexing
(c) Compound indexing
(d) Differential indexing
(e) Angular indexing

(*a*) *Direct or Rapid Indexing*

It is used when a large number of identical pieces are indexed by very small divisions. The operation is performed on plain dividing head and universal dividing head. When using universal dividing, the worm and worm wheel are first disengaged. Rapid index plate having 24 slots is fitted to the frame. While indexing, the pin is taken out and then the spindle is rotated by hand and after the required position is reached, it is again locked by pin. With a rapid index plate of 24 slots, it is possible to divide the work into divisions of 2, 3, 4, 6, 8, 12 and 24.

The formula for indexing is as follows:

$$\text{No. of slots to be moved} = \frac{24}{N}$$

Where N = Number of divisions required.

Example 11.1: Find index movement to mill hexagonal bolt by direct indexing. The plate has 24 slots.

Solution: No. of slots to be moved $= \frac{24}{6} = 4$

After machining one side of the bolt, index plate has to be moved by 4 slots for 5 times the finish work.

(b) Plain or Simple Indexing

This method of indexing is used when the direct method of indexing cannot be employed for obtaining the required number of divisions on the work. Universal dividing head is used for this purpose. This method of indexing involves the use of crank, worm, worm wheel and index plate as shown in Fig. 11.22. The worm wheel carries 40 teeth and the worm is single threaded. With this arrangement 40 turns of the crank are required to rotate the spindle for one revolution i.e. one turn of the crank will cause the worm wheel to make 1/40th of revolution.

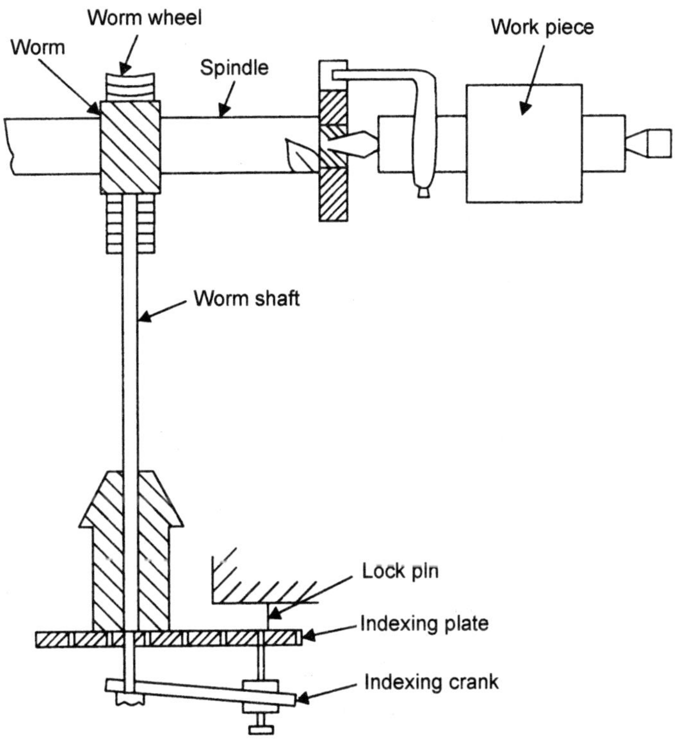

Fig. 11.22 Simple Indexing

To facilitate indexing to fractions of a turn, index plates of different circles are used to cover practically all numbers.

Suppose the work is to be divided into a number of parts, the corresponding crank movement will be as follows:

$$\text{Indexing crank movement} = \frac{40}{N}$$

Where N = Number of divisions required

Example 11.2: Mill 23 teeth on a spur wheel blank.

Solution: Index crank movement = $\frac{40}{23} = 1\frac{17}{23}$

Thus for indexing, one complete revolution and 17 holes of 23 hole circle of the index plate will have to be moved by the index crank.

(c) *Compound Indexing*

This method of indexing is used when the number of divisions required is outside the range of simple indexing. It involves in two stages:

1. By turning the crank a definite amount in one direction in the same way as in simple indexing.

2. By turning the index plate and the crank both in the same or reverse direction, thus adding further movement to or subtracting from that obtained in the first stage.

$$\frac{40}{N} = \frac{n_1}{N_1} \pm \frac{n_2}{N_2}$$

Where N =The number of divisions required

N_1 =The hole circle used by the crank pin

N_2 =The hole circle used by the lock pin

n_1 =Hole spaces moved by the crank pin in N_1 hole circle

n_2 =Hole spaces moved by the plate and crank in N_2 hole circle

Procedure for compound indexing is as follows:

1. Factorise the number of divisions required.
2. Factorise the standard No. 40.
3. Select for trial any two circles on the same plate and on same side. Factorise their difference.
4. Factorise the number of holes of one circle.
5. Factorise the number of holes of the other circle.
6. After obtaining these factors place them as follows:

$$\frac{\text{Factors of divisions required} \times \text{Factors of difference of hole circles}}{\text{Factors of 40} \times \text{Factors of first Circle} \times \text{Factors of 2nd Circle}}$$

Example 11.3: Compound index 69 divisions.

Solution:

1. $69 = 23 \times 3$

2. $40 = 2 \times 2 \times 2 \times 5$

3. Select index circle 23 and 33

Difference = $33 - 23 = 10 = 5 \times 2$

4. $23 = 23 \times 1$

5. $33 = 3 \times 11$

$$\frac{23 \times 3 \times 5 \times 2}{2 \times 2 \times 2 \times 5 \times 23 \times 1 \times 3 \times 11} = \frac{1}{44}$$

as we get one in numerator, the circles selected are correct.

$$\frac{44}{69} = \frac{44}{23} - \frac{44}{33} = 1\frac{21}{23} - 1\frac{11}{33} = \frac{21}{33} - \frac{11}{33}$$

Thus for indexing 69 divisions, the index crank should be moved by 21 holes of 23 hole circle in forward direction and then the plate and the crank together is moved by 11 holes 33 hole circle in the backward direction.

Example 11.4: Compound indexing for 87 divisions.

Solution:

1. $87 = 29 \times 3$

2. $40 = 2 \times 2 \times 2 \times 5$

3. Select indexing circle 29 and 33

   Difference = $33 - 29 = 4 = 2 \times 2$

4. $29 = 1 \times 29$

5. $33 = 3 \times 11$

$$\frac{3 \times 29 \times 2 \times 2}{2 \times 2 \times 2 \times 5 \times 1 \times 29 \times 3 \times 11} = \frac{1}{110}$$

as numerator is one, selected circles are correct.

$$\therefore \quad \frac{40}{87} = \frac{110}{29} - \frac{110}{33} = 3\frac{23}{29} - 3\frac{11}{33}$$

or $\quad 3\frac{11}{33} - 3\frac{23}{29}$

$$\frac{23}{29} - \frac{11}{33} \text{ or } \frac{11}{33} - \frac{23}{29}$$

since we keep the forward motion of the crank as larger than the backward motion, we adopt

$$\frac{23}{29} - \frac{11}{33} \text{ or } \frac{23}{29} - \frac{11}{33}$$

For indexing, the index crank should be moved by 23 holes of 29 circle forward direction and then the plate and the crank together is moved by 11 hole 33 circle in backward direction.

(d) Differential Indexing

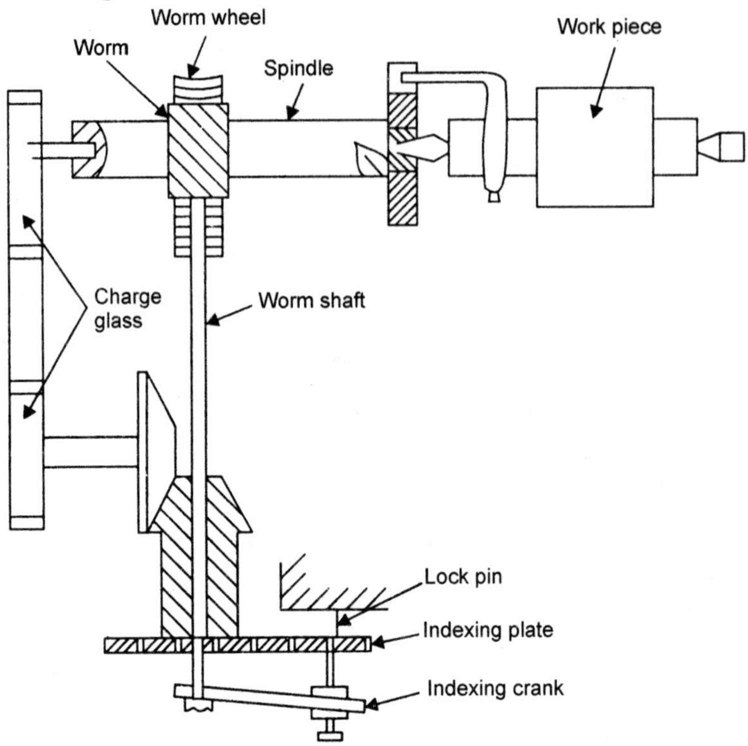

Fig. 11.23 Differential Indexing

Figure 11.23 illustrates the mechanism of differential indexing. Differential indexing greatly resembles compound indexing. This process is also carried out in two stages. In the first operation a crank is moved in a certain direction. In the second phase movement is added or subtracted by moving the plate by means of a gear train.

The rotation of index plate may take place in the same direction as the crank or opposite to it.

Rules for differential indexing.

1. Gear Ratio $= \dfrac{(A - N) \times 40}{A}$

   Where A = Selected number which can be indexed by plain indexing and the number is approximately equal to N.

   N = The required number of divisions to be indexed.

2. In the gear ratio, the numerator indicates driving gears on the index head spindle and the denominator indicates the driven gears on the index plate.

3. Index crank movement $= \dfrac{40}{A}$

The index crank has to be moved for N number of times for complete division of work. If $(A - N)$ is positive, the index plate must rotate in the same direction and if $(A - N)$ is negative, the index plate must rotate in a direction opposite to that of the crank.

To achieve this conditions, the following conditions are used:

(a) If the gear train is simple and $(A - N)$ is positive, only one idle gear is used.

(b) If the gear train is compound and $(A - N)$ is positive, no idle gear is used.

(c) If the gear train is simple and $(A - N)$ is negative, two idle gear are used.

(d) If the gear train is compound and $(A - N)$ is negative, only one idle gear is used.

Example 11.5: Make 83 divisions using differential indexing.

Solution: Assume $A = 86$

1. Gear Ratio $= \dfrac{(A - N) \times 40}{A} = (86 - 83) \times \dfrac{40}{86} = 3 \times \dfrac{40}{86} = \dfrac{72}{24} \times \dfrac{40}{86}$

2. Drivers = 72, 40

   Driven = 24, 86

3. Index crank movement $= \dfrac{40}{86} = \dfrac{20}{43}$

   The index crank have to be moved by 20 holes of 43 hole circle for 83 times.

4. As $(A - N)$ is positive, gear ratio compound, no idle gear is required.

Example 11.6: Make 139 divisions by differential indexing.

Solution: Assume A = 140

N = 139

1. Gear ratio $= (A - N) \times \dfrac{40}{A} = \dfrac{140 - 139}{140} \times 40 = \dfrac{40}{140} = \dfrac{2}{7} = \dfrac{2 \times 3}{7 \times 3} - \dfrac{2 \times 3}{3 \times 7}$

   $= \dfrac{32}{48} \times \dfrac{24}{56}$

2. $\because$ Drivers = 32, 24

   driven = 48, 56

3. Index crank movement $= \dfrac{40}{A} = \dfrac{40}{140} = \dfrac{2}{7} = \dfrac{6}{21}$

   The index crank have to be moved by 6 holes of 21 hole circle for 139 times.

4. As $(A - N)$ is positive, gear ratio compound, no idle gear is required.

(e) Angular Indexing

The angular indexing is used to rotate the job through certain angle, 40 turns of crank makes the work rotate through one complete turn i.e., 40 turns of crank make the work rotate 360°.

For each turn of the crank, the crank will rotate by = $\frac{360}{40} = 9°$

$\therefore$ Crank movement = $\frac{\text{Angular Displacement } (\theta)}{140}$ in degrees

$= \frac{\text{Angular Displacement}}{240}$ in minutes

$= \frac{\text{Angular Displacement of work}}{32400}$ in seconds

Example 11.7: Index angle 60°.

Solution: Crank movement = $\frac{60}{9} = 6\frac{6}{9} = 6\frac{12}{18}$

Crank movement = 6 full turns + 12 holes of 18 holes circle.

Example 11.8: Index angle 15°30'

Solution:

	In degree	*In minutes*
Crank movement	$= \frac{15\frac{1}{2}}{9}$	$15° \times 60 = 900'$
	$= \frac{31}{18}$	$15°30' = 900 + 30 = 930$
	$= 1\frac{13}{18}$	Crank movement = $\frac{930}{540}$
		$= \frac{31}{18} = 1\frac{31}{18}$

Crank movement	*Crank movement*
1 Full turn + 13 holes of 18 holes circle	1 Full turn + 13 holes of 18 holes circle

11.12 MACHINING TIME CALCULATIONS

Time required/cut = $\frac{\text{Length of cut}}{\text{Feed / Min}}$

Since milling cutter is a multipoint cutter, the feed will be as follows:

Feed per rev = Feed per tooth × Number of cutter teeth

Feed per min = (Table Feed) = Feed per rev × rpm of cutter

Length of cut = Length of the job + Added table travel

(*a*) *Slab Milling or Slot Milling Operation*

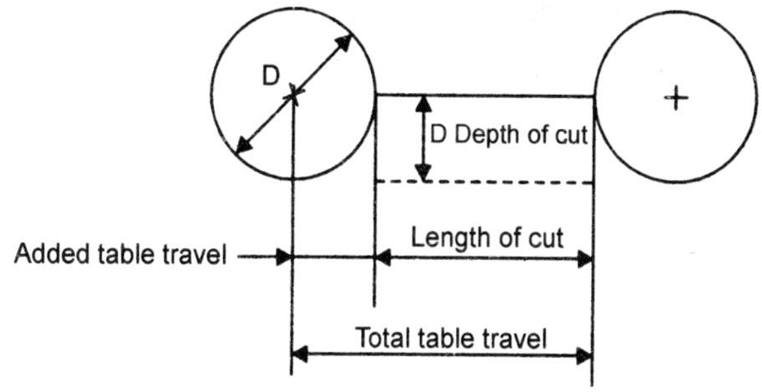

Fig. 11.24 Slab or slot Milling Operation

Fig. 11.24 shows how the cutter is adjusted to depth of cut '*d*' of the job.

Added table travel = $\sqrt{Dd - d^2}$

Where d = Depth of cut

D = Diameter of the cutter

(*b*) *Face Milling Operation*

Fig 11.25 shows face milling operation

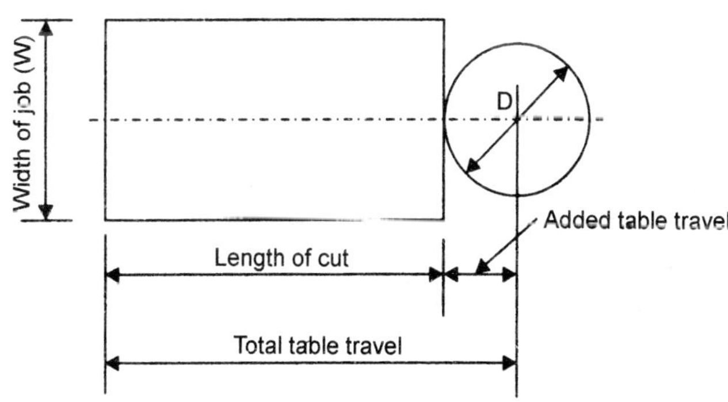

Fig. 11.25 Face Milling Operation

Added travel = $\frac{1}{2}(D - \sqrt{D^2 - W^2})$

Where D = Diameter of the cutter, W = Width of the work

if the cutter diameter is less than the width of the job

Then

$$\text{Added Travel} = \frac{D}{2}$$

Example 11.9: A slot of 25 mm depth is cut in a job 200 mm long with the help of cutter having 150 mm diameter and has 10 teeth. The cutting speed is 50m/min, and feed 0.25 mm per tooth. Determine (i) table feed in mm/min (ii) total travel of the cutter (iii) machining time for machining the slot.

Solution:

$$N = \frac{1000\ V}{\Pi D} = \frac{1000 \times 50}{3.14 \times 150} = 106.16 \text{ rmp}$$

(i) Table feed=0.25 × 10 × 106.16 = 265.4 mm/min

$$\begin{aligned} \text{Added Travel} &= \sqrt{Dd - d^2} \\ &= \sqrt{150 \times 25 - 25^2} \\ &= \sqrt{3750 - 625} \\ &= \sqrt{3125} \\ &= 55.9 \end{aligned}$$

$$\begin{aligned} \text{(ii) Total Travel} &= \text{length of the job + added travel} \\ &= 200 + 55.9 = 255.9 \end{aligned}$$

$$\begin{aligned} \text{(iii) Machining Time} &= \frac{\text{Total travel}}{\text{Feed / Min.}} = \frac{255.9}{265.4} \\ &= 0.964 \text{ min.} \end{aligned}$$

Example 11.10: A 63.5 mm diameter milling cutter having 6 teeth is used to face mill a block of 180 mm long and 30 mm wide. Speed is 1500 rpm and the feed is 0.125 mm per rev. find the machining time.

Solution: Added Travel $= \frac{1}{2}(D - \sqrt{D^2 - W^2})$

$$= \frac{1}{2}(63.5 - \sqrt{63.5^2 - 30^2})$$

$$= 3.7665 = 3.77 \text{ mm}$$

Total Table Travel $= 180 + 3.77 = 183.77$ mm

Feed/Min. (table feed) $= 0.125 \times 6 \times 1500$

$= 1125$ mm/min

Machining time $= \frac{183.77}{1125} = 0.163$ mm.

Example 11.11: A slot is to be made on a milling machine with cutter revolving at 120 rpm. Find the time required to prepare the slot in two cuts, if it uses 20 mm deep and 100 mm long with a cutter diameter 80 mm. Assume the feed as 0.5 mm. sec. (GATE).

$$\text{Added Travel} = \sqrt{Dd - d^2} = \sqrt{80 \times 10 - 10^2} = \sqrt{700}$$

$$= 26.45 \text{ mm}$$

Total Travel = 100 + 26.45 = 126.45 mm

Feed/min. (Table Travel) = 0.5 × 120 = 60 mm/min.

$$\text{Time taken/cut} = \frac{126.45}{60} = 2.1 \text{min}$$

Total time required per two cuts = 2 × 2.1 = 4.2 min.

QUESTIONS

1. How do you classify the milling machines?
2. Draw a neat sketch of horizontal milling machine and label the parts.
3. List the various types of milling cutters. Draw a neat sketch of milling cutter you are familiar with.
4. What are the main attachments used on milling machine?
5. Explain Up milling and Down milling with neat sketches.
6. Draw a neat sketch of a Universal dividing head and explain its working.
7. Explain different types of indexing methods with examples.

12

Shaper, Slotter and Planer

12.1 SHAPER MACHINE

12.1.1 Introduction of Shaper

The shaper is a reciprocating type of machine tool used for producing flat surfaces, which may be horizontal, vertical or inclined. Sometimes irregular or curved surfaces are also produced by shaper. In shaping, tool is given a reciprocating motion. It is generally not used as a production machine. But widely used in machine shop and tool rooms since it is very easy to setup and operate.

12.1.2 Working Principle

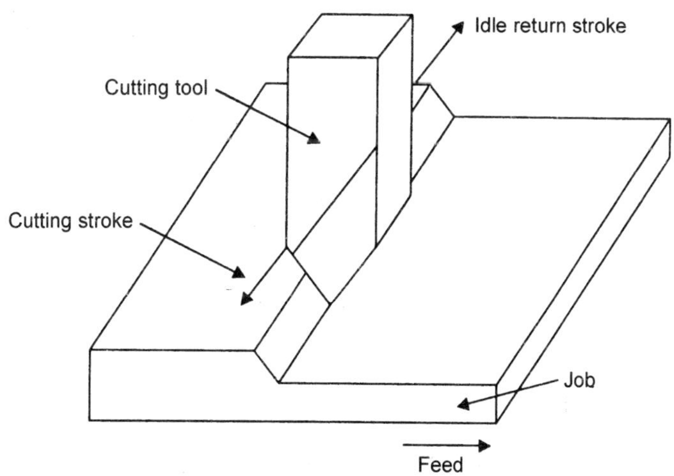

Fig. 12.1 Working Principle of Shaping Machine

Figure 12.1 shows the basic principle involved in shaping. The job is held in a suitable device (vice) clamped rigidly on the machine table. The cutting tool is held in the tool post mounted on the ram of the shaper. The ram reciprocates to and fro and in doing so it cuts the material from the job in cutting stroke. Generally, the cutting action takes place in the forward stroke, which is also known as cutting stroke. No cutting of material takes place during the return stroke of the ram is called idle stroke. At the end of one cycle, consisting of one to and fro motion of the cutting tool, the job is given a feed motion perpendicular to the direction of tool movement. The depth of cut is given by lowering the tool relative to the job.

12.1.3 Principle Parts of a Shaper (see Fig. 12.2)

Base: It consists of a heavy robust cast iron structure, which supports all the other parts of the machine.

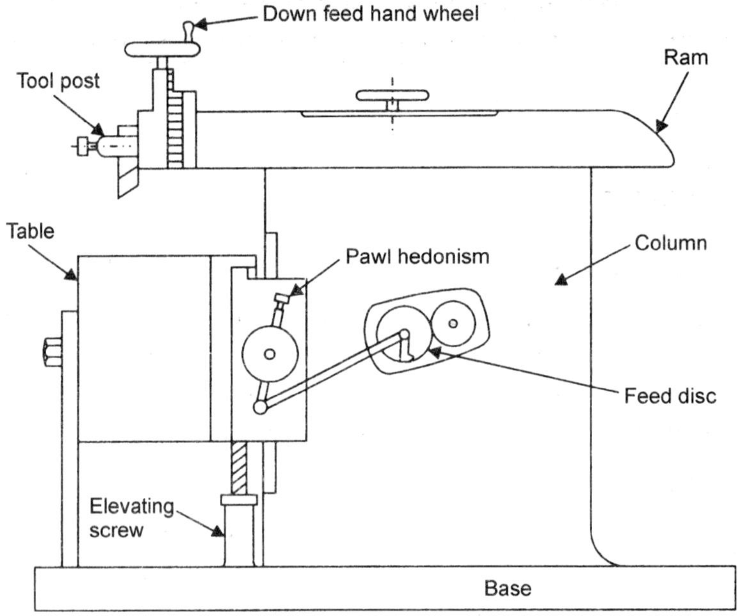

Fig. 12.2(a) Principle Parts of a Shaper

Column: It acts as housing for electrical circuits and operating mechanism of a shaper. It also supports ram, tool head, cross-rail etc. It is a heavy cast iron body attached to the column of the machine. It is used for two purposes — for elevating the table and for traverse of the table.

Table: It is a box type construction with T-slots cut on it to hold the vice and jobs. It slides along the cross-rail to provide feed to the work.

Ram: It is the reciporcating part of the shaper, semi-circular in shape and carries the tool head infront of it. It gets its drive from the quick return mechanism, which is inside the column.

Tool head: The tool head of a shaper is used for holding the tool rigidly. It also provides vertical and angular feed movement of the tool and allow the tool to lift automatically to provide relief during its idle of return stroke. The vertical feed to the tool is provided by rotating the down feed screw handle.

Fig. 12.2(b) 3D View of a Shaper

12.1.4 Classification of Shapers

Shapers are classified in the following ways:

1. According to the length of stroke
   (i) 30 cm shaper
   (ii) 45 cm shaper
   (iii) 60 cm shaper
2. According to movement of ram
   (i) Horizontal type
   (ii) Vertical type
3. According to the type of design of the table
   (i) Standard shaper
   (ii) Universal shaper
4. According to the drive
   (i) Mechanical shapers
      (a) Crank type
      (b) Geared type
   (ii) Hydraulic shaper
5. According to the type of cutting stroke
   (i) Push type shaper
   (ii) Draw type shaper

12.1.5 Types of Shapers

Standard Shaper: It consists of a plain table that may or may not have vertical support at its front. Some machines have a provision for the table to swivel around a horizontal axis parallel to the ram. The material is cut in the forward stroke of the tool and the return stroke is idle.

Horizontal Shaper: It is very popular type of shaper. In this shaper the ram holding the tool head reciprocates in a horizontal axis. Horizontal shapers are mainly used to produce flat surfaces.

Vertical Shaper: In a vertical shaper, the ram holding the tool reciprocates in a vertical axis. In some of the vertical machines provision is made to allow adjustment of the ram to an angle of about 10 degrees from the vertical position. Vertical shapers may be crank driven, rack driven or hydraulic power driven. The work table of a vertical shaper can be given cross, longitudinal and rotary movement. Vertical shapers are used for machining keyways, slots or grooves.

Universal Shaper: This ia also a horizontal shaper, but its table can be swung about a horizontal axis parallel to the ram ways. The top of this table can also be tilted about another horizontal axis, which is normal to the former axis. It is called a universal shaper since the job can be tilted in any direction through the required angle with the help of swivel vice.

Crank Shaper: This is the most common type of shaper in which crank and slotted link mechanism is used to give reciprocating motion to the ram.

Geared Shaper: The reciprocating motion of the ram is given by means of a rack and pinion. The rack teeth that are cut directly below the ram mesh with a spur gear.

Hydraulic Shaper: In this type of shaper, reciprocating movement of the ram is obtained by hydraulic power. Oil under high pressure is pumped into the operating cylinder fitted with a piston. The end of the piston rod is connected to the ram. The high-pressure oil first acts on one side of the piston and then on thc other causing the piston to reciprocate and the motion is transmitted to the ram.

Push Type Shaper: This is the most general type of shaper used in practice. The metal is removed in the forward stroke of the ram.

Draw Type Shaper: In this type of shaper, the metal is removed in the backward stroke of the ram. The tool is set in a reverse direction to that of a standard shaper. Vibrations in these machines are eliminated.

12.1.6 Shaper Size and Specification

The maximum length of cut it can take gives the size of shaper. The usual size range from 175 mm to 900 mm. In addition to this, other details also required to specify a shaper. Specification of a shaper as follows:

1.	Length of stroke	610 mm
2.	Maximum vertical travel of table	475 mm
3.	Maximum horizontal travel of table	450 mm
4.	Size of the side table	473 mm × 330 mm
5.	Power of the motor	3 HP

6. Maximum vertical travel of tool slide 150 mm
7. Ram cycles per minute or strokes per minute 15 to 90
8. Approximate net weight 1750 kg
9. Floor space required 1981mm × 1067 mm

12.1.7 Quick Return Mechanism

In a standard type shaper, metal is removed in the forward cutting stroke, while the return stroke goes idle and no metal is removed during this period. To reduce the total machining time, it is necessary to reduce the time taken by the return stroke. Thus the shaper mechanism should be so designed that the ram should move slowly (slow speed) during forward stroke (cutting stroke) and should move fast in the return stroke.

This mechanism is known as Quick return mechanism. For this purpose, two mechanisms are commonly used:

(a) Crank and slotted link mechanism
(b) Hydraulic mechanism

(a) Crank and Slotted Link Mechanism (see Fig. 12.3)

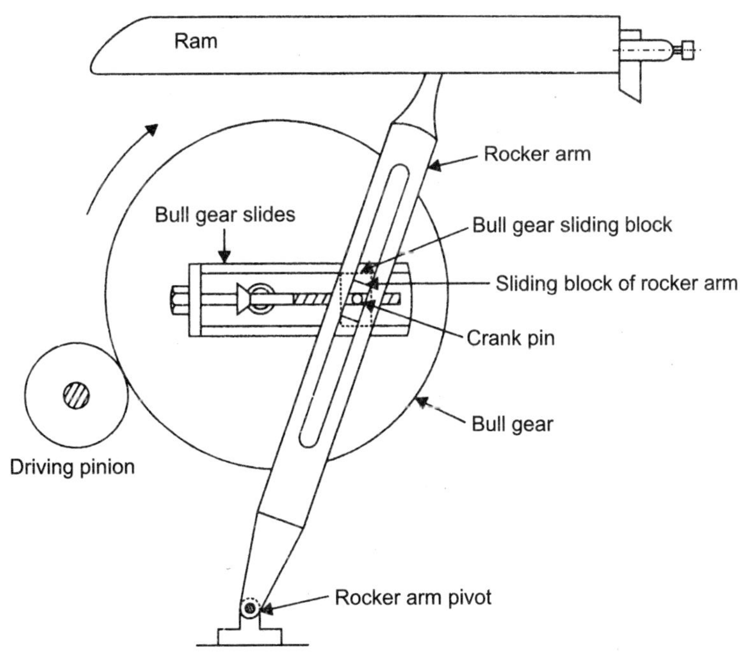

Fig. 12.3 Crank and Slotted Link Mechanism

The drive from the motor is coupled to driving pinion, which is in mesh with the bull gear. Bull gear slide is mounted at the center of the bull gear. There is a sliding block, of rocker arm and a crankpin which passes through the sliding block and is fixed on the bull gear sliding block. The slotted link, which is also known as rocker arm, is pivoted at its bottom end to the frame. The upper end of the rocker arm is forked and connected to the ram. As the bull gear rotates causing the crankpin to rotate, the sliding block of rocker arm fastened to the crankpin will rotate on the crankpin circle, and at the same time will move up and down the slot in the slotted link giving it a rocking movement which is communicated to the ram. Thus the rotary motion of the bull gear is converted to reciprocating motion of the ram.

The principle of quick return motion is illustrated in Fig. 12.4. When the link is in the position PQ the ram will be at the extreme backward position of its stroke and when it is at PR, the extreme forward position of the ram will have been reached. PQ and PR are drawn tangent to the crankpin circle. The forward cutting stroke thererfore, takes place when the crank rotates through the angle SKT ($\alpha = 220°$) and the return stroke takes place when the crank rotates through the angle SLT ($\beta = 140°$).

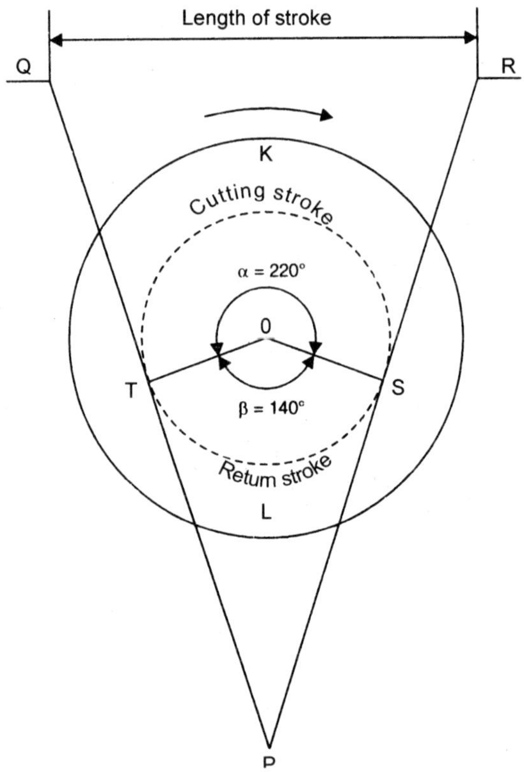

Fig. 12.4 Principle of Quick Return

The ratio between the cutting time and return time may be determined by the formula:

$$\frac{\text{Cutting time}}{\text{Return time}} = \frac{\text{SKT}}{\text{SLT}} = \frac{\alpha}{\beta} = \frac{220}{140} = \frac{22}{14} = \frac{1.6}{1} = \frac{3.2}{2}$$

Cutting time to return time ratio usually varies between 2 : 1 and the practical limit is 3: 2

(*b*) *Hydraulic Mechanism* (see Fig. 12.5)

The machine consists of a constant discharge oil pump. a valve chamber, a cylinder, a piston. The piston rod is connected to the ram body. The oil under high pressure is pumped from the reservoir and made to pass through the valve chamber to the right side of the cylinder exerting pressure on the piston. This causes the ram to perform forward stroke and at the same time oil present on the left side of the cylinder is discharged to the reservoir through the throttle valve. At the end of forward stroke, the shape dog hits the reversing lever causing the valves to alter their positions within the valve chamber. Now the oil pumped to the left side of the piston causing the ram to perform return stroke. Oil present on right side of the piston is discharged to the reservoir. At the end of the return stroke, another dog hits the reversing lever altering the direction of the stroke of the piston. Thus the cycle is repeated.

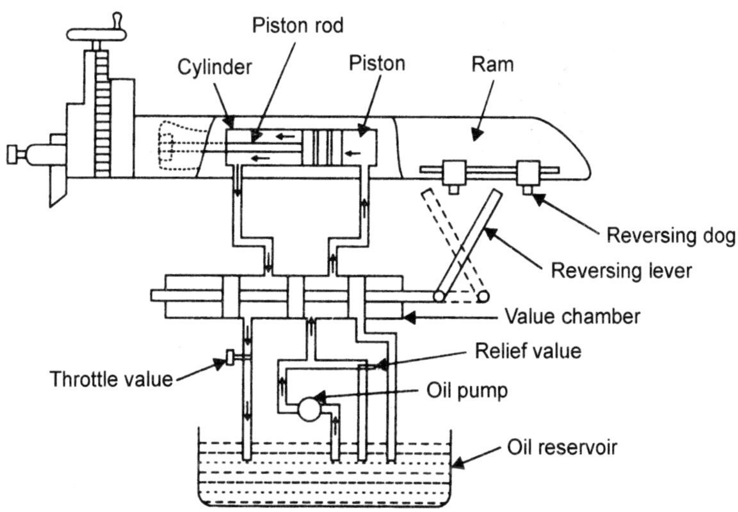

Fig. 12.5 Hydraulic Shaper

The quick return motions is effected due to the difference in the stroke volume of the cylinder at both ends. The left hand end is smaller due to the presence of piston rod. As the pump is a constant discharge one, the same amount of oil is pumped into smaller volume, the pressure is rised automatically and increases the speed during return stroke.

12.1.8 Advantages and Disadvantages of Hydraulic Drive

The advantages of a hydraulic drive can be enumerated as follows:

(a) Cutting speed is constant throughout most of the stroke.

(b) The reversal of the ram is quick without any shock as the oil on the other end of the cylinder provides cushioning effect.

(c) An infinite number of cutting speeds are available and these speeds are easily controlled.

(d) When the cutting tool is overloaded, an overload relief valve may automatically opens, thus prevents the damage of the machine.

However, the disadvantages are:

(a) Initial cost of the hydraulic drive machine tool is high.
(b) Complicated in construction.
(c) Oil leakage.

12.1.9 Feed Mechanism of Shaper

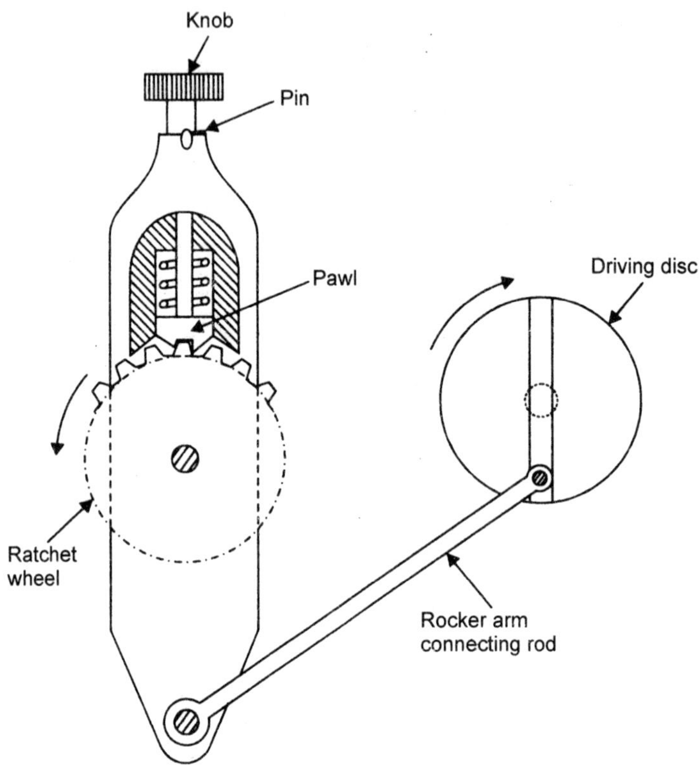

Fig. 12.6 Feed Mechanism of a Shaper

Figure 12.6 shows the automatic cross feed mechanism of a shaper. The rotation of the bull gear causes the driving disc to rotate in particular direction. The driving disc is T-slotted and position of the crankpin attached to the connecting rod may be altered, to give different throw of accentricity. The other end of the connecting rod is attached to the rocking arm by a pin. The rocking arm is fulcrumed at the centre of the ratchet wheel, which is keyed to the cross feed screw. The rocking arm houses a spring loaded pawl, which is straight on one side and bevel on the other side. As the driving disc rotates, the connecting rod starts reciprocating and the rocking arm rocks on the fulcrum. When the driving disc rotates through half of the revolution in the clockwise direction top part of the rocking arm moves in the clockwise direction and the pawl being slant on one side slips over the teeth of the ratchet wheel, imparting no movement. As the driving disc rotates through the other half, the top of the rocking arm now moves in the anticlockwise direction and the straight side of the pawl engages with the teeth of the ratchet wheel causing the wheel to move in anticlockwise direction only. As the driving wheel is connected to the bull gear, the table feed movement is effected, when the bull gear or the driving disc rotates through half of the revolution. To reverse the direction of rotation of the ratchet wheel and consequently the feed, a knob on the top of the pawl after removing the pin is rotated through 180°.

12.1.10 Machining Time Calculations

$$\text{Cutting speed} = \frac{\text{Length of stroke}}{\text{Time required for cutting stroke}}$$

$$\text{Return speed} = \frac{\text{Length of stroke}}{\text{Time required for return stroke}}$$

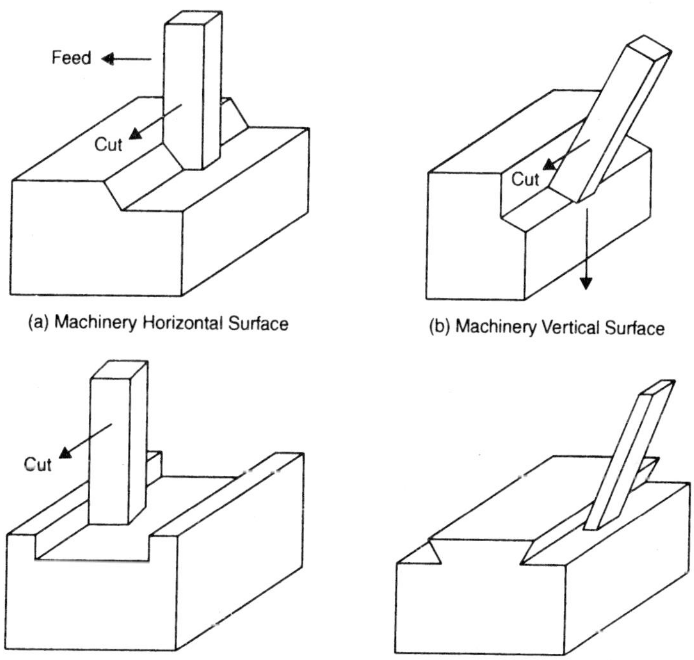

(a) Machinery Horizontal Surface (b) Machinery Vertical Surface

(c) Machinery Slots (d) Machinery Angular Surface (Dovetail)

Fig. 12.7 Operations of Shaper

$$\text{The time required for the forward stroke} = \frac{L}{1000\ V}$$

As the forward and return speeds are different,

$$\text{Time per cycle} = \frac{L}{1000\ V_1} + \frac{L}{1000\ V_2}$$

$$\text{The machining time} = \frac{\text{Width of work}}{\text{Feed /stroke}} \times \text{Time per cycle}$$

$$= \frac{W}{f} \times \frac{L(V_1 + V_2)}{1000\, V_1.V_2}$$

$$\text{Total number of double strokes necessary} = \frac{\text{Width of the job}}{\text{Feed}}$$

Total time required to complete the cut = time for double stroke × No. of double strokes.

Example 12.1: Find the time required for taking a complete cut on a plate 250 mm × 500 mm if the cutting speed is 10 m/min, the ratio of return time to cutting time is 2 : 3 and feed is 2mm. The clearance at each end is 50 mm.

Solution: Length of stroke = approach + length of job + clearance

$$= 50 + 500 + 50 = 600 \text{ mm}$$

$$\text{Cutting time} = \frac{600 \times 60}{1000 \times 10} = 3.6 \text{sec}$$

$$\frac{\text{Return time}}{\text{Cutting time}} = \frac{2}{3}$$

$$\text{Return time} = 3.6 \times \frac{2}{3} = 2.4 \text{ sec}$$

Total time for one complete double stroke = 3.6 + 2.4 = 6 sec

$$\text{Total number of double strokes necessary} = \frac{\text{Width of the job}}{\text{Feed}}$$

$$= \frac{250}{2} = 125$$

Total time required to complete = Time for one double stroke × the cut no. of double strokes

= 6 × 124 = 750 sec = 12.5 min **Ans.**

12.2 SLOTTING MACHINE (SLOTTER)

12.2.1 Introduction

A slotting machine or slotter may be considered as a vertical shaper. The slotter has the vertical ram and a hand or a power operated rotary table. On some machines, the ram may be inclined at 10° to either side of the vertical position when cutting inclined surfaces. A slotter can perform a variety of operations, such as the finishing of the external and internal plain surfaces in addition to slotting.

12.2.2 Principle Parts of a Slotter (Fig. 12.8 (a) and (b))

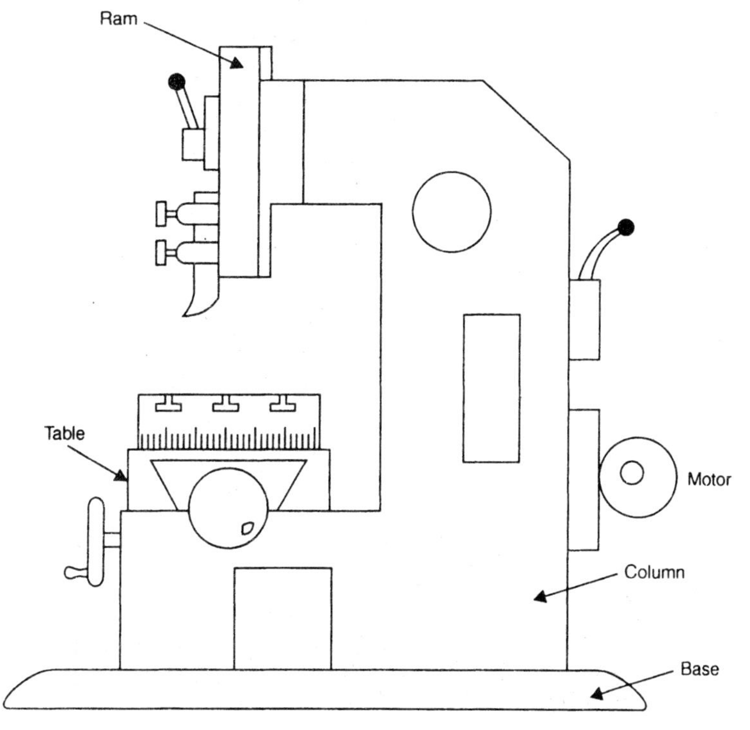

Fig. 9.8(a) Slotting Machine

Base: The base is rigidly built and is cast integral with column. The top of the bed and the front face of the vertical column are accurately machined to provide guideways for the saddle and ram respectively .

Saddle: The saddle is mounted upon the guideways and may be moved towards or away from the column. The top phase of the saddle is accurately finished to provide guideways for cross-slide. These guideways are perpendicular to the guideways on the base.

Cross-Slide: The cross-slide is mounted on the guideways of the saddle and may be moved parallel to the phase of the column.

Rotary Table: The rotary table is a circular table mounted on the top of the cross-slide. This table can be rotated by hand or by automatic device. The angular graduation on the table facilitates work to be machined at angles. The table is also provided with an index plate to one on the dividing head of the milling machine. The use of index plate ensures accurate division on parts.

Fie. 12.8(b): 3D View of Slotting Machine

Ram: Ram reciprocates vertically up and down. At its bottoms it carries the cutting tool. A slot is cut on the body of the ram for changing the position of the stroke. In some machines special types of tool holders are provided to relieve the tool during its return stroke.

12.2.3 Types of Slotting Machines

Slotting machines are mainly of the following three types:

(a) Puncher slotter (b) Production slotter

(c) Tool room slotter

(*a*) *Puncher Slotter*: The puncher slotter is the heavy duty machine and equipped with powerful motor. It is designed to remove large amount of metal from large castings or forgings. The length of the stroke is also large.

(*b*) *Production Slotter*: This is common type of slotter used for general production work. The drive of the ram is by means of slotted disc and connecting rod. The fly wheel is fitted to prevent shock at the end of the stroke.

(*c*) *Tool Room Slotter*: This slotting machine is of precision type and is used for very accurate machining. It is lighter machine and is operated at high speeds. Using special jigs, the machine can handle a number of identical works on a production basis.

12.2.4 Specifications of a Slotter

Slotter is generally specified in terms of the maximum length of the stroke.

1. Maximum stroke 457 mm
2. Diameter of rotary table 915 mm
3. Longitudinal movement 762 mm
4. Cross movement 559 mm
5. H.P. required 7.5 HP

12.2.5 Drive Mechanisms of a Slotter

There are four types of driving mechanisms used in slotter for driving the ram.

(a) Slotted disc mechanism
(b) Slotted link mechanism
(c) Vertical speed reversible motor driving mechanism
(d) Hydraulic drive mechanism

(*a*) ***Slotted Disc Mechanism*** (Fig. 12.9)

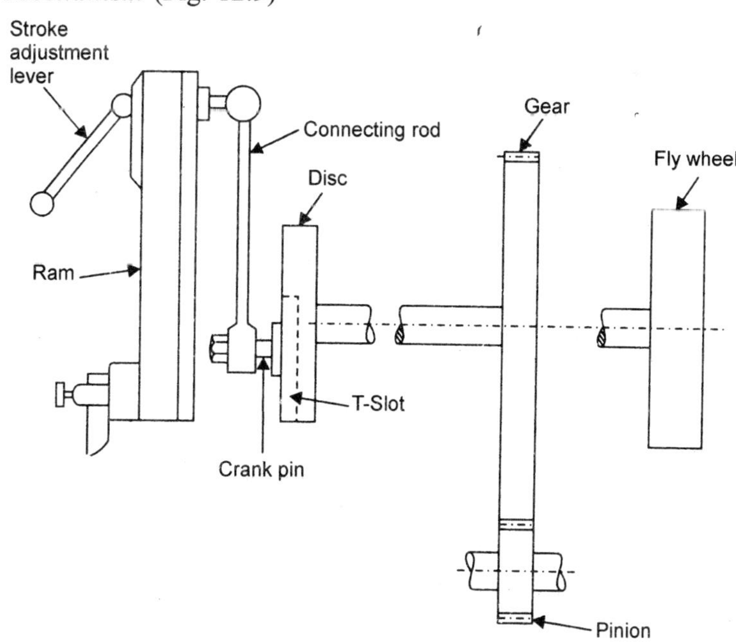

Fig. 12.9 Slotted Disc Mechanism

This is the simplest of all the methods used to drive the ram of a slotter. This mechanism consists of a pinion, a gear, a crank, a slotted disc and crank as shown in Fig. 12.9. Drive comes from the motor to the pinion by V-belt. Then pinion to the drive gear, which is further, connected to the disc. One end of the connecting rod is attached to the disc by means of pin while the other end is to the reciprocating ram. Here the crank and the connecting rod mechanism convert the circular motion of the disc into reciprocation motion of the ram. The stroke can be adjusted by positioning the stroke-adjusting lever in the desired position.

12.2.6 Operation Performed on a Slotter

The operations performed on a slotter are:

(a) Machining flat surfaces
(b) Machining cylindrical surfaces
(c) Machining irregular surfaces
(d) Machining slots, keyways and grooves.

(*a*) *Machining Flat Surfaces*: The external and internal flat surfaces can be machined on a work piece in a slotting machine. The work to be machined is supported be parallels so that the tool will have clearance from the table when it is at the extreme downward position of the stroke. The cross feed is given at the beginning of each cutting stroke. The machining on the work-piece is completed by using a roughing and a finishing tool.

(*b*) *Machining* (*Cylindrical Surfaces*): The external and internal surfaces of a cylinder can also be machined in a slotting machine. The work is clamped on the rotary table. The tool is set radially on the work-piece. While machining, the feeding is done by rotary table-screw, which rotates the table through a small arc at the beginning of each cutting stroke.

(*c*) *Machining Irregular Surfaces*: The work-piece is set on the table. Then by combing cross, longitudinal and rotary feed movements of the table any contoured surfaces can be machined on work pieces.

(*d*) *Machining Slots, Keyways and Grooves*: Internal and external grooves are cut very conveniently on a slotting machine is designed for cutting internal grooves which are to difficult to produce in other machines.

12.3 PLANER

12.3.1 Introduction

Planer is used to produce plane and flat surfaces on work-piece that are too large to be accommodated on a shaper.

12.3.2 Principle Parts of a Planer (see Fig. 12.10 (a) and (b))

Bed: The bed of a planer is a large rigid box-like casting made of cast iron. It supports the column and all the moving parts of a planer. As the table has to make complete stroke on the bed surface, the length of the bed is usually twice the length of the table. V-guide ways should be properly be lubricated.

Table: The table of a planer is a large rectangular thick cast plate and moves over the bed. The upper surface of the table has T-slots to facilitate the clamping of the work-pieces, special fixtures and others with T-belts. Accurate hole drilled on the top surface of the planer table at regular intervals and supporting the poppets and stop pins. At each end of the table a holo space left which acts as a troughs for collecting chips. A groove is cut on the side of the table for fixing dogs and positioning the length of the travel of the table.

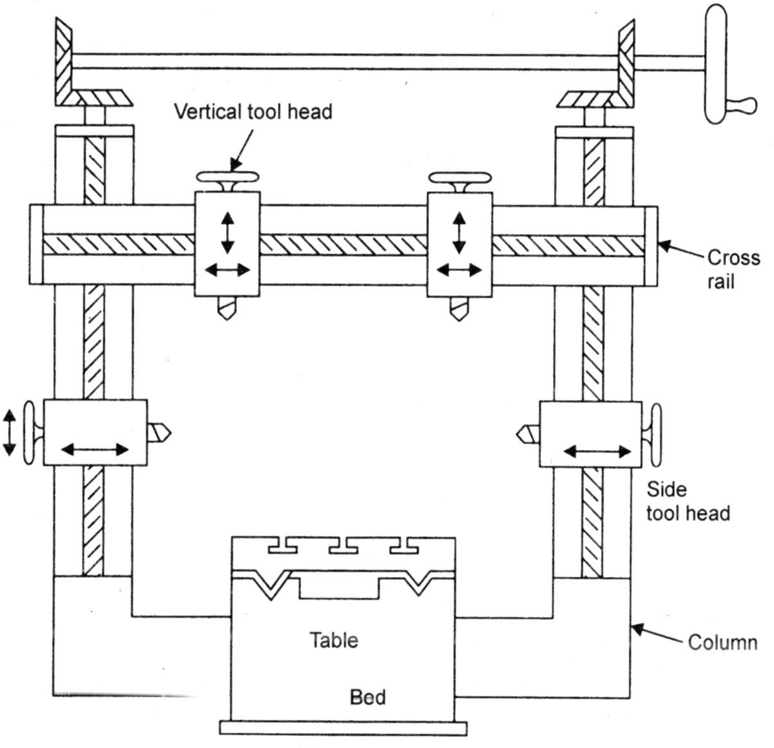

Fig. 12.10(a) Principle Parts of a Planer

Columns: These are rigid box-like vertical structures placed on each side of the bed and are fastened to the sides of the bed. The front space of each column is accurately machined to provide the base on which the cross rail may be made to slide up and down to accommodate different heights of the work-pieces. The cross rail elevating screw, vertical and cross feed screws for tool heads are accommodated within the body of the uprights (columns).

Cross rail: The cross rail is a rigid box like casting connecting the two columns. The cross rail can be moved up and down by means of feed screws. Usually two toolheads are mounted upon the cross rail. The cross rail has screws for vertical and cross feed of the tool heads and a screw for elevating the rail.

Tool head: The tool head of a planer is similar in design and operation to the tool head of a shapper. A planer may be fitted two or more tool heads to perform more than one operation.

Fig. 12.10(b) 3D View of a Planer

12.3.3 Classification of Planing Machines

Planer can be classified in a number of ways

1. According to the drive
   - (a) Gear drive
   - (b) Hydraulic drive
   - (c) Belt drive
   - (d) Variable speed motor drive
2. According to general motor construction
   - (a) Double housing planer
   - (b) Open-side planer
   - (c) Pit type planer
   - (d) Edge of plate planer

(a) Double Housing Planer (see Fig. 12.11)

This is the most widely used planer in the workshops. It consists of a heavy base on which a tablet reciprocate on accurate guide ways and two columns. The columns support the cross rail. The cross rail carries one or two vertical tool heads. Fig. 12.11 shows a double housing planer.

(b) Open-Side Planer (see Fig. 12.12)

An open side planer has a column on one side only, which permits the machining of wide and large jobs. The cross rail is suspended from the column as a cantilever. As the single column has to take up the entire load, it is made extra massive to resist the forces.

(c) Pit Type Planer

This is a massive planer capable of a holding very big work-pieces. It differs from ordinary planers the bed is stationery and the tools is move over the job.

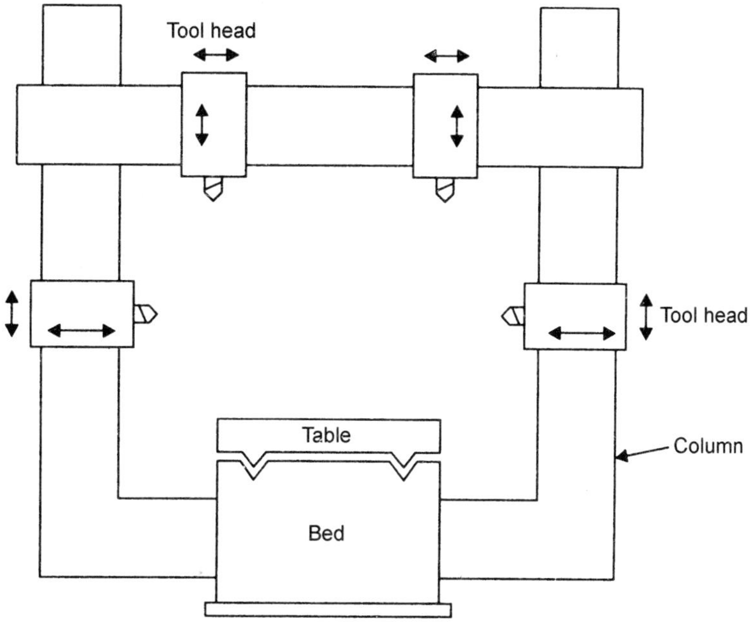

Fig. 12.11 Double Housing Planer

(d) Edge or Plate Planer

This is a specially designed planer used for the edges of having heavy steel plates, pressure vessels and ship building works. The job to be machined is mounted rigidly on the bed and the carries supporting the tool is moved back and forth along the edge.

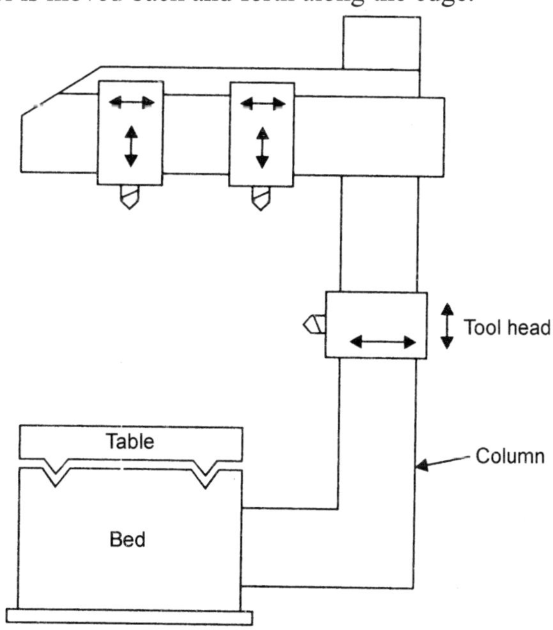

Fig. 12.12 Open-Side Planer

12.3.4 Planer Size and Specifications

Planers are made different sizes and they are specified by the following:

1. Length of bed 2910 mm
2. Width of bed 550 mm
3. Table size (working surface) 1830 × 686 mm
4. Height under cross rail 914 mm
   (vertical traverse)
5. Table surface speed
   (a) Forward stroke 9.5 metres per minute
   (b) Return stroke 14 metres per minute
6. Main drive
   (a) Main drive motor (kW) 2.2 kW
   (b) Cross rail elevating motor 0.7 kW
7. Overall dimensions
   (a) Length 2190 mm
   (b) Width 2134 mm
   (c) Height 2362 mm
   (d) Weight approximately 4200 kg

12.3.5 Difference between Shaper and Planer

S.No.	Shaper	Planer
1.	It is a comparatively light machine.	It is a heavy-duty machine.
2.	It requires lesser floor space area.	It requires more floor area.
3.	Cutting takes place by moving the cutting tool over the job.	Cutting takes place by moving the work under the tool.
4.	Shaper is used for small sized and light work pieces.	Planer is used for large sized and heavy work-pieces.
5.	Shaper tools are normally simple.	Planer tools are quite massive and can sustain large cutting forces.
6.	Generally a single tool is used.	More than one tool can be used simultaneously.
7.	Setting of the work requires less time and skill.	Setting of the work requires more time and skill.
8.	In shaper due to over hanging of the ram during the cutting stroke the accuracy can't be expected up the mark.	The tool is rigidly supported and hence obtained maximum accuracy on the machined surface.

contd...

9.	Rate of production is less.	Rate of production is more as number of Jobs can be fixed on the table and machined simultaneously.
10.	A shaper consumes 15-20 HP.	A planer consumes 120 HP.
11.	Cost of the shaping machine is less.	Cost of the planing machine is more.
12.	Stroke length is small.	Stroke length is substantially greater than that of a shaper.

12.3.6 Table Drive Mechanism

Different methods are used for driving the table of planer. They're as follows:

(a) Open and cross belt drive (fast and loose pulleys drive)

(b) Reversible motor drive

(c) Hydraulic drive

(a) Open and Cross Belt Drive (see Fig. 12.13)

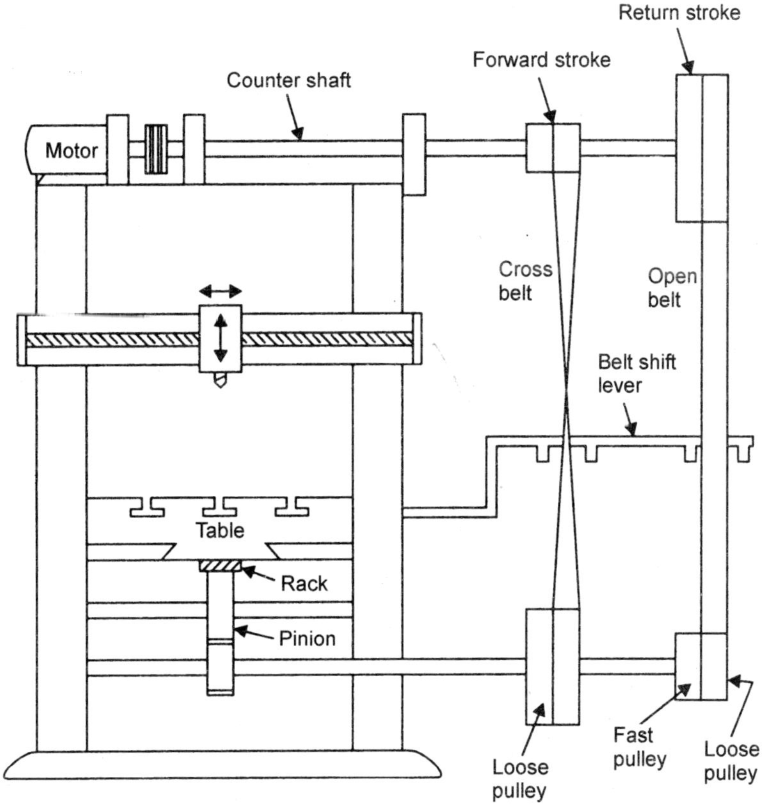

Fig. 12.13 Open and Cross Belt Drive

In small planers, the open and cross belt drive is used for the quick return of the table. Fig. 12.13 shows the mechanism. The motor drives the counter shaft, which carries two driving pulleys; one for open belt and other for cross belt.

The main driving shaft is provided below the bed. One end of it passed through the housing and carries a pinion, which meshes with the rake provided under the table of the machine. The other end of this shaft carries two pairs of pulleys–each pair consists of fast pulley and loose pulley. One of these pairs is connected to one of the drive pulleys by means of an open belt and the other to the second driving pulleys by means of a cross-belt.

In the given diagram the crossed-belt will be used for forward stroke and open belt for return stroke. The driving pulleys on the counter shaft for the crossed belt is smaller than the pair of fast and loose pulleys for the same. The driving pulleys on the counter shaft for open belt is bigger than the pair of fast and loose pulleys for the same. Consequently for the same speed of counter shaft the main driving shaft will run faster when connected by open belt than when the cross-belt is used. It is obvious, that the return stroke is faster than the return stroke.

The pulleys are so arranged that when the cross-belt is on the fast pulley i.e. in forward stroke, the open belt will be on loose pulley and it's reverse will take place during the return stroke. The shifting of the belt may take place automatically at the end of each stroke, without stopping the machine, with belt shifter and it's operating lever. Trip dogs are mounted, one each at both the ends on the table. At the end of the stroke these dogs strike against the operating lever alternatively and the belt shifted accordingly. Thus the table movement reversed automatically.

(*b*) *Reversible Motor Drive*

Number of modern planers used this system. In this a DC reversible motor is directly coupled to the driving shaft. The direction of rotation of this motor can be instantaneously changed by reversing the polarity. This is done by operating two different switches which are actuated by means of trip dogs provided at each end of the table. Also the speed of this motor can be controlled by varying the supply of the electric current in the field.

(*c*) *Hydraulic Drive*

These are becoming increasingly popular these days. They provide uniform speed throughout the cutting stroke. The oil is pumped into the piston by variable delivery electric pump. The speed of the piston is controlled by regulating the delivery of the oil into the piston with adjusting valves.

12.3.7 Work Holding Methods and Devices (see Fig. 12.14 (a) and (b))

Planing is high power cutting operation during which the work-piece is subjected to extremely high cutting force. Hence a number of clamps, stops and other anchoring devices are used to prevent the work-piece to move. Thin flat work pieces are often held by magnetic chuck. An end stop is used to prevent the longitudinal movement of the plate. Fig. 12.14(a) illustrates another method of holding the plate by means of chisel points, T-slot stop block, stop-pin and guide stops for initial adjustments.

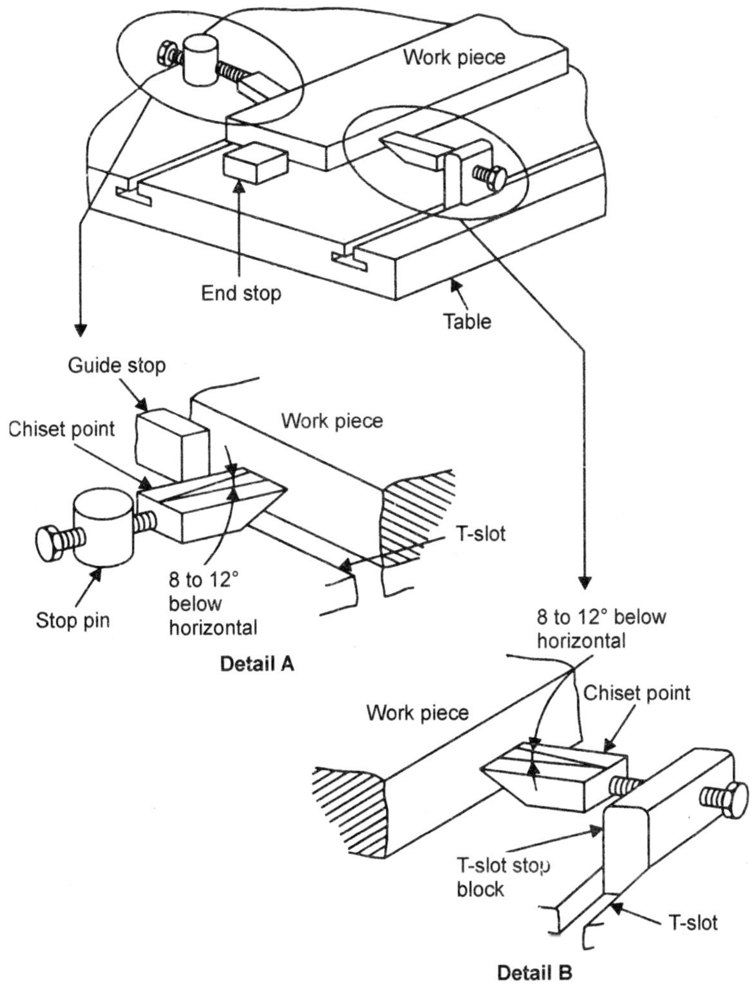

Fig. 12.14(a) A Common Method for Securing a Long Flat Plate to Planer Table

Heavy, irregular work-pieces require additional care in setting up. First it should be ascertained that the work-pieces rest securely on the table. If there is any wobble, shim jacking must be used. A typical illustration of securing a large irregular casting to a planer table is shown in Fig. 12.14(b).

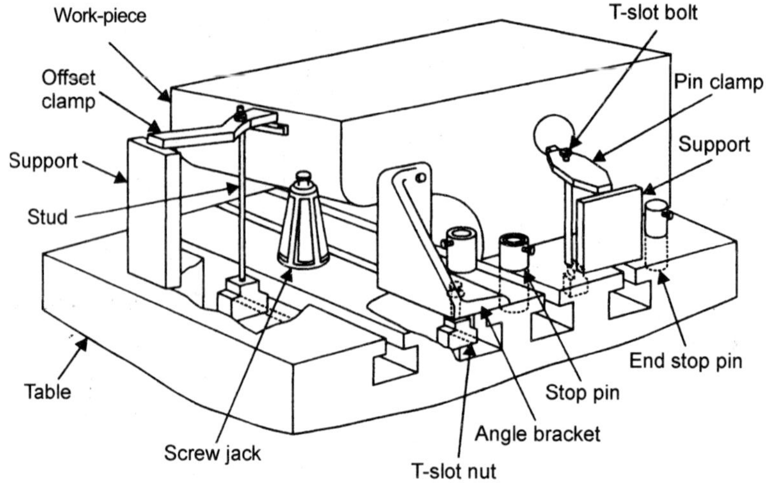

Fig. 12.14(b) Typical arrangement of securing and supporting devices for mounting a large irregular work-piece on a planer table

QUESTIONS

1. Sketch and explain the working principle of a shaper.
2. How shapers are classified?
3. Explain Quick-return mechanism used in a shaper with neat sketch.
4. Draw a linen diagram of a slotter and indicate its main parts.
5. Draw a neat sketch of planer and explain the parts.
6. How planers are classified?
7. Describe with the help of a neat sketch how the quick return motion on a planer is obtained with open and cross belt drive.
8. State the main differences between shaper and planer.

13

Grinding and Grinding Machines

13.1 INTRODUCTION

Grinding is a process of removing material in the form of small chips by the abrasive action of revolving wheel on the surface of a work piece. The wheel used for performing the grinding operation is called grinding wheel. It is basically a finishing process used for producing close dimensional and smooth surface finish.

13.2 GRINDING WHEELS

The grinding wheel is composed of two main elements:

(a) Abrasives, (b) Bonding agents.

The abrasives are intended for actual cutting action and hence they are hard substances. The bonding agent holds the abrasives during operation and hence they should have good binding properties.

(*a*) *Abrasives*: An abrasive is a hard material used for making the grinding wheels. Abrasives are small particles bonded together in different shapes.

Types of abrasives: The abrasives are of two types:

(i) Natural,

(ii) Artificial or manufactured.

(*i*) *Natural Abrasives*: These are produced by uncontrolled forces of nature. These are obtained from mines. The following are the natural abrasives:

(a) Sand abrasives, (b) Emery,

(c) Corundum, (d) Diamond.

(*ii*) *Artificial Abrasives*: These are manufactured under controlled conditions in closed electric furnace in order to avoid the introduction of impurities and to achieve necessary temperature for the chemical reaction to take place. These possesses better cutting properties and higher efficiency than natural abrasives. Most commonly used manufactured abrasives are (1) Silicon carbide (2) Aluminium oxide.

13.3 MANUFACTURING OF ARTIFICIAL ABRASIVES

1. *Silicon Carbide* (*SIC*): It is available in variety of colours. Bluish-green is very suitable for grinding tipped tools.

Ingredients:

1. Silicon sand = 50 parts
2. Petrolium coke = 34 parts
3. Common salts = 4 parts
4. Saw dust = 12 parts

The ingredients are thoroughly mixed and heated in an electric furnace at about 2000°C for 34 hours. The whole solid mass is crushed, washed and treated with alkalis. It is again washed and finally ground into small particles. These are sieved in different number of sieves. For grinding, particles of 180-200 mesh numbers are taken. The green grit and black grit abrasives are used.

2. *Aluminium Oxide* (Al_2O_3): It looks like brilliant white crystal. It is manufactured by fusing mineral Bauxite mixed with coke and iron scrap. This is fused in an electric furnace. After fusing, it is crushed, washed, treated with alkalis. Again washed and finally ground. It is used for grinding materials of high-speed steels, wrought iron etc.

13.4 BONDS AND BONDING PROCESSES

A bond is a material that holds the abrasive grains together in the form of wheel. The most commonly used bonds for manufacturing of grinding wheels are :

(i) Vitrified bond (denoted by V)
(ii) Silicate bond (denoted by S)
(iii) Shellac bond (denoted by E)
(iv) Rubber bond (denoted by R)
(v) Resinoid bond (denoted by B)
(vi) Oxychloride bond (denoted by O)

(i) Vitrified Bonding Process

In this process, the abrasive grains are mixed with clay together with sufficient water to make the mixture uniform. The fluid mixture is then poured into the moulds and allowed to dry. After it is cut and trimmed to perfect size and shape, it is then heated or burned in a kiln at 715°C for 12 to 14 days. When the burning proceeds, the clay vitrifies i.e., it fuses and connects the abrasive grains. About 75% of all wheels are made by vitrified process.

Advantages

1. The grinding wheels are porous. Due to the porosity, the metal removal rate is high.
2. These wheels are not affected by water, oil and acids.
3. The bond itself is very hard and acts as an abrasive.

Disadvantages

1. The process is very slow.
2. Cracks may develop in the large wheels during fusion.
3. Proper control during fusion becomes difficult.
4. High temperature in the kiln tends to make the abrasive grains week.

(ii) Silicate Bonding Process

Grinding wheels are produced by this process by mixing abrasive grains with sodium silicate. The mixture is moulded in the moulds and allowed to dry for several hours. Finally, baking is carried out at a temperature of 260°C to 280°C for 25 to 80 hours.

Advantage

1. It is more rapid process than vitrified bond.
2. Large wheels, up to 1500 mm diameter can easily be produced.
3. Since it is processed at low temperature, there is no tendency to weaken the grains.

Disadvantage

1. Wear of the wheel is high.
2. Extra hard wheels can not be produced with this bond.

(iii) Shellac Bonding Process

Abrasive grain particle and shellac are mixed thoroughly to give a uniform mixture. The mixture is then rolled and pressed to desired shapes. Since this mixture is very sticky, it can not be moulded. Finally, the wheels are baked a few hours at a temperature around 150°C. This bond is adopted to thin wheels, as it is very strong and has some elasticity. Shellac bond wheels are used for grinding cam shafts, bearing rollers etc.

Advantages

1. Because of high elasticity, these wheels are used for grinding under severe working conditions.

Disadvantages

1. Wheels of bigger diameters can not be produced.

(iv) Rubber Bonding Process

Rubber bonded wheels are manufactured by mixing abrasive grains with pure rubber with some amount of sulphur which acts as vulcanizing agent. The abrasive grains are spread and finally vulcanized. By vulcanization, the whole thing becomes joined and acts as a solid wheel while rubber acting as the bond.

Advantages

1. The wheels are very hard and tough.
2. Wheels as thick as 0.125 mm can be made.

(v) *Resinoid Bonding Process*

In this process, the abrasive grains are mixed with powdered synthetic resin and a liquid solvent which dissolves resin. The mixture is rolled or pressed to the desired shape and baked in an electric oven for a few hours at a temperature of 205°C to 260°C. These are used for general-purpose grinding and widely used in foundries.

Advantage

1. This bond is very hard and strong.

(vi) *Oxychloride Bonding Process*

This process using abrasive grains with magnesium oxide and magnesium chloride produces abrasive wheels. The process of mixing is similar to that for vitrified bonding. These wheels are used for disc grinding.

13.5 GRIT, GRADE AND STRUCTURE OF GRINDING WHEELS

(*a*) *Grit*: Grit number indicates the size of the abrasive grains used in making a wheel. The following Table 13.1 shows the grain sizes ranging from coarse to very fine which are used in the manufacture of grinding wheels.

In general, coarse wheels are used for fast removal of material. Fine grained wheels are used where finish is an important consideration. Coarse wheels are used for soft, ductile material, but generally a fine grain should be used to grind hard and brittle material.

Table 13.1

Type of Grit	*Grain Size of Grit Number*						
Coarse	8	10	12	16	20	24	
Medium	30	36	46	54	60		
Fine	80	100	120	150	180		
Very Fine	220	240	280	320	400	500	600

(*b*) *Grade*: The grade indicates the strength of the bond in the wheel.
Table 13.2 shows the various grades.

Table 13.2

Soft	A, B, C D, E, F, G, H
Medium	I, J, K, L, M, N, O, P
Hard	Q, R, S, T, U, V, W, X, Y, Z

(c) *Structure*: This term denotes the spacing between the abrasive grains, or in other words the density words the density of the wheel. The structure commonly used is denoted by numbers as given in table 13.3.

Table 13.3

Dense	1, 2, 3, 4, 5, 6, 7, 8
Open	9, 10,11,12,13,14,15 or Higher

The structure of grinding wheel depends on the hardness of the material being cut. Soft, ductile materials and heavy cuts require an open structure, whereas brittle materials and finishing cuts require a dense structure.

13.6 TYPES OF WHEELS

Grinding wheels are available in a large number of shapes. The various shapes of grinding wheels are shown in Fig. 13.1.

The straight wheels shown at (a), (b) and (c) are used for cylindrical, internal and surface grinding operations. The recesses are provided to accommodate the mounting flanges.

The tapered wheel, shown as (d) is used for grinding thread, gear teeth etc. straight cup wheel (e) is mainly used for surface grinding on both the horizontal and vertical spindle grinders.

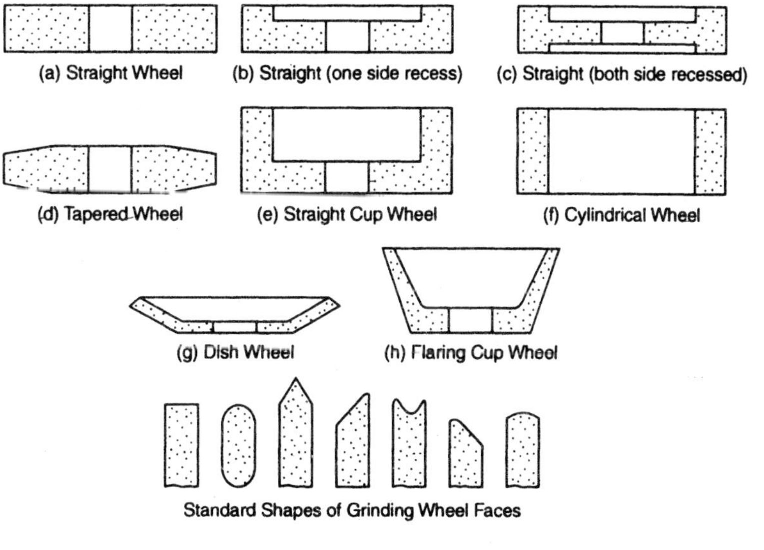

Fig. 13.1 Types of Grinding Wheels

Cylindrical wheel (f) has the same application as straight cup wheel.

Dish wheel (g) finds application on tool and cutter grinder, for grinding teeth of various cutting tools like milling cutters etc.

Flaring cup wheel (h) is also mainly used on tool and cutter grinders.

13.7 METHOD OF SPECIFYING A GRINDING WHEEL

The Indian standard marking system for ginding wheels (IS 551-1954) has been designed to designate various characteristics in a systematic manner. The standard marking system is shown in Fig. 13.2.

13.8 SELECTION OF GRINDING WHEELS

The proper selection of a grinding wheel is very important for getting best results in the grinding work. The Indian Standard (IS 1249-1958) gives recommendations for the selection of the grinding wheels for various applications. In selecting the grinding wheels, there are four constant factors and four variable factors as described below:

Constant Factors

(*i*) *Physical properties of material to be ground*: Materials of high tensile strength such as steel, tough varieties of bronze and other materials which are hard yet tough and strong are best ground with aluminium oxide wheel. Materials of low tensile strength which are penetrated easily such as soft bronze, case and chilled iron and aluminium are ground with silicon carbide wheels. Hard wheel is used for soft materials and soft wheel for hard materials. Close spacing is required for hard and brittle materials and wide for soft and ductile materials.

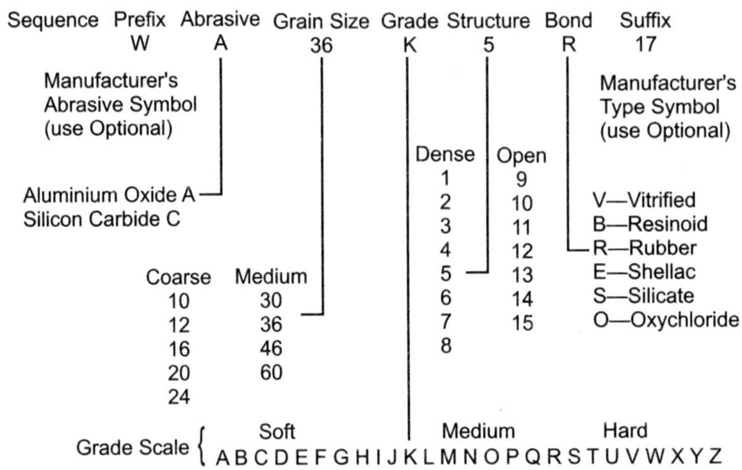

Fig. 13.2 Indian Standard Marking System

(*ii*) *Amount of stock to be removed*: This involves accuracy and finish coarse grain is used for fast cutting and fine grain for fine finish.

(*iii*) *Area of contact*: The area between the wheel and the work effects the pressure over the number of cutting points and therefore influence the selection of the wheel. Small area of contact calls for a wheel of fine grain, closed spaced structure so that the pressure is distributed over a number of cutting points. The grade should be therefore be medium hard. Cylindrical grinding work is an example of the small area of contact. Surface grinding with the rim of a cup shaped wheel is an example of larger contact area. Thus the wheel used in this case will be coarse grain, with widely spaced structure and soft grade.

(*iv*) *Type of Grinding Machine*: Type of grinding machine determines the grade of the wheel. Heavy rigidly constructed machines take softer wheels than the lighter and more flexible types. The combination of feeds and speeds on precision machines may affect the grade of the wheel desirable for best results. Surface grinding machines using cup wheels requires soft wheels of more open structure than similar machines using a straight wheel.

Variable Factors

(*i*) *Wheel Speed*: The speed of grinding wheel is influenced by the grade and bond. The higher the speed of a grinding wheel the softer it is. Vitrified bond is specified for speeds up to 2000 m/min. Rubber, shellac or resinoid bonds over 2000 m/min. surface speed.

(*ii*) *Work Speed*: The speed at which the work piece traverse across the wheel face is known as the work speed, the greater is the wear and tear of the wheel. If the work speed is low, the wheel wear is also low. However, low speed results in local overheating, produces deformation and lowers the hardness of work pieces. Most grinding machines are provided with variable speed mechanisms. As the diameter of the wheel decreases, the work speed needs to be increased accordingly to provide optimum working conditions.

(*iii*) *Machine Condition*: In the selection of grinding wheel, due consideration should be given to the condition of the machine. The grinding wheel cannot work properly if the machine is in poor condition or improperly set.

(*iv*) *Personal Factor*: The skill of workman is another variable factor which should be considered in selecting the wheel. An unskilled worker cannot handle soft wheels and he is likely to break the wheel. Thus unskilled worker should be allowed to work on hard wheels.

13.9 DRESSING AND TRUING OF GRINDING WHEELS

Grinding wheel slowly wears out during use and in addition it loses its efficiency due to loading and glazing. During the operation, the chips formed get entrapped in the inner granular space of abrasive particles. This is called loading of the wheel and it results in inefficient cutting operation.

When the band of the abrasive wheel is very hard, it doesn't dislodge an abrasive particle which has become blunt. This results in this process get a shining appearance. This is known as glazing of the wheel. The ineffectiveness in the cutting action of grinding wheel by loading and glazing is removed and dressing and truing of the wheel expose the new sharp edges.

Dressing: The process of the loading and breaking away the glazed surface so that new sharp abrasive particles are again present to work for efficient cutting is called dressing. This is done with various types of dressing tools namely:

(a) Star dressing tool
(b) Round abrasive stick
(c) Diamond dressing tool

(*a*) *Star dressing tool*: A common type of star-dresser is illustrated in the Fig. 13.3.

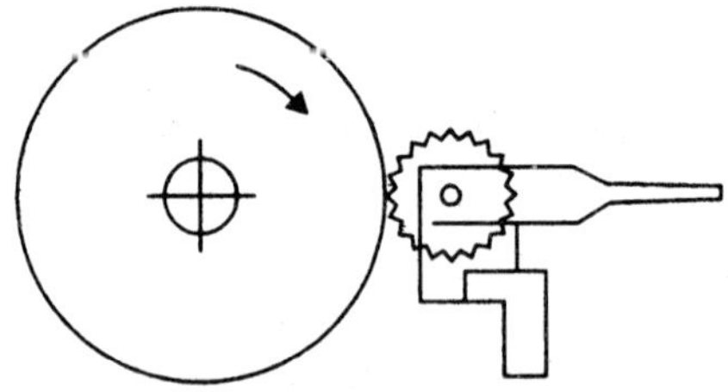

(a) Star dressing tool

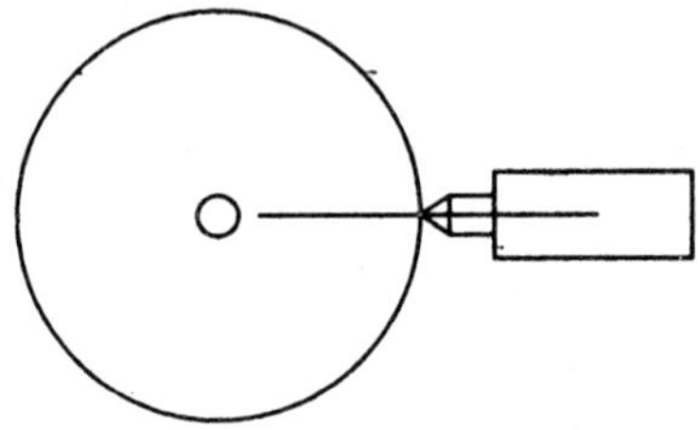

(b) Abrasive stick

Fig. 13.3 Dressing a Grinding Wheel

It consists of a number of hardened steel wheel with points on their periphery. The dresser is held against the face of the revolving wheel and moved across the face to dress the surface.

(*b*) *Round abrasive stick* (Fig. 13.3(b)): This type of dressing tool consists of a steel filled with a bonded abrasive. The end of the tube is held against the wheel and moves across the face.

(*c*) *Diamond dressing tool*: For precision and high finish grinding, small industrial diamonds are used.

Truing : Truing is the process of changing the shape of the grinding wheel as it becomes worm from the original shape, due to the breaking away of the abrasive and bond. This is done to make the wheel true and concentric with the bore or to change the contour for form grinding. Truing and dressing are done with the same tool.

13.10 BALANCING OF GRINDING WHEEL

If the centre of gravity of a grinding wheel and its axis of rotation coincide, the grinding wheel is said to be balanced. Since the grinding wheel speeds are high, straight out of balance condition may give use to large forces. This may results in execssive vibrations poor surface finish, and faster wheel breakdown and may even dangerous to the operator. Therefore, particular attention should be given to the balancing of the wheel.

The commonly used procedure for the balancing of grinding wheels is to use a balancing bench. The following steps are involved in this process.

(a) Thoroughly clean and inspect the wheel for cracks.

(b) Place the balancing stand on the flat surface and align it horizontally with an accurate level.

(c) Place the grinding wheel on the balancing stand (Fig. 13.4)

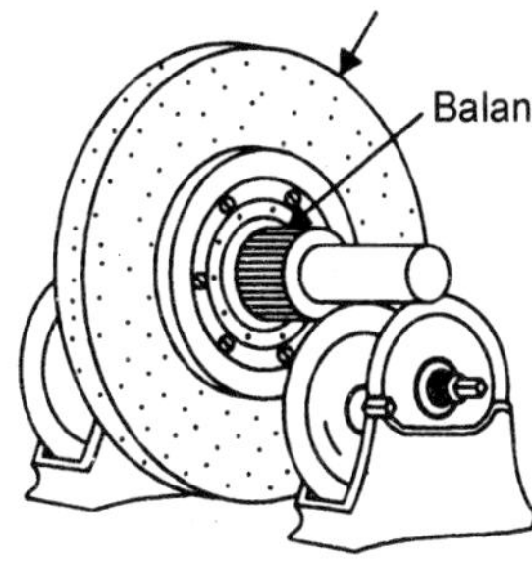

Fig. 13.4 Revolving Wheel Balancing Stand

(d) Set the wheel in any arbitrary position. Balanced wheel set in any position. If the wheel is not properly balanced, the heavier part will move downwards.

(e) Now bring the wheel to static position by moving the balance weights.

13.11 TYPES OF GRINDING MACHINES

Metal working machines in which the cutting of metal is by abrasive action is known as grinding machines.

Grinding machines can be divided into three groups:

1. Rough grinders
2. Precision grinders
3. Special purpose grinding machines.

13.11.1 Rough Grinders

These include floor stand grinder, bench grinder, portable and flexible shaft grinder, abrasive bell grinders. Rough grinding is preferred when large amount of stock is to be removed and accuracy is of secondary consideration.

(*a*) *Floor Stand Grinder*: It is mounted on a base and consists of a horizontal spindle with grinding wheels mounted at each end of the motor shaft extensions. The work is held by the operator in hand and pressed against the wheel to remove the material.

(*b*) *Bench Grinder*: It is similar to a floor grinder except for the size. It is fitted on the bench. These machines are used for grinding of tools and miscellaneous parts.

(*c*) *Portable and Flexible Grinder*: The portable grinders resembles a portable electric drill. These are used for finishing castings, welded joints in a structural work etc.

(*d*) *Swing Frame Grinder*: It is used to remove material from the objects which are heavy and inconvenient to handle. It has a horizontal frame 2 to 4 meters long, suspended at the centre of gravity and having a grinding wheel on one end. The operator guides the frame and applies the wheel to the job. This is used for snagging castings which are too heavy and large.

(*e*) *Abrasive Belt Grinder*: These machines are designed to use an endless abrasive belt for grinding instead of a regular type of grinding wheel. The belt runs round the pulley or rollers and work is fed against the revolving abrasive bell.

13.11.2 Precision Grinding Machines

These machines are used to manufacture parts of accurate dimensions and good surface finish. The cylindrical grinding machines, surface grinding machines and internal grinding machines comes under this category.

13.11.2.1 Cylindrical Grinder

The principal of cylindrical grinder is illustrated in the Fig. 13.5. In this work piece is held between the dead centres and rotated by a dog and driver on the face plate. There are four movements in a cylindrical centre type grinding.

(i) The work must revolve, (iii) The work must pass the wheel,
(ii) The wheel must revolve, (iv) The wheel must pass the work.

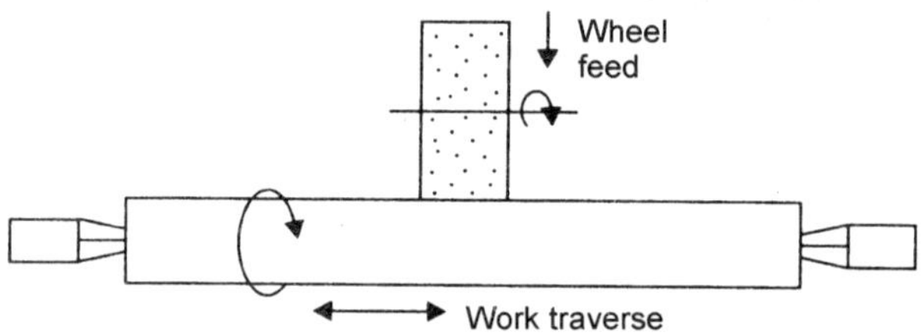

Fig. 13.5 Cylindrical Grinder

These grinding machines are used for grinding plain cylindrical parts, although they can also be used for grinding contoured cylinders, tapers, shoulders etc.

In cylindrical grinding, two types of grinding operations are done:

(i) Traverse grinding, (ii) Plunge grinding.

Traverse Grinding (see Fig. 13.6)

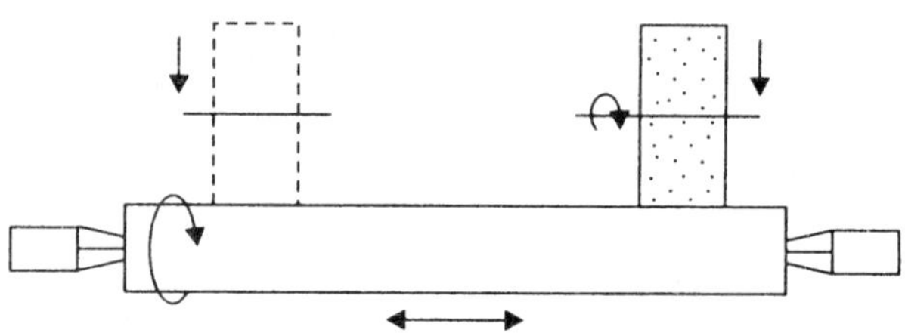

Fig. 13.6 Traverse Grinding Operation

In this work is reciprocated as the wheel feeds to produce cylinders longer than the width of the wheel.

Plunge Grinding (see Fig. 13.7)

In plunge grinding, the work rotates in fixed position as the wheel feeds to produce cylinders of a length equal to or shorter than the width of the wheel.

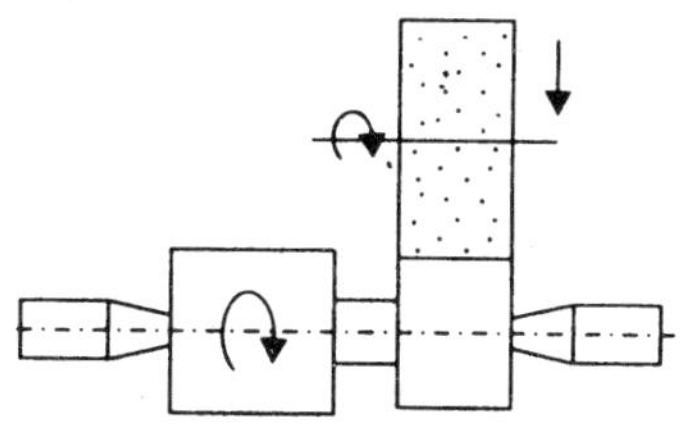

Fig. 13.7 Plunge Grinding Operation

(*a*) *Plain centre type cylindrical grinding machine*

A plain centre type grinding machine and cylindrical grinding machine are shown in Fig. 13.8(a) and (b). It consists of the various parts.

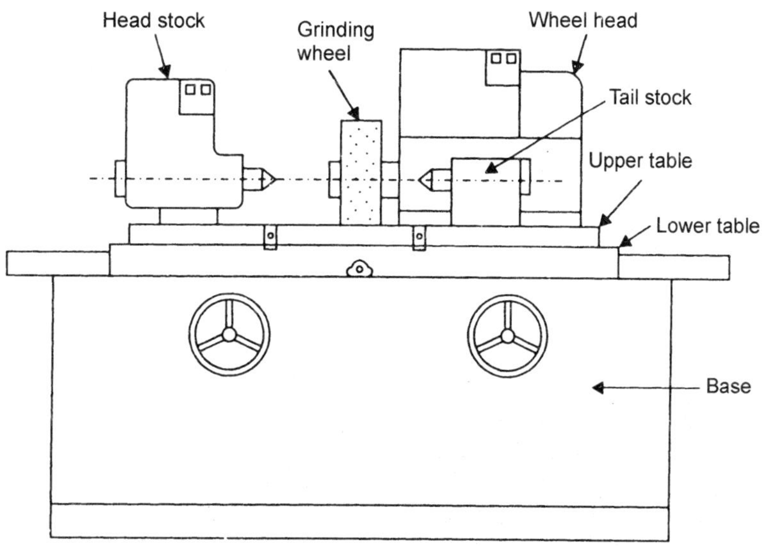

Fig. 13.8(a) Plane Centre Type Grinder

Base: The base is the main casting that rests on the floor and supports the parts mounted on it. On the top of the base horizontal ways are set on which the table slides to give traverse motion to the work piece. The table drive mechanism is incorporated in the base itself.

Tables: There are two tables — Lower table and Upper table. The lower table slides on the ways of the bed and provides traverse of the work past the grinding wheel. It can be moved by hand or power within the limits. Headstock and tailstock are mounted on the upper table. The upper table can be swivelled up to 10° relative to the main table traverse.

Headstock: The headstock supports the work-piece by means of a dead centre and drives it by means of a dog or it may hold and drive the work piece in a chuck. The work piece is rotated by separate motor housed in the headstock.

Tailstock: The tailstock can be adjusted and clampled to accommodate different lengths of work pieces.

Wheel Head: The wheel head carries a grinding wheel and rotated by a motor housed in the headstock. The wheel head can be moved perpendicular to the table ways by hand or power to feed the wheel to the work.

Fig. 13.8(b) Cylindrical Grinder

(b) Universal centre type grinders

Universal grinders are widely used in tool room for grinding tools etc. The features of this machine are similar to those of plain grinders, but in addition it is provided with a swivelling headstock and swivelling wheel head.

(c) Centreless grinders

Centreless grinding is a method of grinding exterior cylindrical, tapered, and formed surfaces on work pieces that are not held and rotated on centres.

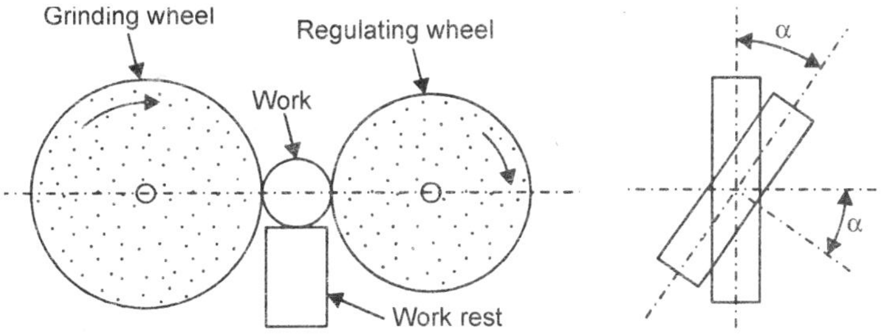

Fig. 13.9 External Centreless Grinding

The principle elements of an external centreless grinder are shown in Fig. 13.9. The grinder has two wheels, a larger grinding wheel revolving at a high speed and a small regulating or controlling wheel revolving at a slow speed. Work rest is located between the wheels. The work is placed on the work rest. The regulating wheel is fed forward forcing the work against the grinding wheel.

The axial movement of the work past the grinding wheel is obtained by tilting the regulating wheel at a slight angle from the horizontal.

Methods of Centreless Grinding: Basically there are three different methods by which centreless grinding can be done on different following types of jobs.

(1) Through feed, (2) Infeed, (3) End feed.

These are illustrated in Fig. 13.10.

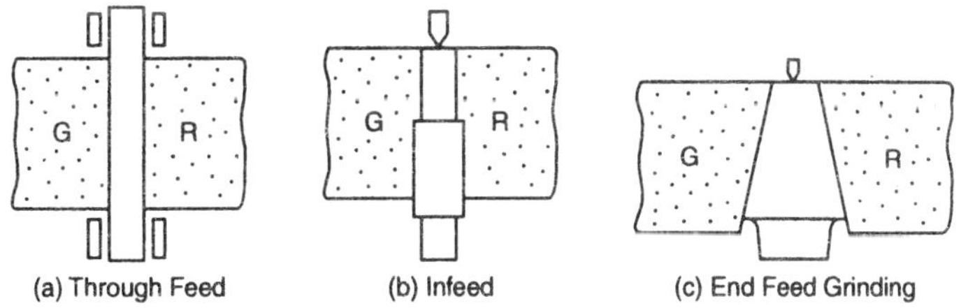

(a) Through Feed (b) Infeed (c) End Feed Grinding

Fig. 13.10 Methods of Centreless Grinding

Through Feed Grinding (Fig. 13.10 (a)): Through feed grinding is used for straight cylindrical work like long shafts or bars. In this method, the work enters from one side of the machine and comes out from the other side with guides at the both ends.

Infeed Grinding (Fig. 13.10(b)): It is similar to plunge or form grinding. The regulating wheel is drawn back so that the work-pieces may be placed on the work rest table. Then it is moved into feed the work against the grinding wheel. This method is useful to grind shoulders, and formed surfaces.

End Feed Grinding (Fig. 13.10(c)): It is used to produce taper, either the grinding wheel or regulating wheel or both are formed to a taper. The work is fed lengthwise between wheels and is ground as it advances until it reaches the end stop.

Advantages

1. The work is supported throughout its entire length, so there is no chatter or deflection.
2. No centre holes, no chucking or other holding devices are required.
3. As a true floating condition exists during the grinding process, less metal needs to be removed.
4. The process is continuous, so it is used for production work.
5. The size of the work is easily controlled.

Disadvantages

1. Work having multiple diameters is not easily handled.
2. In hollow work, there is no certainly that the outside diameter will be concentric with the inside diameter.

13.11.2.2 Internal Grinder

Internal grinders are used to finish straight, tapered or formed holes to the correct size, shape and finish. According to the construction features, there are three types of internal grinders:

(a) Chucking (b) Planetary (c) Centreless

(a) Chucking grinders

In this, the work piece is chucked and rotated about its axis to bring the surface to be ground in contact with the grinding wheel. The grinding wheel is rotated and at the same time reciprocated back and forth through the length of the hole as shown in Fig. 13.11.

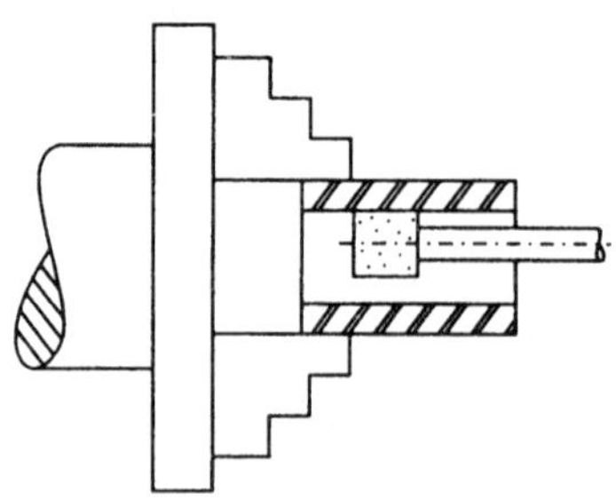

Fig. 13.11 Chucking Grinder

(b) Planetary type internal grinders

The work remains stationary and the rotated wheel spindle is given an eccentric motion, according to the diameter of the hole to be ground. Such type of operation is used where the work is difficult to be rotated. Since in this operation, the motion of the grinding wheel is in the form of planet and hence it is called planetary grinding.

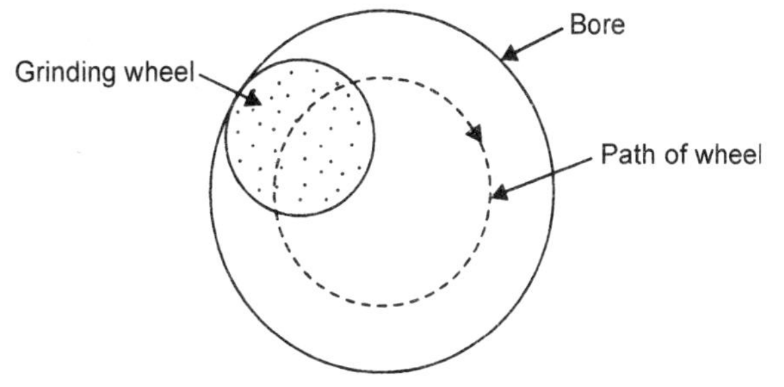

Fig. 13.12 Planetary Internal Grinder

(c) *Centreless grinding*

The external centreless grinding principle is also applied to internal grinding. In internal centreless grinding, the work is supported by three rolls. One is the regulating wheel, second one is a supporting roll and other is pressure roll to hold the work-piece against the support and regulating rolls. The process is illustrated in Fig. 13.13.

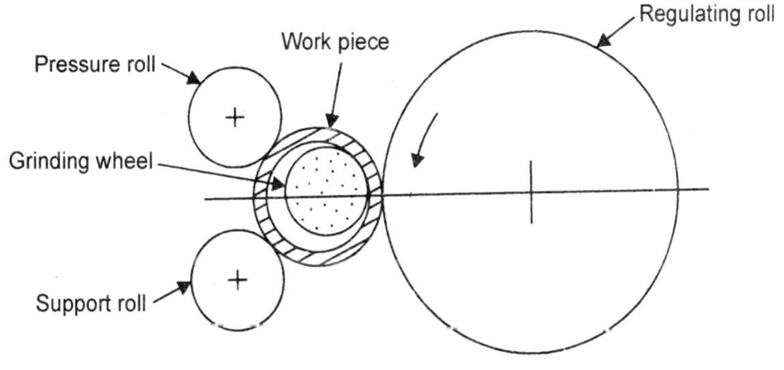

Fig. 13.13 Internal Centreless Grinding

The grinding wheel contact the inside diameter of the work-piece directly opposite the regulating roll, thus assuring a part of absolutely uniform wall thickness and concentricity. The pressure roll is mounted to swing aside to permit loading and unloading.

13.11.2.3 Surface Grinding Machines (see Fig. 13.14)

Surface grinding machines are used to produce and finish flat surfaces. With special fixtures and form dressing devices, angular and formed surfaces can also be ground. The surface grinders are classified depending on the construction, design and other features as follows:

1. According to the table movement they are classified as
   (a) Reciprocating table type, (b) Rotary table type.
2. According to the position of wheel spindle, they are classified as
   (a) Vertical spindle type, (b) Horizontal spindle type.

Four different types of surface grinders:

(a) Horizontal spindle reciprocating table, (b) Horizontal spindle rotary table,
(c) Vertical spindle reciprocation table, (d) Vertical spindle rotary table.

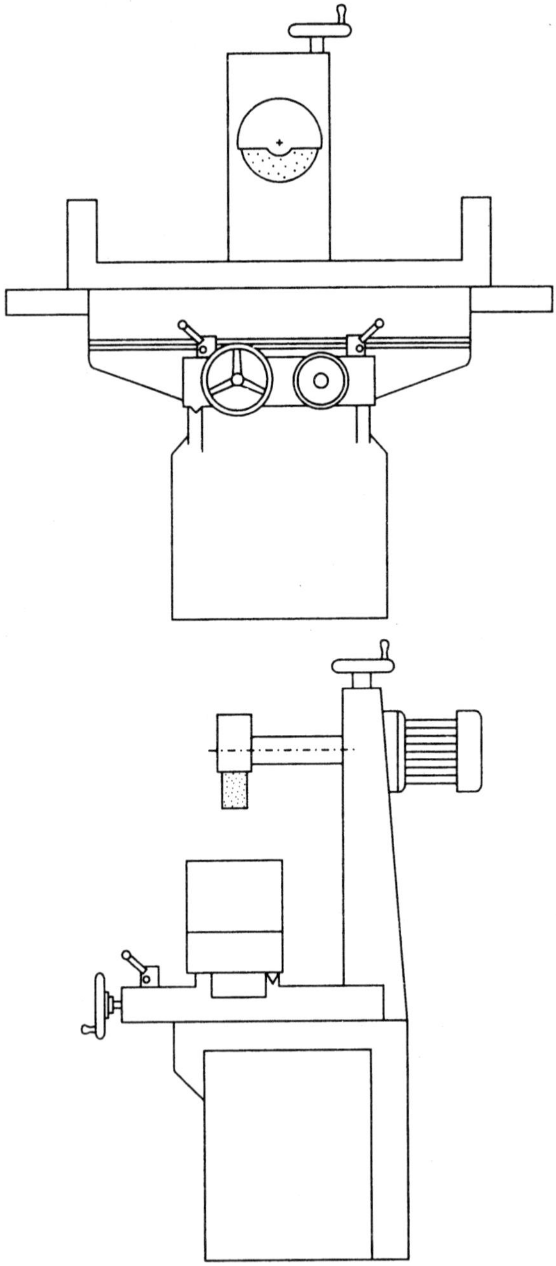

Fig. 13.14 Surface Grinding Machine

(*a*) *Horizontal spindle reciprocating table* (see Fig. 13.15)

It consists of a horizontal spindle carrying the grinding wheel and rectangular work table. The table is provided with a longitudinal feed movement. The table top has T-slots for mounting the magnetic chuck, vices, fixtures etc. These are used where smooth finish and close tolerances are required.

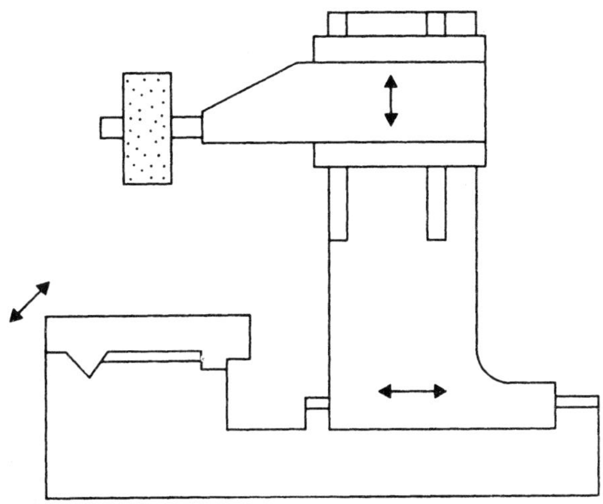

Fig. 13.15 Horizontal Spindle, Reciprocating Table

(*b*) *Horizontal spindle rotary table* (see Fig. 13.16)

These surface grinders are generally used for precision grinding. The grinding wheel spindle is carried on a wheel slide and can be traversed across the work which is mounted on a revolving horizontal axis table.

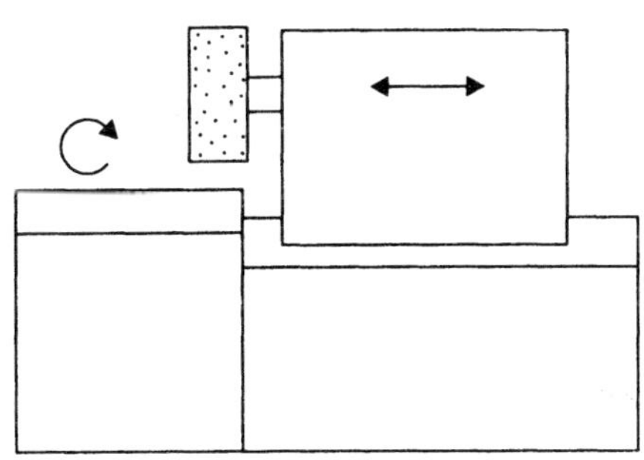

Fig. 13.16 Horizontal Spindle, RotaryTable

(*c*) *Vertical spindle reciprocation table* (see Fig. 13.17)

These machines use cup, cylinder wheels for faster stock removal and where accuracy's are not stringent. These type of machines are usually if higher capacity and are used as production machines.

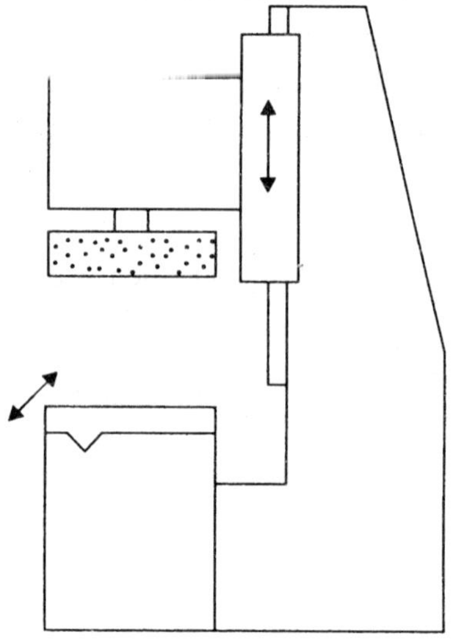

Fig. 13.17 Vertical Spindle, Reciprocating Table

(*d*) *Vertical spindle rotary table* (see Fig. 13.18)

The grinding spindle is mounted vertically on the face of a column and rotates in fixed position, feeding only along its axis. The *rotary* table travels beneath the wheel as it rotates.

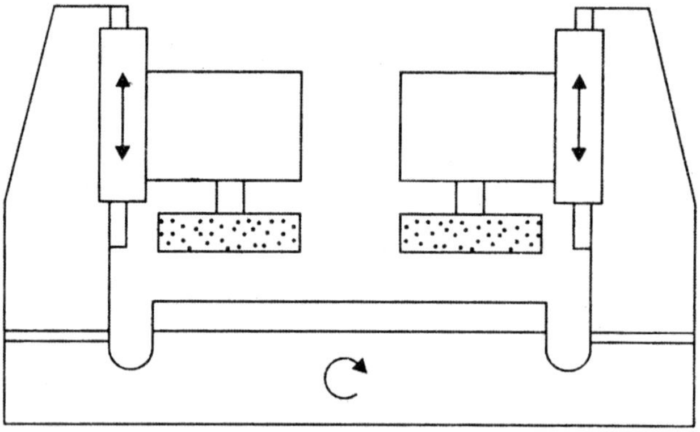

Fig. 13.18 Vertical Spindle, RotaryTable

13.11.2.4 Special Purpose Grinding Machines

Some grinders are designed for highly specialized work:

(*a*) *Tool and cutter grinder*

Tool and cutter grinders are used to sharpen and recondition multiple tooth cutters like reamers, milling cutters, drills, taps, hobs and other type of tools used in the shop. With various attachments they can also do surface grinding, cylindrical and internal grinding operations. They are classified according to the purpose of grinding into two groups:

(i) Universal tool and cutter grinders,

(ii) Single purpose tool and cutter grinders.

Universal tool and cutter grinders are particularly intended for sharpening of miscellaneous cutters.

Single purpose grinders are used for grinding toos such as drills, tool bits etc.

(i) Universal tool and cutter grinders

The Fig. 13.19(a) shows the principle parts of an universal tool and cutter grinder. The parts and their function are described below.

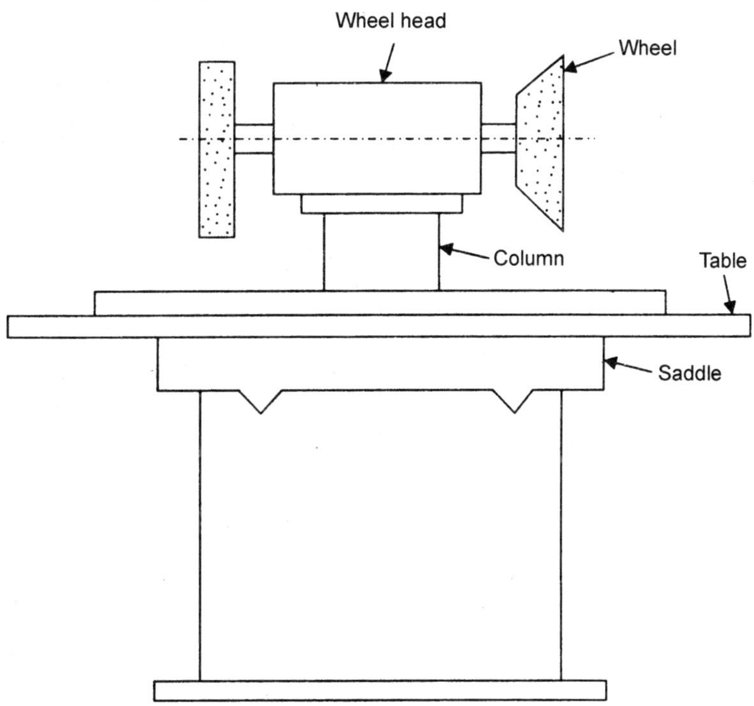

Fig. 13.19 (a) Universal Tool and Cutter Grinder

Base: The base gives rigidity and stability to the machine. It is heavy and box type.

Saddle: The saddle is mounted on the top of the base. It moves on anti-friction bearings on hardened ways. The column supporting the wheel head is mounted on the saddle and it can be moved up or down and swivelled to either side.

Table: The table rests and moves on the top base which is mounted over the saddle.

Headstock and Tailstock: The headstock and tailstock are mounted on either side of the table similar to those on a cylindrical grinder.

Wheel Head: The wheel head is mounted on a column on the back of the machine. It can be swivelled and positioned on the base for various setups.

Grinding Wheels: Three different types of grinding wheels are used. (1) The straight or disc shaped wheel, (2) The cup type, (3) The dish type.

Fig. 13.19(b) Block Diagram of Universal Tool and Cutter Grinder

(b) Crankshaft grinders

A crankshaft grinder is basically a cylindrical grinder using the principle of plunge grinding. In plunge grinding, the work rotates in a fixed position and the wheel is fed to produce cylinders. These grinders are used for grinding crankshafts of automobile engines, aircraft engines, diesel engines, compressors etc.

(c) Cam grinders

These machines are basically cylindrical grinding machines with additional feeding and withdrawal mechanisms for the work-piece. It consists of a separate base that carries the headstock and tailstock. The complete unit can oscillate about a centre below the work-piece. Before carrying out the operation a small template is mounted on the headstock. A hardened steel roller in conjunction with a template actuates the movement of the whole unit to produce the desired shape. Modern cam shaft grinders are provided with automatic feed mechanisms for rapid production.

(d) NC grinding machines

Numerical control has been applied to various types of grinding machines—surface grinder, rotary as well as reciprocating table type, cylindrical grinders, centreless grinders and tool and cutter grinders.

Digital readouts, solid state programmable controllers, numerical controls, computer numerical controls and adaptive controls have been applied to grinding machines to achieve different degrees of automation. Microprocessor based controls are finding many applications in grinding not only for accurate positioning of slides during metal removal but also vital auxiliary functions such as wheel dressing, wheel compensation and in-process gauging.

Size control is provided by automatic in-process gauging systems interfaced with the controls or by dressing the wheel with reference to a fixed datum position before every grinding cycle. Electrical or Electro-hydraulic stepper motors or dc servo drives actuates the slides through ball screws and nuts. Servo drivers are now becoming increasingly popular.

QUESTIONS

1. What are Natural and Artificial abrasives?
2. What are the different types of bonds in the manufacturing of grinding wheels?
3. Explain the factors to be kept in mind in selecting a grinding wheel.
4. Explain dressing and truing of a grinding wheel.
5. Describe the centreless grinding process. What are the various feeding methods used in centreless grinding?
6. Sketch and explain plain type cylindrical grinder.
7. Sketch and explain tool and cutter grinder.

14

Jigs and Fixtures

14.1 INTRODUCTION

Over the centuries, man has been finding easier and better ways to produce goods and services to satisfy the customer needs and wants. Now fully automatic machines make it possible to remove much of the burden of the manual work from the operator.

When the articles are to be produced in large quantities with a high degree of accuracy and interchangeability at a competitive cost, some specially designed tooling is required. These are consists of special tools, Jigs and Fixtures etc. The Jigs and Fixtures are the most economical means to produce repetition type of works.

14.2 USES OF JIGS AND FIXTURES (ADVANTAGES)

1. It eliminates the marking, measuring and other setting methods before machining.
2. It increases the machining accuracy because the workpiece is automatically located and the tool is guided without making any manual adjustment.
3. It enables production of identical parts.
4. It increases the production capacity by enabling a number of workpieces to be machined in the single set up and in some cases a number of tools may be made to operate simultaneously.
5. Handling time is greatly reduced due to quick setting and locating the work.
6. It enables semi skilled operator to perform operations. This saves the labour cost.
7. It reduces the expenditure on the quality control of the finished products.
8. It reduces the overall cost of machining.

(a) Jig

It may be defined as a device which holds and locates a workpiece and guides and control one or more cutting tools. In construction, a Jig comprises of a plate, a structure or box made of metal or non metallic and then guiding the tool in correct position on the work with reference to the production drawing.

(b) Fixture

It may be defined as a device which holds and locate the workpiece during machining or inspection. In construction, it comprises of designed workholding devices which are clamped on the machine table.

14.3 DIFFERENCES BETWEEN A FIXTURE AND JIG

Fixture

1. A fixture holds and positions the work, but does not guide the tool.
2. A fixture is bolted to the machine table.
3. The fixtures are heavier in the construction.
4. Fixtures are used for milling, grinding, shaping and welding operations.

Jig

1. Jig holds, locates and as well as guide the tool.
2. Usually Jig is not fixed to machine Table.
3. Jigs are made lighter for quicker handling.
4. Jigs are used for drilling, reaming and boring operations.

14.4 PRINCIPLES OF JIGS AND FIXTURES DESIGN

Jigs and Fixtures have the following components.

1. Location

(a) Ensure that the workpiece is given the desired constraint.
(b) The locators must be positioned in such a way that swarf will not cause misalignment.
(c) Make the location points adjustable if a rough casting or forging is being machined.
(d) Make all location points visible to the operator from his working position.

2. Clamping

(a) The clamps should be positioned to give best resistance to the cutting force.
(b) Position the clamps so that they should not cause deformation of workpiece.
(c) If possible, make clamps integral with the fixture body.
(d) Make all clamping and location motions easy and natural to perform.

3. Clearance

(a) Allow enough clearance to allow for variation of workpiece size.
(b) Allow ample clearance for the operator's hands.
(c) Ensure that there is ample swarf clearance.

4. Stability and Rigidity

(a) Provide four feet so that uneven seating will be avoided.
(b) Make the equipment as rigid as is necessary for the operation.
(c) Provide means of positioning and bolting the equipment to the machine table or spindle if required.

5. Handling

(a) Make the equipment as light as possible and easy to handle.
(b) Ensure that no sharp corners are present.
(c) Provide lifting points if it is heavy.

6. General

(a) Keep the design simple in order to minimize the cost.
(b) Utilize standard parts as much as possible.

14.5 TYPES OF JIGS AND FIXTURES

(a) Jigs and Fixtures used with Machine tools, Drilling fixtures, Milling fixtures, Broaching fixtures etc.
(b) Devices for locating and clamping the process tools: Arbors, Holders etc.
(c) Assembly fixtures
(d) Inspection fixtures
(e) Special Jigs and Fixtures.

14.6 PRINCIPLES OF LOCATION

Assume a workpiece in a space as shown in Fig. 14.1. Consider the possible movements of the free body shown with respect to the three mutually perpendicular axes x-x, y-y, z-z.

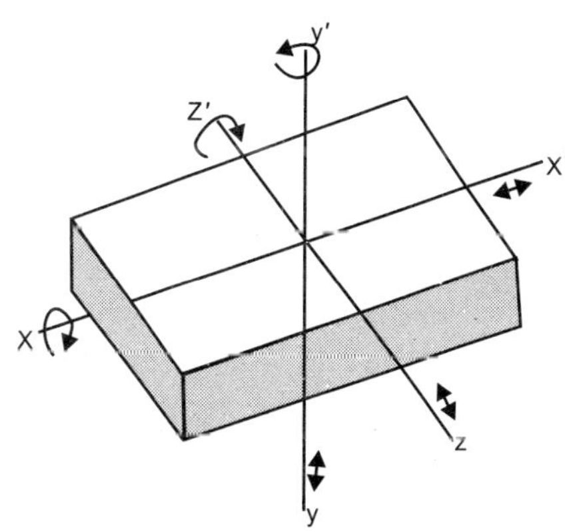

Fig. 14.1 Six Degrees of Freedom

It can

1. Move along x-x
2. Move along y-y
3. Move along z-z

} Three freedoms of TRANSLATION

4. Rotate about x-x
5. Rotate about y-y
6. Rotate about z-z

} Three freedoms of ROTATION

In order to locate the block correctly within a Jig, all these six movements must be restrained by arranging suitable locating points and then clamping the block in position.

14.7 SIX POINT LOCATION OF A RECTANGULAR BLOCK (See Fig. 14.2)

The bottom of the block is supported by three points, the rear face of the block bears against two points, the side of the block rests against a single point. The downward movement along y-y axis is restrained by three supporting points (1, 2, 3). The movements along z-z axis and x-x axis are restrained by the double points (4, 5) and the single point (6) respectively. The rotary movements of the block about x-x, y-y and z-z axis are also restrained. This method is called 3-2-1 location or six point location.

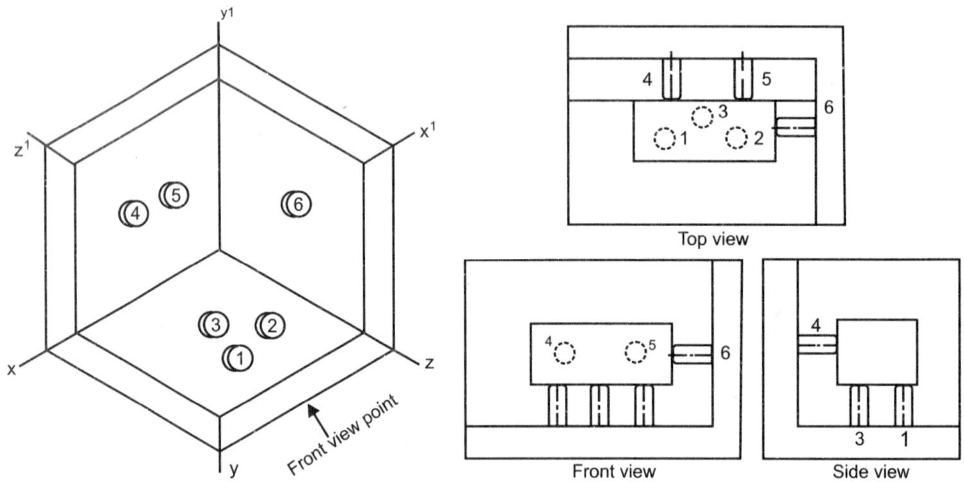

Fig. 14.2 Six Point Location of a Rectangular Block

The reasons for placing 3 points in the first plane are: If a body rests on two points only, it can lean over to one side. It is also not advisable to use more than three fixed points in the first plane, as any extra point of support become redundant. This is clear from the fact that three legged chair can be easily placed on any type of floor, whereas all the legs of a four legged chair will contact the floor only if the floor is absolutely plane. The three points of support in the first plane should be so selected that the weight of the workpiece is evenly distributed on them and the center of gravity of the work-piece is positioned properly with respect to them.

In the second plane, if only one point is provided, it would be possible to swivel the work about this point.

In the third plane, there is only one direction of the movement and therefore one point is sufficient there.

14.8 LOCATING DEVICES

The locating devices locate the workpiece inside a Jig or Fixtures. The type of locating system to be used depends on the operation to be performed on the workpiece. Various types are:

(a) Flat Locator (b) Cylindrical Locator

(c) Conical locator (d) V-locator
(e) Drill bush locator

(a) Flat Locator

This type of locators are used for locating flat machined faces of the component. Pins are used for location from rough surfaces and they support the work-pieces exactly beneath the clamps. These may be fixed pin and adjustable pin.

(b) Cylindrical Locator

These are employed for locating components having drilled holes. The cylindrical locator is fitted on the Jig body is inserted in the drill hole of the component to locate it in position.

(c) Conical Locator

This is used for locating workpieces having different drilled holes. The conical locator is superior to pin locator due to its capacity to accommodation a slight variation in the hole diameter of the workpiece.

(d) V-Locator

It is used to locate workpieces having circular or semicircular profiles.

(e) Drill Bush Locator

It is used for locating cylindrical workpieces. The bush has conical opening for locating purpose. It also serves the purpose of guiding the tool.

14.9 TYPES OF CLAMPING DEVICES

The clamping devices should full fil the following points

(i) The clamp devices must hold the workpiece rigidly against the cutting forces
(ii) The time required to loosen and tighten the clamp on the workpiece should be minimum.
(iii) While clamping, the clamp should not damage the surface of workpiece.
(iv) The movement of the screw, lever or cam of the clamping device whether it is rotary or reciprocating type should be strictly limited to make the device quick acting.

Some of the clamping devices are shown in Fig. 14.3.

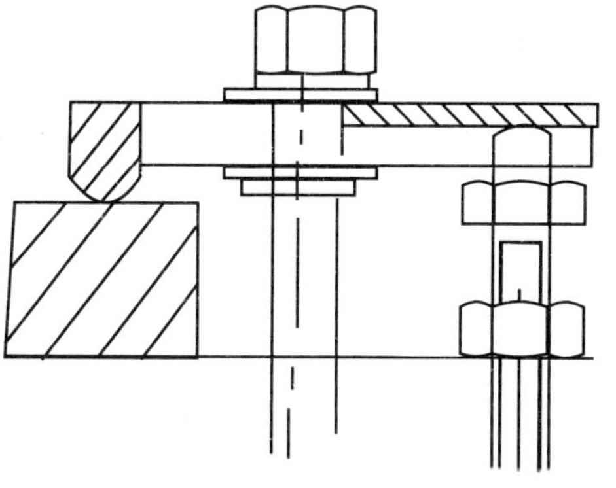

Fig. 14.3 (a) Clamp with Adjustable Heal Pin

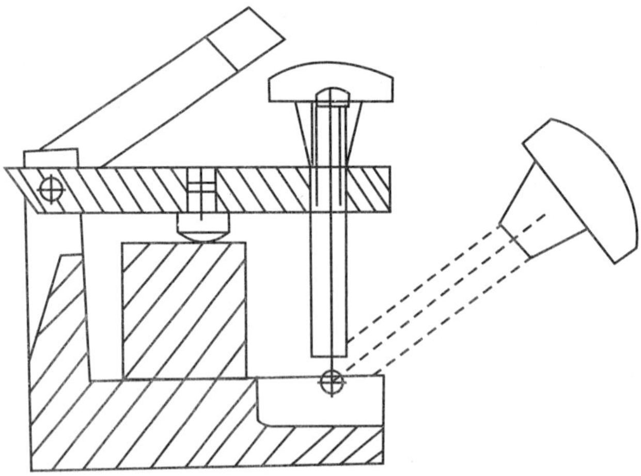

Fig. 14. 3 (b) Latch-Type Clamp

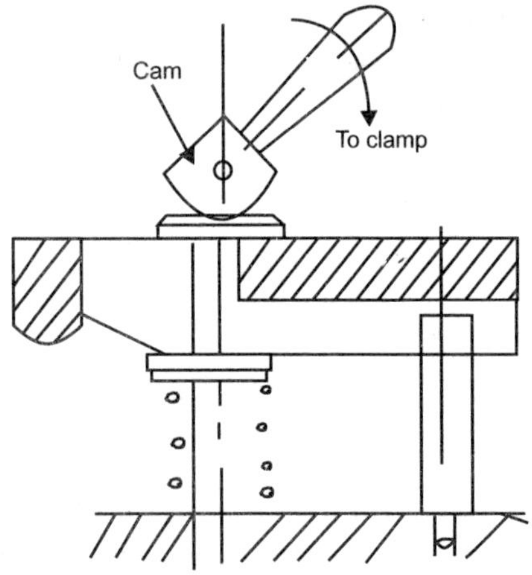

Fig. 14. 3 (c) Cam Operated Clamp

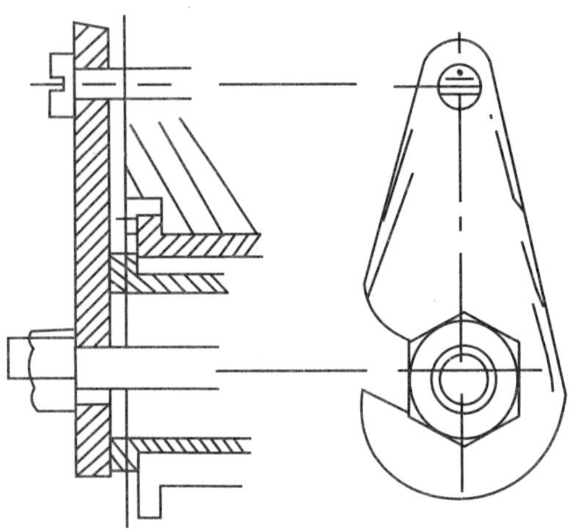

Fig. 14.3 (d) C-Clamp (Swing Washer)

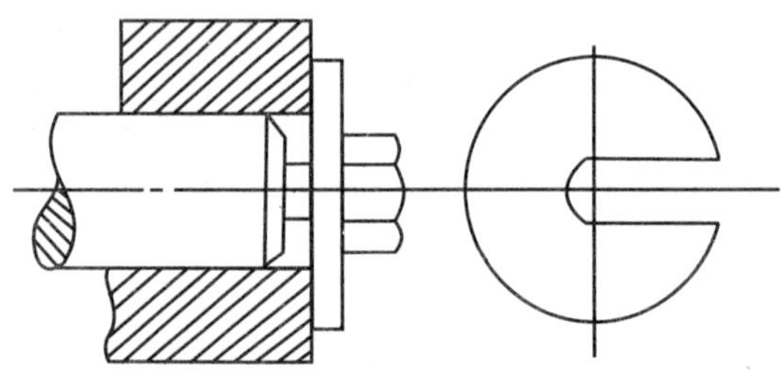

Fig. 14.3 (e) C-Clamp (C-Washer)

14.10 JIG BUSHES

Drill bushes are generally used only with Jigs. These are used to locate and guide cutting tools such as drills, reamers, boring bars etc. Made of M.S. but are case hardened to minimize wear due to use. Classified as follows:

(i) Fixed Bushes: (a) Plain type, (b) Headed type

(ii) Liner Bushes: (a) Plain type, (b) Head or Flange type

(iii) Renewable Bushes (iv) Slip Bushes

(v) Screw or Clamp Bushes (vi) Special Bushes

14.10.1 Fixed Bushes (see Fig. 14.4)

These are pressed into the Jig plate. The cost of plain type bush is Low. There is possibility of the plain bush to come out through Jig plate either by pressure of the drill itself or because of blows from the drilling machine spindle. Therefore the headed bushes are commonly used.

Press fit bushes are also used as liners for renewable and slip bushes.

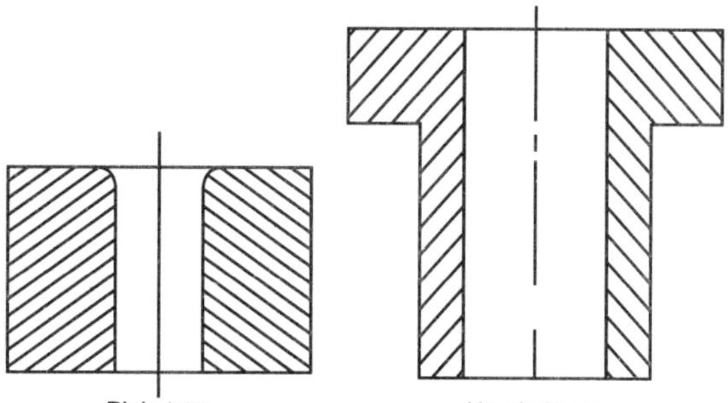

Fig. 14.4 Fixed Bushes

14.10.2 Liner Bushes (see Fig. 14.5)

These act as a hardened guide for both slip and renewal type bushes and sometimes used as guide to tools. In this way, the Jig plate of soft metal could be used. Liners has to be replaced and not the whole of the Jig plate.

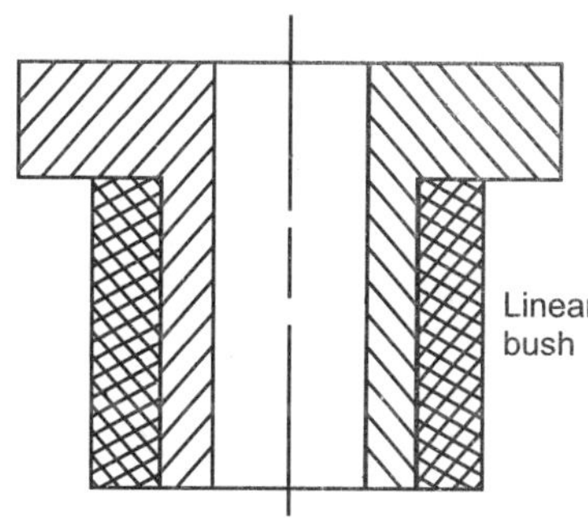

Fig. 14.5 Liner Bush

14.10.3 Renewable Bushes (see Fig. 14.6)

These are special type of fixed bushes. When they require to be replaced due to wear, a retaining screw is removed and worn out bush is taken out.

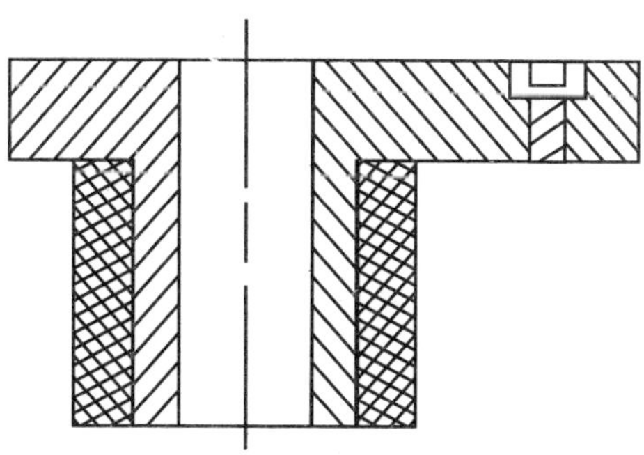

Fig. 14.6 Renewable Bush

14.10.4 Slip Bushes (see Fig. 14.7)

When a hole is to be drilled in stages, the best practice is to use slip bushes rather than using separate Jig for each operation. Retaining screw is used to prevent rotation and uplift when the tool is withdrawn from the hole.

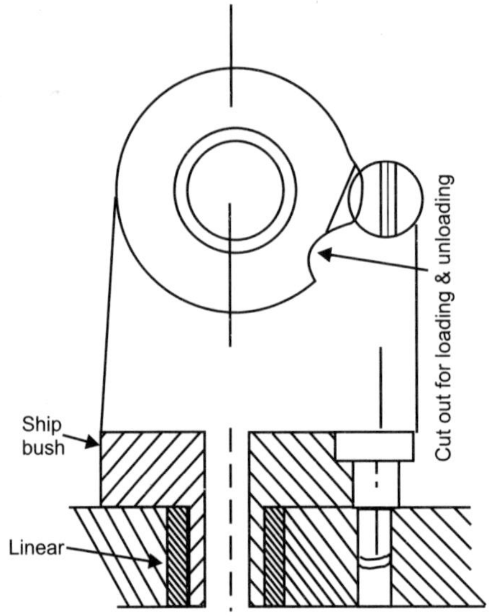

Fig. 14.7 Slip Bush

14.10.5 Screw or Clamp Bushes (see Fig. 14.8)

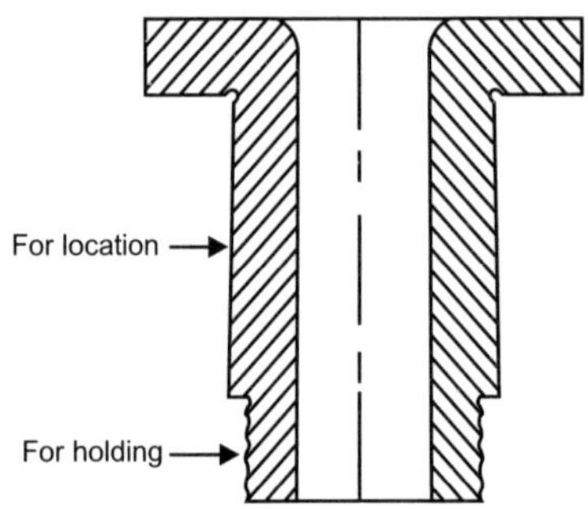

Fig. 14.8 Screw Bush

In this type, threaded portion is used for holding purposes and the plain portion of the bush is used for location.

14.10.6 Special Bushes (see Fig. 14.9)

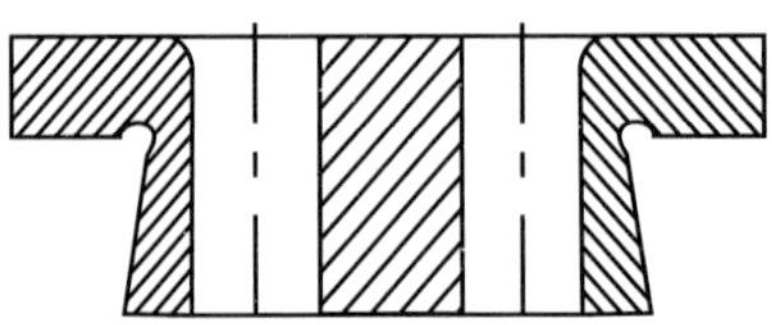

Fig. 14.9 Special Bush

When two holes are close together heads and walls of the two standard bushes would interfere with each other. A special bush with two holes are designed to meet the conditions.

14.11 JIGS AND FIXTURES

Jigs and fixtures used depend on the type of work to be machined. These are so many types of drill Jigs in use. However type of Jigs which have become common to most industries are:

(a) Template Jig (b) Plate Jig
(c) Channel Jig (d) Leaf Jig
(e) Pot Jig

(a) Template Jig (see Fig. 14.10)

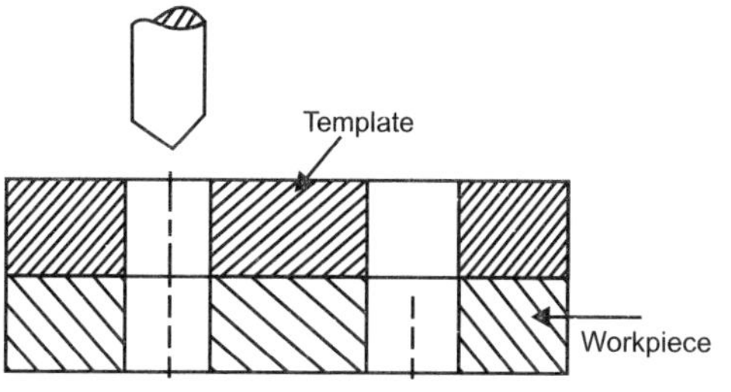

Fig. 14.10 Template Jig

Template Jig is very simple type. A plate having holes at desired position serves as a template which is fixed on to the workpiece to be drilled.

(b) Plate Jig (see Fig. 14.11)

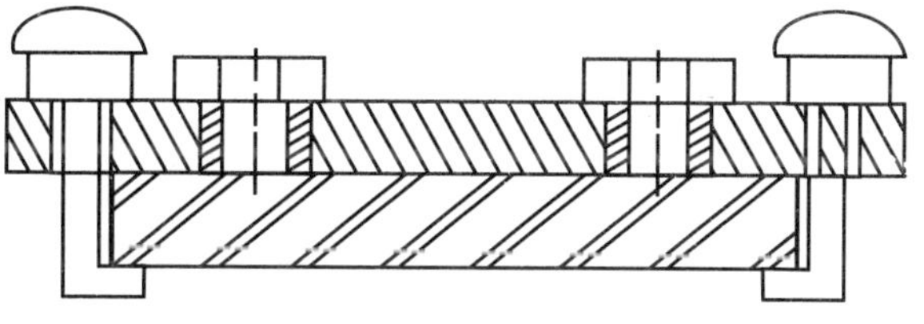

Fig. 14.11 Plate Jig

Plate Jig is the improvement of the template Jig by incorporating drill bushes on the template.

(c) Channel Jig (see Fig. 14.12)

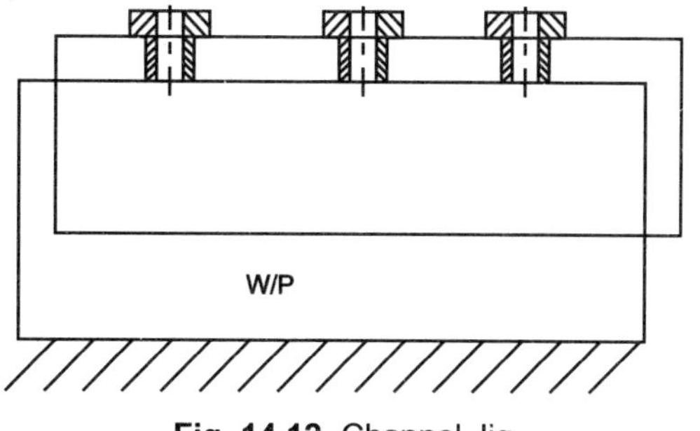

Fig. 14.12 Channel Jig

It is a simple type of Jig having channel like cross-section. The component is fitted within the channel and is located by rotating knurling knob. The tool is guided through the drill bushes.

(d) Leaf Jig (see Fig. 14.13)

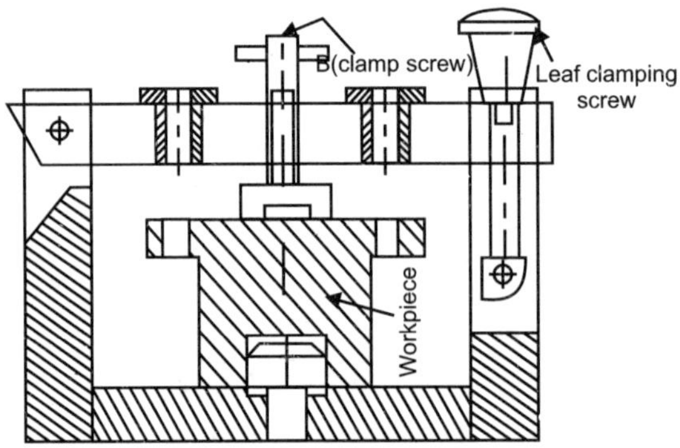

Fig. 14.13 Leaf Jig

Leaf Jig is shown in above figure. The leaf or plate may be swung open or closed on the work for loading and unloading purpose. Workpiece is clamped by screw.

(e) Pot Jig (see Fig. 14.14)

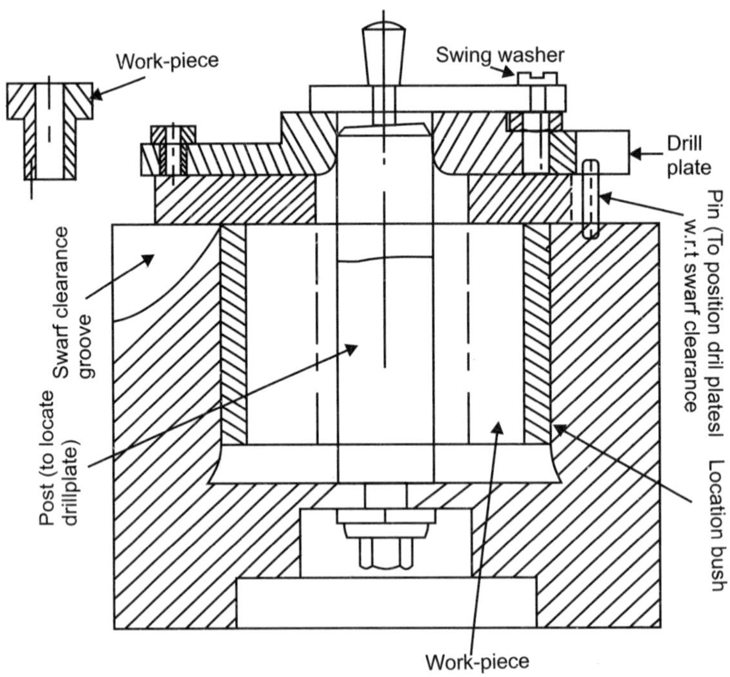

Fig. 14.14 Pot Jig

14.12 MILLING FIXTURES

A milling fixtures holds the part in correct position relative to the milling cutter as the table movement carries the part through the cutters.

Design Principles of Milling Fixture

(i) The fixture should be designed so as to permit rapid loading and unloading the work.
(ii) It should have sufficient chip clearance.
(iii) The fixture should be as strong and rigid as possible.
(iv) The design of fixture should be as simple as possible.

14.13 MILLING METHODS

(a) Straddle Milling
(b) Gand Milling
(c) String or Line Milling
(d) Pendulum Milling
(e) Profile Milling

14.14 ELEMENTS OF A MILLING FIXTURE

(*a*) *Base*: The base of a fixture is about 20 to 25 mm thick and is rigid enough not to deflect upwards during upcut milling. Base is provided with lugs on each side for fixing the base to machine table.

(*b*) *Tenon Strip*: The position of the base on the table is accurately located by means of tenon strips. The tenons are identical in width with the slot in the machine table and are fixed below the base. The length of a tenon strip is twice the width and made of steel.

(*c*) *Setting Block*: Milling fixtures are provided with a setting block so that a feeler gauge may be used for setting the fixture relative to the cutter. The setting block is fixed to the base of fixture by means of screw and dowels.

(*d*) *The bolts*: The fixture is bolted to the machine table with tee bolts suitable for the slots provided in the machine.

(*e*) *Clamping Devices*: Clamping devices are used to clamp workpiece on the fixture.

14.15 TYPES OF MILLING FIXTURES

1. Special Vice Jaw Fixtures
2. String Milling Fixtures
3. Gang Milling Fixtures
4. Continuous Milling Fixtures
5. Index Milling Fixtures
6. Air Operated or Pneumatic Milling Fixtures
7. Duplex Milling Fixtures
8. Some Other Special Milling Fixtures for Various Jobs.

In the above Milling Fixtures, some will be explained below. Special Vice Jaw Fixtures: (See Fig. 14.15).

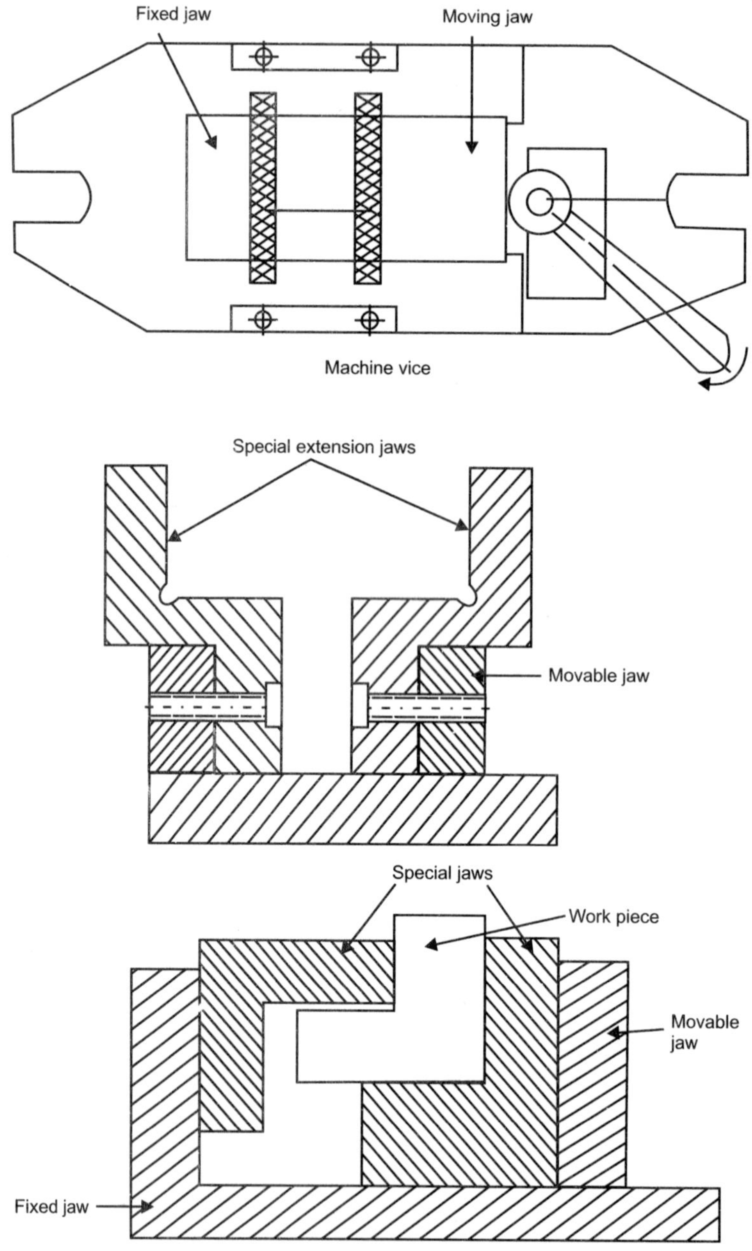

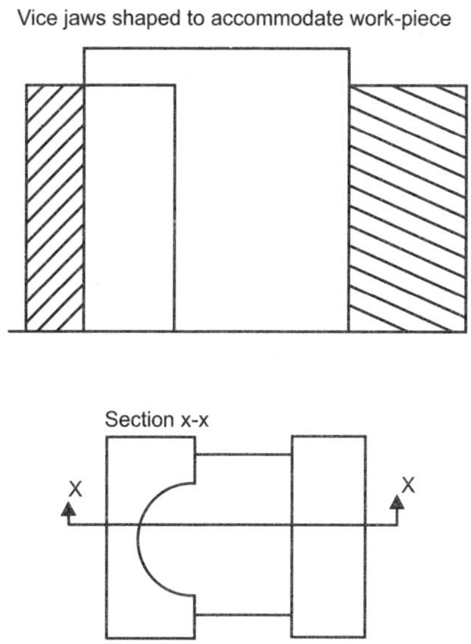

Fig. 14.15 Special Vice Jaw Fixtures

A commonly used work holding device for milling machine is Machine Vice. Provision is made for attaching special Jaw inserts to the fixed and movable vice Jaws. Expenditure on special milling fixtures can often avoided by carefully adopting special vice Jaws.

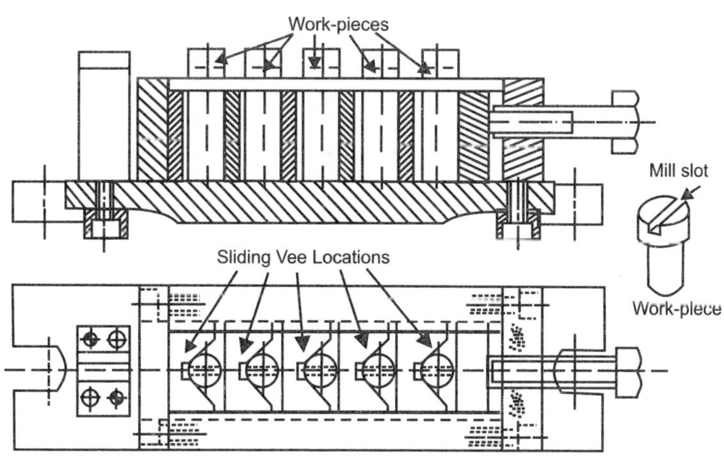

Fig. 14.16 String Milling Fixture

When a number of components are held in a line, the fixture is called a string or line milling fixture.

14.16 PNEUMATIC MILLING FIXTURE (See Fig. 14.17)

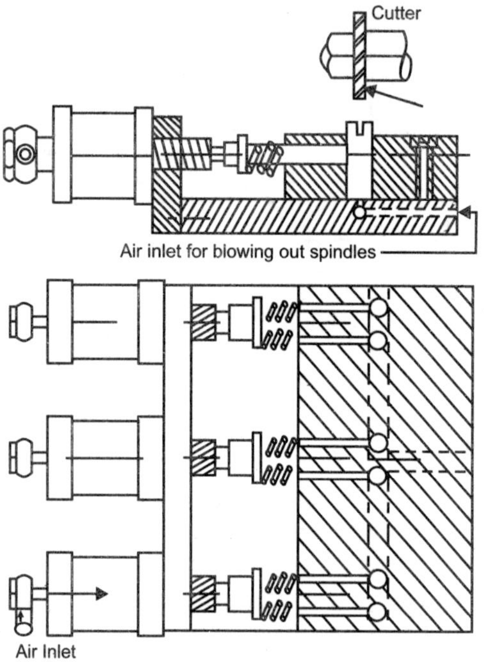

Fig. 14.17 Pneumatic Milling Fixture

This fixture is designed for a slot milling operation on a small cylindrical part. Six workpieces are located by spring loaded plungers. These plungers are operated by a separate air cylinder. When the operating lever is actuated all sin plungers are driven by air to hold work-pieces. After milling operation is over, the pressure is released and the plungers return to their initial position by spring pressure. Now air is admitted under the workpiece to eject them.

14.17 GRINDING FIXTURES

Surface Grinding Fixtures (see Fig. 14.18)

Rectangular plates can be ground under magnetic clamping alone, and no other fixtures are necessary. However for light works (cylinder parts) having less resisting area tend to tilt and fly off from the magnetic table due to high speed of grinding wheel and high feeds used in grinders. Hence it is necessary to provide additional support by nesting the workpiece (Fig. 14.18). This can be done by placing solid plates around the workpiece.

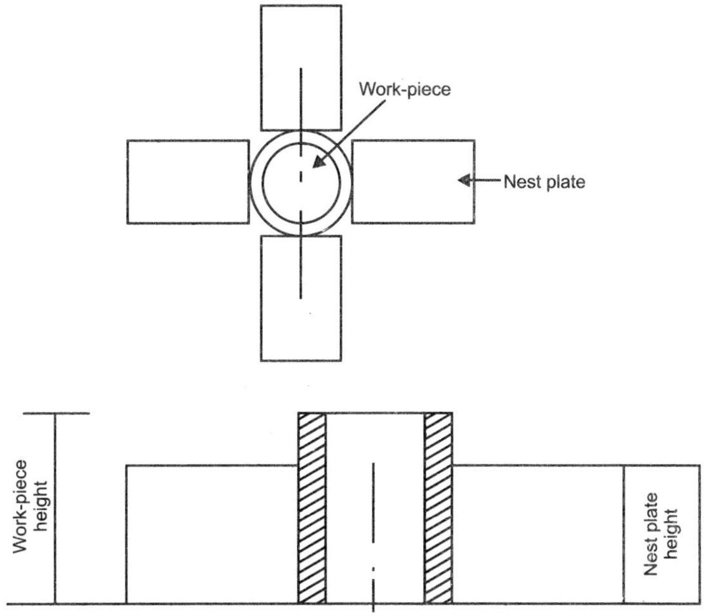

Fig. 14.18 Single Piece Nesting on Surface Grinder

In mass production, the capacity of machines should be utilized to the maximum possible extent. Consequently maximum possible area of magnetic table should be utilized to grind as many workpieces as possible in a single batch.

For round spacers, workpieces are arranged in rows with common supporting nest plates around. The thickness of the nest plate should be lesser than the finish height of the workpieces to prevent obstruction of the grinding wheel. The arrangement is shown in Fig. 14.19.

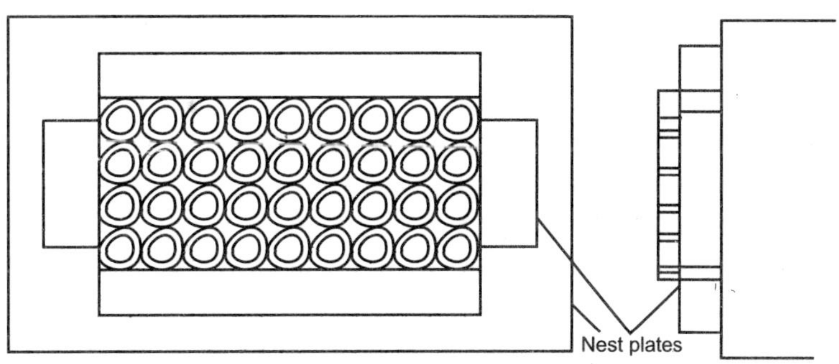

Fig. 14.19 Round Spacers Nesting on Surface Grinder

For odd-shaped workpieces with little variation in size an epoxy resin nest can be used. The nest prevents tilting and sliding of workpieces during grinding operation. The arrangement can be seen in Fig. 14.20.

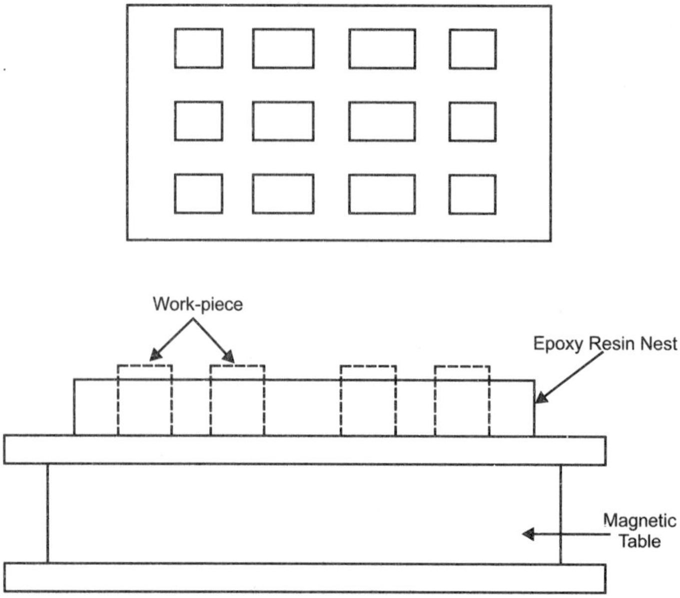

Fig. 14.20 Epoxy Resin Nest for Odd Shaped Parts on Surface Grinder

14.18 TURNING FIXTURES

Holding the workpieces for the lathe operations is successfully achieved with the help of numerous types of equipment available such as chucks, mandrels and collets. However, there are components which can not effectively hold in the above work holding devices or equipment. Such components require the design of turning fixture which can be mounted on face plates with the help of dowels and screws.

The following points require while designing turning fixture.

(a) Grip the rotating workpieces securely to the fixture to with- stand torsional forces.

(b) The fixture should be rigid and over hang should be minimum possible.

(c) Locate the workpiece on critical surfaces from which all major dimensions are taken.

(d) Provide adequate support for frail sections.

(e) Fixtures should be accurately balanced to avoid vibrations at high spindle speeds.

(f) Fixture should be free from projections which are liable to injure operator.

Turning Fixture

The base plate is used to locate and clamp turning fixture. Turning fixtures are provided with clamping studs which are inserted in the clamping holes in base plate and secured by hexagonal nuts. Fixing of the turning fixture with base plate is shown in the Figure 14.21.

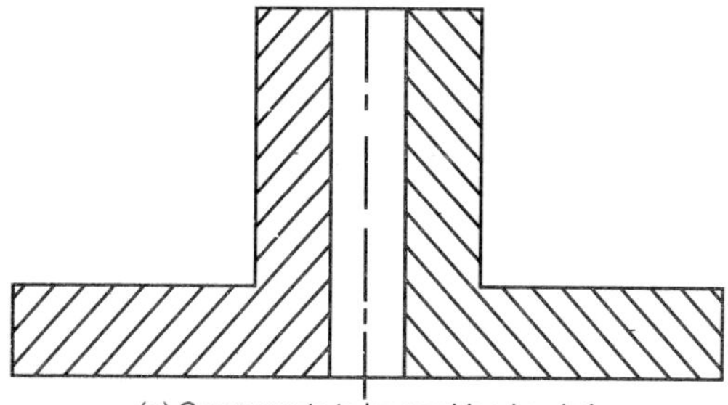

(a) Components to be machined on lathe

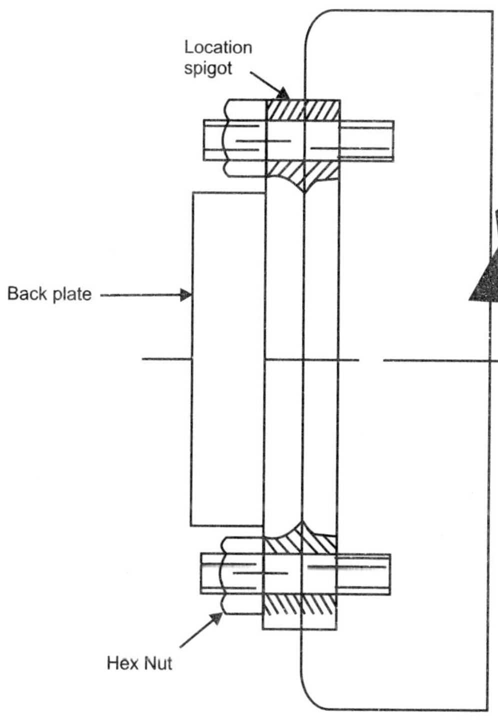

(b) Location of clamping of turning fixture

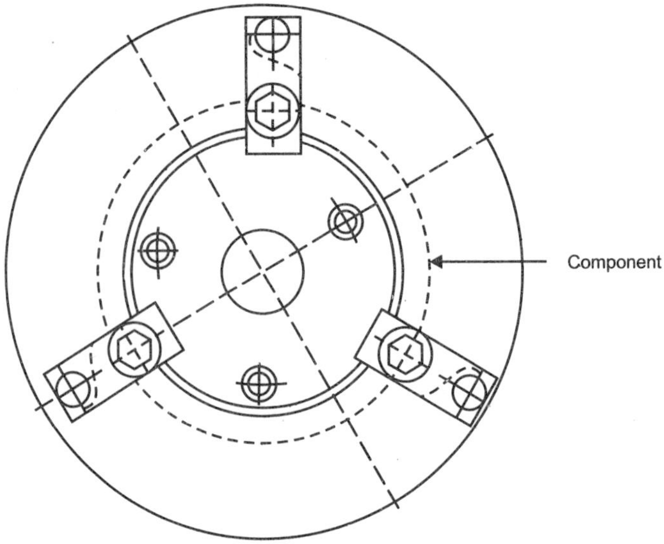

(c) Front view of the fixture

Fig. 14.21 Turning Fixture

QUESTIONS

1. Differentiate between a Jig and Fixture.
2. Explain the principle of 3-2-1 location.
3. List the kind of Locators commonly used and explain each with the help of a sketch.
4. With the help of sketches explain various clamping devices in use.
5. Describe different types of Drill Jig Bushes and their applications.
6. What are the various types of Milling fixtures? Sketch and explain any one.
7. Sketch string milling fixture to make a slot on the top of cylindrical workpiece.
8. Explain grinding fixture on surface grinding machine.
9. Explain points of consideration while designing a turning fixture.
10. Sketch of explain different types of Jaws used on a Lathe.

15

Broaching

15.1 INTRODUCTION

Broaching is a machining operation in which a tool having a series of cutting teeth, called broach is either pulled or pushed by the broaching machine past the surface of a workpiece. In doing so, each tooth of the tool takes a small cut. Most of the cutting is done by the first and intermediate teeth, where as the last few teeth finish the surface to the required size.

15.2 PRINCIPAL PARTS OF A BROACH (See Fig. 15.1)

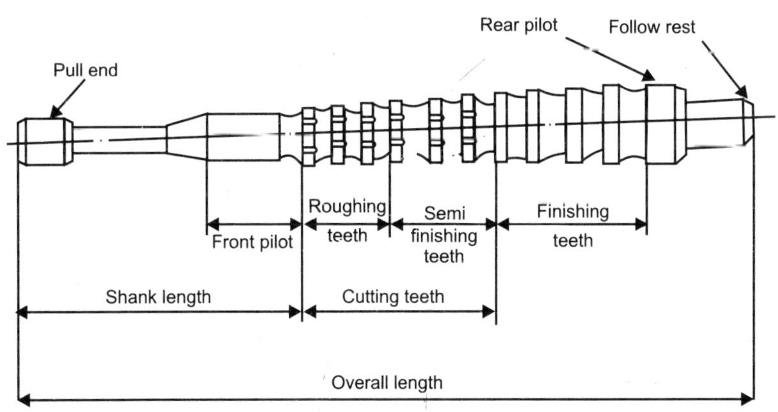

Fig. 15.1 Principal Parts of a Broach

***Pull End*:** This is used to permit engagement of broach with the broaching machine.

***Front Pilot*:** This centres the broach in the hole before the teeth being cut.

***Rough and Semi Finishing Teeth*:** They remove the most of the stock in the hole.

***Finishing Teeth*:** They used for sizing the hole.

***Rear Pilot and Follow Rest*:** They support the broach after the last tooth leaves the hole.

15.3 BROACH CLASSIFICATION

(a) Method of operation

(i) Internal pull type (ii) Internal push type

(iii) Surface type

(b) Method of Construction

(i) Solid type (ii) Built up type

(c) Character of Operation

(i) Hole (ii) Spline

(iii) Serration (iv) Spiral

(v) Surface (vi) Keyway

Besides the above classification, there may be broaches of special applications.

15.4 GEOMETRY OF THE BROACH TEETH (See Fig. 15.2)

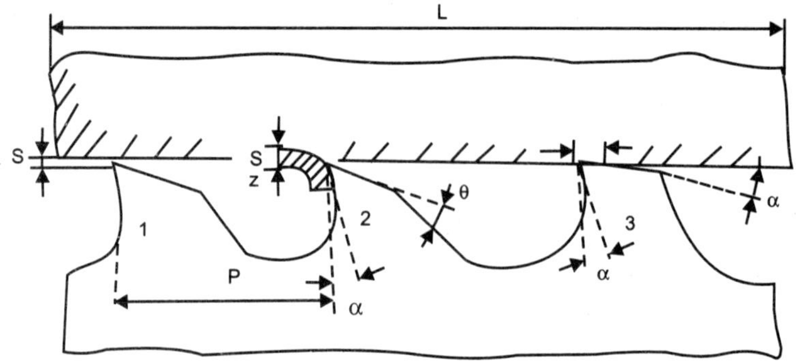

Fig. 15.2 Geometry of the Broach Teeth

P = Pitch

Sz = Feed for tooth-difference between two successive teeth

Tooth: Tooth 1 and 2 are cutting teeth

Tooth 3 is a sizing teeth

Relief angle (Back-off angle θ)

The back off or relief angle is measured in the direction of cutting motion.

It is independent of work material, but ensure good condition of cutting by reducing friction between the tooth flank and machined surface. The relief angle is provided based on the operation.

Relief angle—For roughing operation, it is 3° to 4°.

For semifinishing operation, it is 2° to 3°

For sizing or finishing operation , it is –1° to 2°

Finishing teeth are provided with a narrow wear land (f) 0.05 to 0.2 mm.

The back off angle of internal broaches are made as small as possible from the point of view of the strength of the section.

Rake angle (α)

Rake angle (α) depend on the type of operation and the work material.

15.5 CONSIDERATIONS IN BROACH DESIGN

While designing a broach, considerations should be given to the type of operation, quality of manufacture, feed per tooth, depending upon material, force and power consumption, velocity of approach and tool geometry.

In addition geometry of the workpiece, type of broach, depth to tooth, strength, pitch and length of broach should also be given due considerations.

In practice, design of broach is much more complicated than the design of any other multipoint tool.

The Pull End: It is used to fix the broach to the broaching machine at the machine puller head. The diameter of the pull end is to be 0.5 to 1 mm less than the diameter of the hole to be broached.

The Front Pilot: It guides the broach at the beginning of the cut. The length of the front pilot is made equal to the length of the hole to be broached. Its diameter is made equal to the minimum diameter of the hole in the work-piece before broaching.

Rear Pilot: The purpose of the rear pilot is to guide the broach and maintain proper alignment as it passes out of the workpiece after cutting. Further it prevents any damage of the broach teeth and the finished teeth. The diameter of the rear pilot is made equal to the minimum permissible diameter of the machined hole. The length of the rear pilot should be 0.5 to 0.7 L (Where L is length of hole to be broached). But it should not be less than 20 mm.

Broaching Allowance: It is defined as the total thickness of the material to be removed by broaching is the stock left for machining.

Under normal circumstances, the broaching allowance left on the workpiece to be broached is

$$A = 0.005D + 0.2\sqrt{L}$$

where, D = Diameter of the hole, mm

L = Length of the hole to be broached

In designing a broach, the cut per tooth or thickness of the undeformed chip is to be assigned for successive teeth to fix the incremental rise in the diameter of the broach in the cutting teeth range.

The thickness of the undeformed chip (a) is the cut per tooth depends upon factors like the material being machined, the type of broach, rigidity of machine, tool life and cutting force on the broach.

Table 15.1: Cut per tooth in broaching, mm

Material	Type of Broach		
	Round	Spline	Keyway/Surface
Steel	0.02–0.05	0.04–0.08	0.03–0.20
CI	0.03–0.08	0.04–0.10	0.06–0.15
Brass	0.05–0.10	0.05–0.12	0.06–0.20
Aluminium	0.02–0.05	0.02–0.10	0.05–0.20

Table 15.2: Broadh hook or rake angle

Work Material	Rake angle. deg.	
	Roughing teeth	Finishing teeth
Steel	10–18	5
Cast Iron	10	5
Aluminium, Copper	20	20
Brass, Bronze	5	–10

Number of teeth: The total number of teeth (Roughing and finishing) is found from the formula

$$Z = \frac{A}{2a} + (2 \text{ or } 3)$$

where, A = Broaching allowance to be removed by teeth

a = cut per teeth or thickness of the undeformed chip

Pitch of tooth (P): Pitch P is given by

$$P = 1.25 \text{ to } 1.5\sqrt{L} \quad \text{— For plain Broaches}$$

$$= 1.45 \text{ to } 1.9\sqrt{L} \quad \text{— For progressive cutting broaches}$$

where L = Length of the job to be machined.

Height of the tooth (h) Fig. 15.3

The height 'h' of the tooth is given by

$$h = 1.13\sqrt{KaL}$$

K = Factor depending upon type of broach

= 3 to 5 for Surface Broaches

= 6 to 10 for Internal Broaches

a = Difference in height between two successive cutting teeth and depends on work material

= 0.02 to 0.12 mm

r = (0.2 to 0.3) P

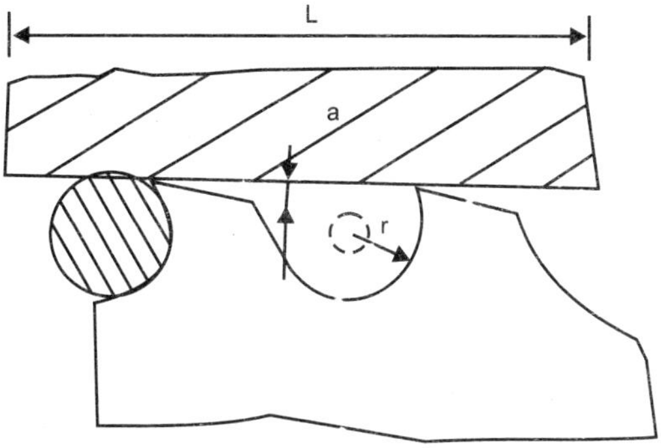

Fig. 15.3 Height of Tooth

The other consideration is that the pitch should be such that it is possible to provide sufficient gullet space for chips. The chips do not fill the entire gullet space.

15.6 CUTTING SPEED AND POWER REQUIREMENTS

The cutting speed in broaching in the speed of the broach movement

$$V = \frac{CK}{T^x S^y}\ m/min$$

where C is coefficient characterizing the process conditions

T = Broach life in minutes

S = Cut per tooth in mm

K = Coefficient depending upon broach material

For Broaching Round holes in steel

C = 12, T = 100 minutes; $x = 0.62$, $y = 0.62$

K = 1 to 1.45 in accordance with broach material

Generally, the value of cutting speed in Broaching Ranges 1-18 m/min.

The power required for Broaching is given by

$$P = \frac{F_2 V}{60 \times 120}\ \text{Kilowatts}$$

The machining time $T_v = \dfrac{hLA}{1000}$

where, h = Machining allowance on each side in mm
L = Length of broached surface in mm
A = Ratio of return stroke to working stroke speed.

15.7 BROACH TOOL MATERIALS

High speed steel is the most commonly used material for making broaches. H.S.S. broaches produce a good surface finish. These are used for broaching Mild Steel and Cast Iron components. Inserted tooth carbide broaches are used for broaching Cast Iron in the automotive industry. Carbide tools produce good surface finish and are capable of working at thrice the speed of high speed steels.

15.8 BROACHING MACHINES

These broaching machines consists of a work holding fixture, a broach, a drive mechanism and a suitable supporting frame. These machines are usually pull or push, broach through or past the workpiece. Most broaching machines are hydraulically operated to secure a smooth uniform actions.

Generally broaching machines are classified as

(a) Horizontal broaching machines
(b) Vertical broaching machines
(c) Continuous broaching machines
    (i) Rotary type
    (ii) Chain type

(a) Horizontal Broaching Machines (See Fig. 15.4)

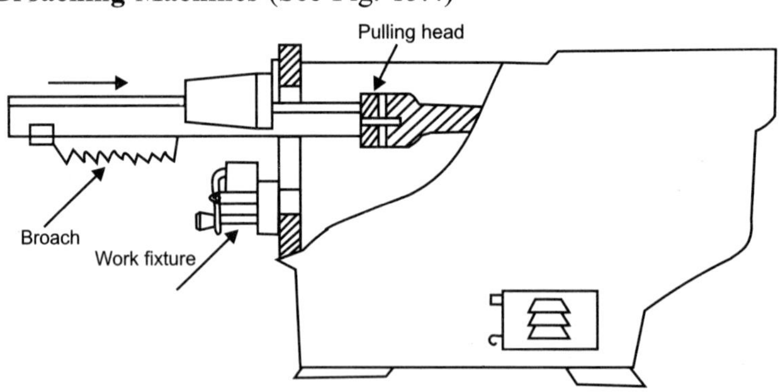

Fig. 15.4 Horizontal Broaching Machine

All horizontal machines are of the pull type. These machines are shown in Fig. 15.4. These may be used either internal or external broaching. These are mostly used for internal work. These machines are used for broaching key ways, splines, slots etc. These machines occupy more floor space. Available with the capacities upto 100 tonnes and stroke upto 9 mm.

(b) Vertical Broaching Machines (see Fig. 15.5)

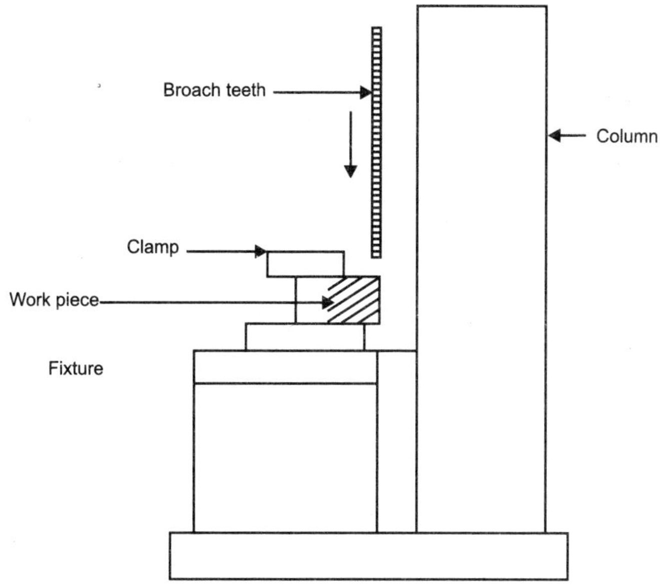

Fig. 15.5 Vertical Broaching Machine

In vertical machines, the travel of the broach is vertical. These machines can be further classified as vertical pull down machines, vertical pull up machines and vertical surface broaching machine.

(c) Continuous Broaching Machines

(*i*) ***Rotary Type Continuous Broaching Machine*** (see Fig. 15.6)

In these machines, the work-pieces are loaded on a series of work holding fixtures which are mounted on rotary table. These machines are used only for surface broaching of small parts.

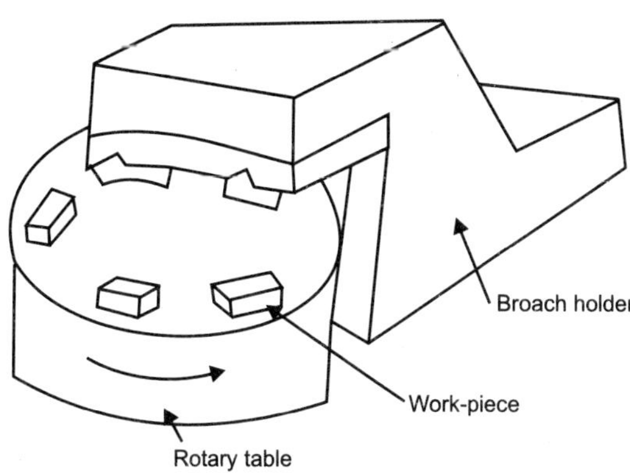

Fig. 15.6 Rotary Type Continuous Broaching Machine

(*ii*) ***Chain Type Continuous Broaching Machine*** (See Fig. 15.7)

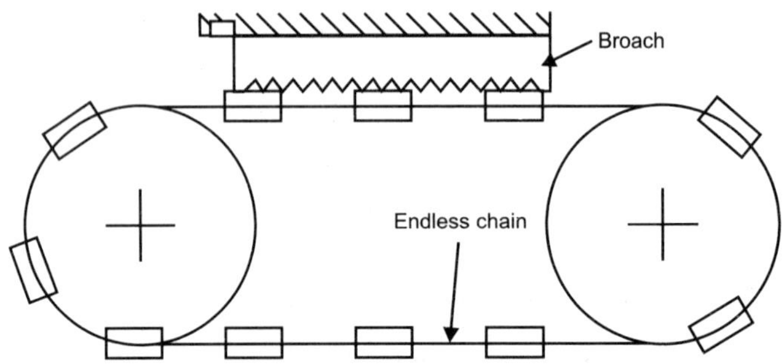

Fig. 15.7 Chain Type Continuous Broaching Machine

These machines have continuous chain travelling in a horizontal position over sprockets. Fixtures for holding the workpieces are mounted on the chain itself at intervals.

15.9 APPLICATIONS OF BROACHING

Broaching is used for small lot jobs as well as mass production. The main application of broaching process lies in machining irregular shaped holes of considerable length. Keyways, straight and spiral splined holes, square, hexagonal and other odd shaps are produced by broaching.

15.10 ADVANTAGES OF BROACHING

(a) Rate of production is very high.
(b) As each tooth of the broach takes a small cut only once in one operation, the broach has a longer life.
(c) The broaching performs both roughing and finishing operation.
(d) Highly skilled operators are not required.
(e) The process can be used for either internal or external surface finishing.

15.11 LIMITATIONS OF BROACHING

(a) A broach is a costly tool.
(b) A surface having obstruction in the way of broach travel cannot be machined.
(c) Large amount of metal cannot be removed.
(d) Parts to be broached must be capable of being rigidly supported and must be able to withstand the forces, during cutting operation.

QUESTIONS

1. Sketch and explain principal parts of a Broach.
2. Describe the various broaching machines used in industry.
3. Explain clearly the advantages and limitations of Broaching.

16

Super Finishing Processes

16.1 INTRODUCTION

If a better finish is desired, for looks, for accuracy or for any other reasons, one of the microfinishes that include lapping, honing, polishing, buffing may be employed.

16.2 LAPPING

It is an abrading process employed for improving the surface finish by reducing roughness, waviness and other irregularities on the surface. Very thin layer of metal (0.005 to 0.01 mm) are removed in lapping.

16.3 ABRASIVES AND LAP MATERIALS

Abrasive powders such as emery, corundum, Iron oxide, Chromium oxide etc. mixed with oil or special pastes with some carrier are used in lapping. Aluminium oxide is preferred for lapping soft ferrous and non-ferrous materials. Silicon and natural corundum are used for hardened steel parts.

Most lapping is done by means of lapping shoes or quills, called 'laps' that are rubbed against the work. The face of a lap charged with abrasive particles. Charging a lap means embedded the abrasive particles into its surface.

Laps are made of soft Cast Iron, Brass, Copper, Lead or soft material. Cast Iron is the best lap material.

16.4 LAPPING METHODS AND MACHINES

Lapping is done in the following two ways:

1. By hand — Called Hand Lapping
2. By machine — Called Machine Lapping

16.4.1 Hand Lapping

Lapping is done by hand held tools for both flat work and external cylindrical work.

(a) Flat work hand lapping

(b) External cylindrical hand lapping (Ring lapping)

(a) Flat Work Hand Lapping (see Fig. 16.1)

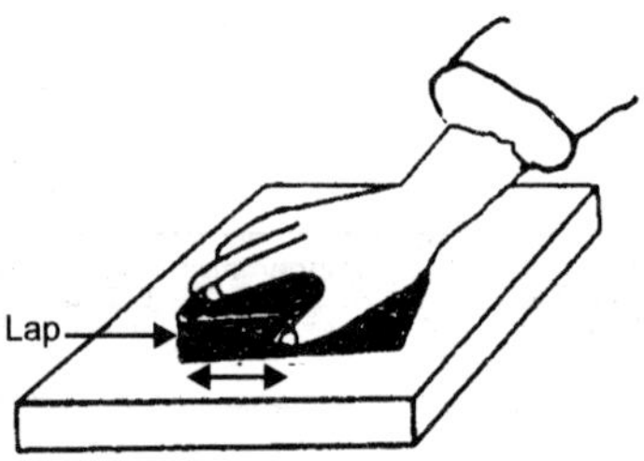

Fig. 16.1 Hand Lapping Method

In hand lapping, either the lap or the work-piece is held by hand and the motion of the other, enables the rubbing of the two surfaces in contact. This method is widely used in lapping dies and moulds for casting and limit gauges etc.

(b) External Cylindrical Hand Lapping (Ring Lapping) (see Fig. 16.2)

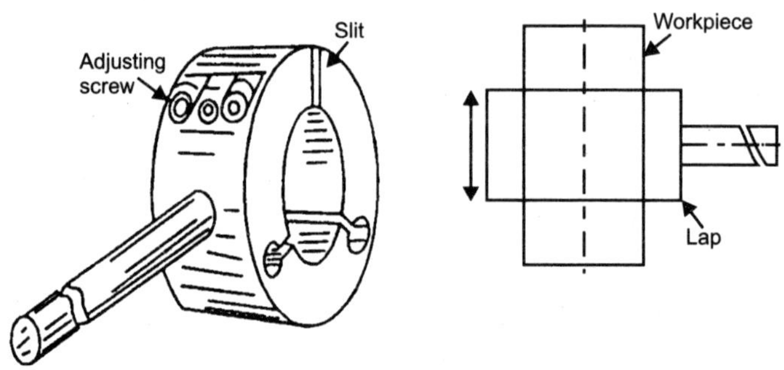

Fig. 16.2 Adjustable Ring Lap

Ring lapping is done for finishing external cylindrical surfaces. These laps are made of cast Iron. The ring lap has slots through. Screws are provided for precise adjustment. The size of the ring lap should be slightly shorter than the work. The ring lap reciprocates over the work-piece surface. The abrasive and vehicle are fed through the slot.

16.4.2 Machine Lapping

This is performed for obtaining highly finished surface on races of ball and roller bearings, crank shafts, various automobile parts like spray nozzle, injector pump parts etc. Many different types of machines are used.

(a) Vertical spindle lapping machine
(b) Abrasive belt lapping machine
(c) Centreless lapping machine
(d) Spherical lapping machine.

(a)Vertical Spindle Lapping Machine (See Fig. 16.3)

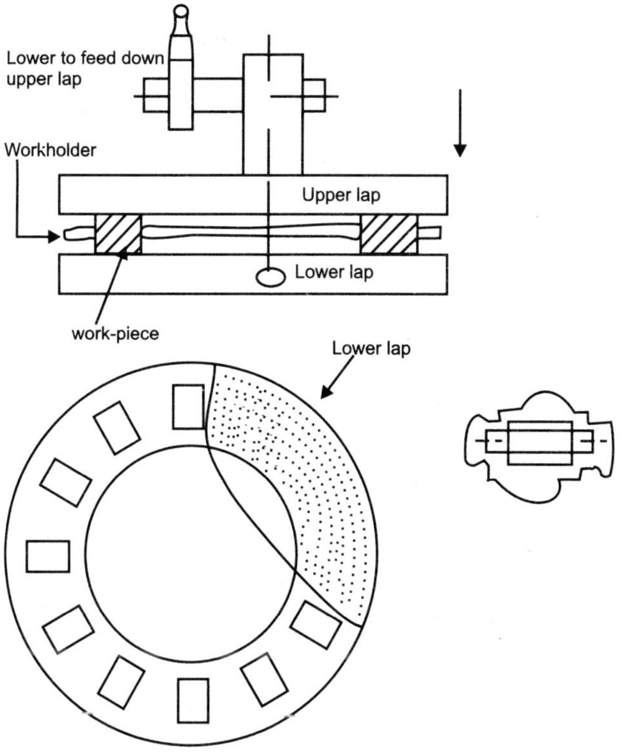

Fig. 16.3 Vertical Spindle Lapping Machine

The vertical spindle lapping machine laps flat or round surfaces between two opposed laps on vertical spindles. The upper lap is free to float and rest upon the work which rides upon the face of the lower lap. Pressure is applied by gravity. The work is held loosely in a work guide or holder.

(b) Abrasive Belt Lapping Machine

In this machine a continuous moving belt with an abrasive is used for lapping. These are employed for lapping crank shaft and pints etc.

(*c*) ***Centreless Lapping Machine*** (see Fig. 16.4)

This machine is similar to centreless grinding machine except that extra-long grinding wheel and regulating wheels are used to allow the work-piece to remain in abrading contact for a longer time.

The spindles of lapping wheel and regulating wheel are swivelled in the vertical plane and are not parallel. Due to this work-piece comes in contact with the wheel. This machine is used to lap piston pins, shafts and bearing races.

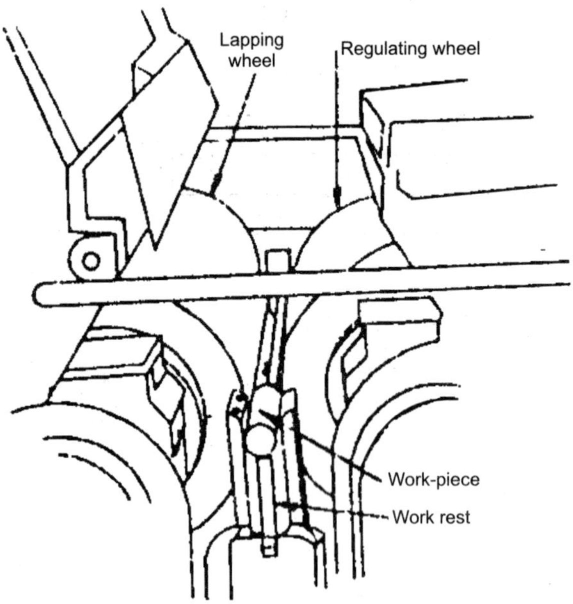

Fig. 16.4 Centreless Lapping Machine

(*d*) ***Spherical Lapping Machine*** (see Fig. 16.5)

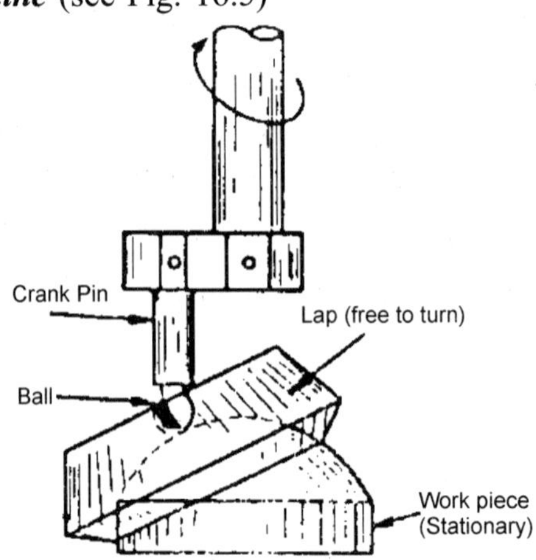

Fig. 16.5 Spherical Lapping

Spherical surfaces are lapped on a machine similar to drill press. A cast Iron Lap is used which is counterpart of the work surface to be lapped. A crank is held in the spindle and crankpin is provided with a ball that enters freely into a blind hole in the back of the lap. The work-piece axis is aligned with spindle axis and the spindle is then rotated which gyrates the lap.

16.5 HONING

It is also an abrading process used for finishing previously machined surfaces. It is mostly used for finishing internal cylindrical surfaces such as drilled or bored holes. The tool used called a "hone". Honing removes maximum stock out of all surface finishing operations (0.1mm to 0.75 mm).

16.6 MATERIAL OF HONING STONES

Honing stones are made from common abrasive and bonding materials often impregnated with sulphur, resin or wax to improve cutting action and for long tool life. The various abrasives used to make honing stones are silicon carbide, aluminium oxide, diamond or cubic nitride. Silicon carbide is used for honing cast iron and non ferrous materials, whereas aluminium oxide is used to hone steel parts. Diamond is also used as an abrasive to hone parts of ceramics or hard carbides. The abrasive grain size ranges from 80 to 600 grit. Mostly honing is done on internal cylindrical surfaces such as automobile cylinder walls.

16.7 HONING PROCESS

The honing tool works more or less in the same way as an expanding reamer.

It is a 'wet' process and it is necessary to use coolant in ample quantity during the operation.

16.8 FIXTURE FOR HONING TOOLS

The honing stones may be loosely held in holders cemented into metal shells which are clamped into the holders, or cemented directly on holders. Some stones are spaced at regular intervals around the holder, while others are interlocked so as to present continuous surface to bore.

The stones may be expanded against the bore by

(a) Mechanical type
(b) Hydraulic type

(a) Method of Applying Pressure in Mechanical Type

(i) Caged spring: Spring is compressed as the tool enters the bore. Pressure of compressed spring pushes the connecting rod and expands the tool through a cone (see Fig. 16.6).
(ii) The movement of the cone is controlled by threads that move the cone rod in relation to drive shaft.

(b) Hydraulic Type

In this, sizing arrangements are provided which contact the hone when the correct size has been reached.

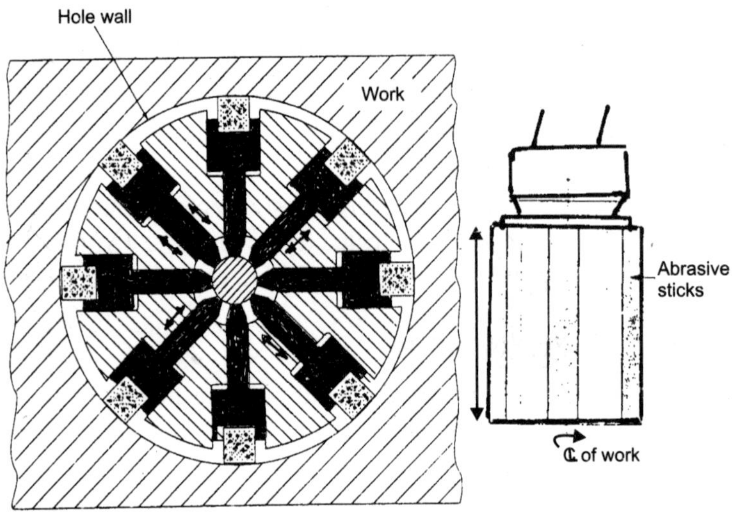

Fig. 16.6 Honing Tool

16.9 HONING MACHINES

Honing is done in the following ways

(a) Manual Honing

(b) Machine Honing

(a) Manual Honing

When honing is done manually the tool is rotated and the work-piece is passed back and forth over the tool. For precision honing, the tool is given a slow reciprocating motion as it rotates.

(b) Machine Honing

Honing is done on general purpose machine tools. These machines are classified as

(i) Horizontal Honing Machines

(ii) Vertical Honing Machines

(b)(i) Horizontal Honing Machines

In some machines, the work is held in horizontal position and rotated about its own axis. The honing tool is rotated and reciprocates. Horizontal honing machines are used for honing long gun barrels and large bores.

(b)(ii) Vertical Honing Machines (see Fig. 16.7)

A general arrangement of a vertical honing machine is shown in Fig. 16.7. The honing tool will follow the axis of the hole of the work-piece, therefore the honing tool or fixture must be free to float. This is done by using universal joint as shown in figure. Due to this, the honing tool becomes selfcentering and it is not necessary to line up the hole and hone axes precisely. These machines are best suited for shorter jobs. Vertical machines are designed for work up to 500 mm diameter. In appearance, these machines resemble the drilling machine.

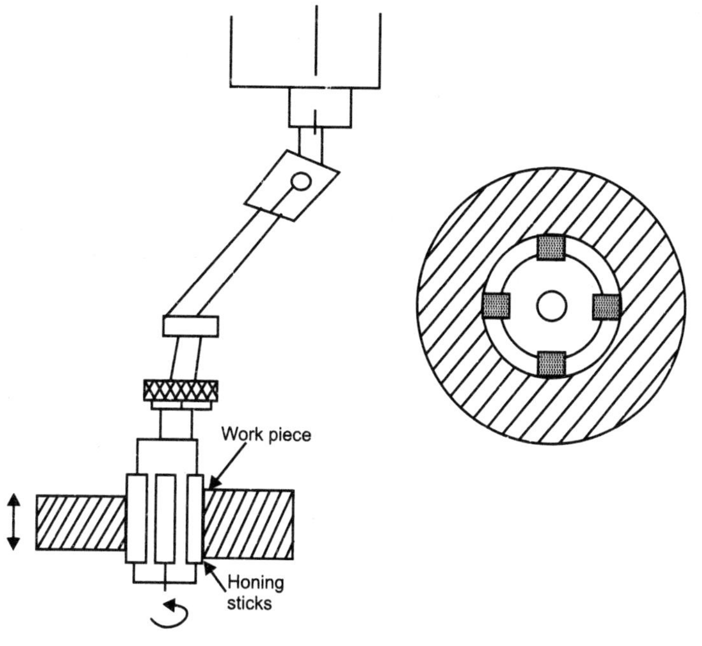

Fig. 16.7 Vertical Honing Machine

16.10 ADVANTAGES OF HONING PROCESS

(a) This process produces highly accurate holes.

(b) Many holes can be honed simultaneously on multiple spindle machines.

(c) Hole of any dimension can be honed.

16.11 DISADVANTAGES OF HONING PROCESS

(a) It is not possible to improve lack of straightness in holes.

(b) It is difficult to machine tough non-ferrous metals.

16.12 POLISHING

Polishing is an intermediate abrading operation which follows grinding and precedes buffing. In polishing operation, the smoothness on a surface is produced by cutting action of abrasive particles adhering to the surface of resilient wheel. In polishing deep scratches, nicks, discoloration and other surface imperfections occurring due to grinding are removed.

The polishing processes uses abrasive grains which are firmly attached to flexible belt or flexible wheel. Coated abrasive belts become production, cutting tool for polishing of metals. The polishing wheels are generally made of canvas, felt of leather.

The abrasive used for polishing are aluminium oxide or silicon carbide. Aluminium oxide is used for most carbon steel, alloy steel, high speed steel and non-ferrous metals. Silicon carbide is recommended for finishing low tensile strength materials such as brass, copper, cast iron and aluminium. Natural abrasives like emery and corundum are sometimes used in specialised operations for getting the finest quality finishes. The bonding agent used on polishing wheel is either hot glue or cold silicate based cement.

16.13 BUFFING

It is the smoothing and brightening process of a surface by the rubbing action of fine abrasive in a lubricating binder applied intermittently to a moving wheel of wood, cotton, fabric, felt or a cloth or a felt belt. Buffing wheels are made more or less firm by the amount of stiching used to fasten the layers of the cloth together.

Buffing can be divided into two operations—cutting down and coloring. Cutting down is done to refine a surface by removing scratch lines from polishing, die marks or other imperfections. Coloring refines the cut down surface and produces a high finish or luster.

16.13.1 Selection of Buffing Wheels

Wheels are selected by considering the following:

(a) Select the wheel to give correct surface metre pre minute when operated.
(b) Select the wheel of largest practical diameter.
(c) Select wheel face slightly wider than the work.
(d) Select hardness of wheel and wheel material based on the work to be buffed.

16.13.2 Method of Buffing

Buffing can be done manually or automatically. The manual buffing machine uses a double ended shaft having wheels on both ends. If buffing is done manually, it becomes expensive due to the labour cost. In the automatic machine, conveyors and workholders are incorporated. Buffing wheel speeds are 30 to 40 m/s.

16.13.3 Applications

Buffing process produce mirror-like finish. Objects used in automobiles, motor-cycles, boats, bicycles, and household utensils and appliances.

16.14 SUPER FINISHING (See Fig. 16.8)

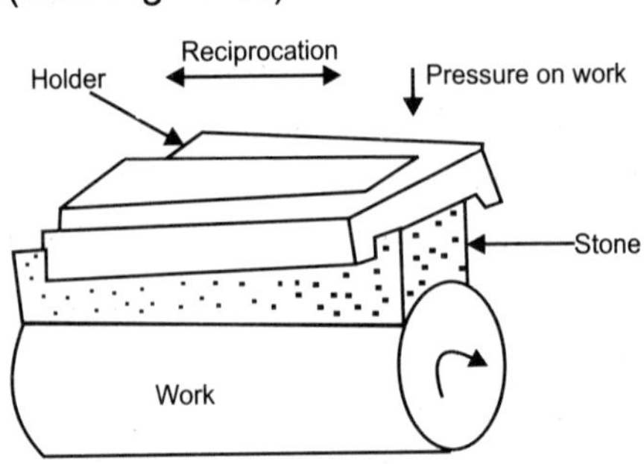

Fig. 16.8 Super-Finishing

Super finishing is a microfinishing process that produces a controlled surface condition on parts which is not obtainable by any other method. It consists in scrubbing a stoned against a surface to produce a fine quality of metal finish. Super finish is mainly used for removing chattering marks, feed spirals and other imperfections left by grinding.

The method is performed by rapidly reciprocating a fine grit stone with a soft bond and pressing it against a revolving cylindrical workpiece.

16.14.1 Applications

Bearing surfaces, automotive cylinders, piston, clutch plates, guide pins, computer memory drums, brake drums etc.

(a) Equalising lapping

When work and lap mutually improve their shape and surface, for example : When gears are run together with some abrasive, or tapered valves are seated in seats.

(b) Form Lapping

Shape of lap is imparted to work.

(c) Advantages of lapping

(i) Increase the work life.
(ii) Provides superfine surface finish.
(iii) Provides liquid and gas tight seals without using gaskets.
(iv) Removes errors in gears, thereby reduces noise and wear.

QUESTIONS

1. Briefly explain the process of lapping.
2. What are the abrasives used for lapping operation?
3. What is having and explain various machines used?
4. What are the advantages and limitations of honing?
5. Write short notes on:
   (a) Lapping (b) Honing
   (c) Polishing (d) Buffing
   (e) Super finishing operation.

Appendix II
Objective Type Questions

1. In orthogonal cutting of metals
   (a) The cutting edge of the tool is perpendicular to the direction of tool travel
   (b) The cutting edge is inclined to the workpiece
   (c) The cutting edge is wider
   (d) All of the above
2. In oblique cutting, of metals
   (a) The cutting edge of the tool is perpendicular to the direction of tool
   (b) Parallel to the direction of tool travel
   (c) Inclined at an angle
   (d) Perpendicular to the workpiece
3. Chips produced when machining cast iron
   (a) Continuous chips
   (b) Discontinuous chips
   (c) Continuous chips with built up edge
   (d) None of the above
4. While machining ductile material, the chips produced are:
   (a) Continuous chips
   (b) Continuous chips with BUE
   (c) Discontinuous chips
   (d) All of the above
5. The factors responsible for the production of continuous chip with BUE
   (a) Low cutting speed and large rake angle
   (b) Low cutting speed and small rake angle
   (c) High cutting speed and large rake angle
   (d) None of the above
6. Discontinuous chips are produced while machining–
   (a) Brittle material (b) Ductile material
   (c) Hand material (d) Soft material
7. Continuous chips are produced while machining speed is–
   (a) High (b) Low
   (c) Medium (d) Design value
8. Segmental chips are produced while machining–
   (a) Mild steel (b) High speed steel
   (c) Tungsten carbide (d) Cast iron
9. The type of tool used on a Lathe
   (a) Single point tool
   (b) Two point cutting tool
   (c) Three point cutting tool
   (d) Multi point cutting tool

10. Drill bit belongs to
    (a) Single point utility tool
    (b) Two point cutting tool
    (c) Three point cutting tool
    (d) Multipoint cutting tool
11. Grinding wheel belongs to
    (a) Two point cutting tool
    (b) Single point cutting tool
    (c) Multipoint cutting tool
    (d) Three point cutting tool
12. In metal cutting, chips break due to
    (a) Plasticity (b) Ductility
    (c) Toughness (d) Work hardening
13. In Metal cutting, the shear angle is the angle made by shear plane with
    (a) Direction of tool travel
    (b) Direction of tool axis
    (c) Central plane of work-piece
    (d) Perpendicular to tool axis
14. While machining Mild Steel, the chips produced are–
    (a) Segmental chips
    (b) Discontinuous chips
    (c) Continuous chips
    (d) Continuous chips with BUE
15. The velocity of the tool relative to workpiece
    (a) Cutting velocity (b) Shear velocity
    (c) Chip velocity (d) Mean velocity
16. The velocity of chip relative to tool
    (a) Cutting velocity
    (b) Shear velocity
    (c) Mean velocity
    (d) Chip flow velocity
17. The strength of the tool depends on
    (a) Rake angle (b) Cutting angle
    (c) Clearance angle (d) Lip angle
18. The negative rake angle is provided on
    (a) High carbon steel tools
    (b) High speed steel tools
    (c) Cemented carbide tools
    (d) All of the above
19. Chip Breaker is used to
    (a) Increase tool life
    (b) Break the chips
    (c) Remove chips
    (d) To minimize heat generation
20. Tool life equation is
    (a) $VT^n = c$ (b) $V^nT = c$
    (c) $V^n/T = c$ (d) $V/T^n = c$
21. Machinability depends on
    (a) Physical and Mechanical properties of workpiece
    (b) Cutting force
    (c) Type of chip
    (d) Tool life
22. Lathe is used for Machining
    (a) Flat surfaces
    (b) Convex surfaces
    (c) Cylindrical surfaces
23. Lead screw of a Lathe is used for
    (a) Thread cutting
    (b) Taper turning
    (c) Cylindrical surfaces
24. Lathe bed is made of
    (a) M.S. (b) Cast Iron
    (c) Tool Steel (d) Ceramics
25. Compound rest is used in
    (a) Turning short taper
    (b) Turning Long tapers
    (c) Chamfering the Job
26. In a lathe the carriage and tailstock are guided on
    (a) Same guide ways
    (b) Different guide ways
    (c) Not guided on guide ways

27. Lathe spindle got
    (a) Internal threads
    (b) Taper threads
    (c) External threads
28. The lathe centers are provided with standard taper known as–
    (a) Morse taper (b) Seller's taper
    (c) Metric taper (d) Chapman taper
29. The angle between lathe centers is
    (a) 30° (b) 45°
    (c) 60° (d) 90°
30. The taper on the Lathe spindle is
    (a) 1 : 10 (b) 1 : 12
    (c) 1 : 15 (d) 1 : 20
31. The chuck used for setting heavy and irregular shaped work is
    (a) Magnetic chuck (b) Four Jaw chuck
    (c) Three Jaw chuck (d) Drill chuck
32. The easiest way of centering a cylindrical job on a lathe is to use
    (a) 4-Jaw chuck
    (b) Face plate
    (c) Self centering chuck
33. The tail stock set over method is used for taper turning of
    (a) Internal tapers
    (b) Long slander tapers
    (c) Small tapers
    (d) Steep taper
34. The spindle speeds are in
    (a) Arithmetical progression
    (b) Geometrical progression
    (c) Harmonic progression
35. Face plate is used to hold
    (a) Odd jobs
    (b) Cylindrical jobs
    (c) Pre drilled Jobs
36. The Mandrel is used for holding
    (a) Cylindrical jobs
    (b) Bar stock
    (c) Jobs with predrilled holes or bored jobs
37. Cutting speed is expressed in
    (a) mm/sec. (b) m/min
    (c) mm/min
38. Feed is expressed in
    (a) mm/sec (b) m/min
    (c) both
39. Collets are used for holding
    (a) Bar stock of various shapes
    (b) Irregular
    (c) Cylindrical
40. When the tool moves parallel to the lathe axis, movement is turned as
    (a) Cross feed
    (b) Angular feed
    (c) Longitudinal feed
41. The chamfering is essential operation after
    (a) Knurling (b) Thread cutting
    (c) Boring (d) Rough turning
42. The average cutting speed for turning M. S with HSS tool
    (a) 15 to 20 m/min (b) 25 to 30 m/min
    (c) 60 to 100 m/min (d) 120 m/min
43. The cutting speed is maximum while machine with HSS tool is
    (a) Cast Iron (b) M.S.
    (c) Brass (d) Aluminium
44. Drilling machine is used for
    (a) Originating hole (b) Boring
    (c) Slotting
45. In drilling operation, the metal is removed by
    (a) Shearing (b) Extrusion
    (c) Shearing and extrusion
    (d) Shearing and compression

46. Size of the lathe is specified by
    (a) Weight of the lathe
    (b) Swing over the bed
    (c) Diameter of the Job
    (d) Height of centers from ground
47. Steady Rest is
    (a) Support for long jobs
    (b) Used while turning symmetrical jobs
    (c) Workpiece subjected to vibration
48. Determine the angle at which compound rest swivel when D = 45; d= 30; L = 200
    (a) 5° (b) 10°
    (c) 2° 9′ (d) 3° 5′
49. Follower rest is mounted on
    (a) Lathe bed guidways
    (b) On the saddle of carriage
    (c) On the tailstock
50. The purpose of tumbler gears in lathe is
    (a) Cut gears
    (b) Cut threads
    (c) Reduce spindle speeds
    (d) Give direction of movement to the lathe carriage
51. The slowest speed in lathe is used for
    (a) Normal turning
    (b) Taper turning
    (c) Turning big diameters
    (d) Thread cutting
52. In a lathe center height is 15 cm, then swing over bed is
    (a) 30 cm (b) 7.5 cm
    (c) 15 cm (d) 45 cm
53. Capstan and turret lathes are used for
    (a) Make small components
    (b) Mass production
    (c) Large components
    (d) For ordinary work
54. In the capstan and turret lathes, parting tool is fixed
    (a) In collet (b) Turret
    (c) Rear tool post (d) Chuck
55. In capstan of turret lathes, threads are cut by
    (a) Thread cutting tool
    (b) Die head and taps
    (c) Special tool
56. In capstan lathe, the turret is mounted on
    (a) A short slide of ram sliding on the saddle
    (b) Compound rest
    (c) Back tool post
57. In the head of turret of a capstan lathe, no. of tools to be fixed
    (a) Single tool only
    (b) Any number of tools or tool holders
    (c) Sin tools
58. On bar type turret lathe, work is held by
    (a) Three jaw chuck (b) Four jaw chuck
    (c) Pneumatic chuck (d) Collet
59. The drill spindle is provide taper as
    (a) More taper
    (b) Seller's taper
    (c) Chapman taper
    (d) Brown and Shaper taper
60. Lip angle of the drill
    (a) 120° (b) 118°
    (c) 130°
61. Drill chucks are used for holding
    (a) Taper shank drills
    (b) Straight shank drills
    (c) Square shank drills
62. In Gang drilling machine, the spindles are driven by
    (a) Individual motor (b) A single motor
    (c) Universal motor
63. A twist drill is
    (a) Side cutting tool (b) Front cutting tool
    (c) End cutting tool

64. The operation of accurate way of sizing and finishing is
   (a) Drilling (b) Reaming
   (c) Boring (d) Tapping
65. The operation of making cone-shaped enlargement of hole is
   (a) Counter-sinking (b) Counter-boring
   (c) Trepaning
66. The operation of smoothing and squaring the surface around hole is
   (a) Counter-sinking (b) Counter-boring
   (c) Spot facing
67. The helix angle of a drill bit
   (a) 20° (b) 30°
   (c) 45° (d) 60°
68. Milling cutter is mounted on
   (a) Shaft (b) Arbor
   (c) Dividing head
69. Down milling is also called
   (a) Climb milling (b) End milling
   (c) Face milling
70. Up milling is also called
   (a) Climb milling
   (b) Conventional milling
   (c) Face milling
71. Gears are produced on Mass production by
   (a) Shaping (b) Casting
   (c) Hobbing (d) Milling
72. UDH in milling machine for
   (a) Plain indexing
   (b) Differential indexing
   (c) Direct indexing
73. In which milling, the thickness is minimum at the beginning and reaches maximum at the end
   (a) Up Milling (b) Down Milling
   (c) Face milling
74. Any number of equal divisions can be made on milling machine by
   (a) Simple indexing (b) Direct indexing
   (c) Differential indexing
75. Dovetail milling cutter is a
   (a) Plain milling cutter
   (b) End Milling cutter
   (c) Side milling cutter
76. Helical gears can cut on
   (a) Vertical milling M/c
   (b) Horizontal milling M/c
   (c) Universal milling M/c
77. Shaping machine is used for
   (a) Cutting gears
   (b) Machining flat surfaces
   (c) Cylindrical surfaces
78. Cutting motion in a shaper is obtained by
   (a) Downward motion of tool
   (b) Reciprocating of the ram
   (c) Horizontal movement of table
79. Quick return motion is obtained by
   (a) Crank and slotted lever mechanism
   (b) Gear train
   (c) Crank and connecting rod mechanism
80. In slotting machine the ram moves
   (a) Horizontal (b) Vertical
   (c) Both Vertical and Horizontal
81. In a planner, while machining
   (a) The job is stationary
   (b) Tool is stationary
   (c) Both alternatively change in motion
82. Grinding operation is used for
   (a) Forming (b) Shaping
   (c) Finishing (d) Dressing
83. The highest cutting speed used in
   (a) Surface grinding machine
   (b) Centreless grinding machine
   (c) Internal grinding machine
   (d) Cylindrical grinding machine

84. For grinding flat surfaces use
   (a) Surface grinding machine
   (b) Internal grinding machine
   (c) Cylindrical grinding machine
85. The workpiece advanced in centreless grinding due to
   (a) Machine drive
   (b) Force exerted by grinding wheel
   (c) Force exerted by regulating wheel
86. Artificial abrasives are
   (a) Sandstone, Diamond
   (b) Silicon carbide, Aluminium oxide
   (c) Corundum
87. Majority of grinding wheels uses
   (a) Silicate bond (b) Rubber bond
   (c) Vitrified bond
88. For soft material, the grain of abrasives used is
   (a) Coarse grains (b) Fine grains
   (c) Medium grains
89. For hard material, the grains of abrasives used as
   (a) Coarse grains (b) Fine grains
   (c) Medium grains
90. Surface speed of the grinding wheel in centreless grinding is
   (a) 1500–1800 (b) 10000–1500
   (c) 100–500
91. Truing of grinding wheel is done by
   (a) Balancing the wheel
   (b) Dressing the wheel
   (c) Glazing the wheel
92. A Jig is defined as
   (a) Holds and locates the workpiece and also guide the tool
   (b) Used to check the accuracy of workpiece
   (c) Hold and locates the workpiece
93. Fixture is defined as
   (a) Holds and locates and guide the tool
   (b) Used to check accuracy of workpiece
   (c) Holds and locates the workpiece
94. If the diameter of hole is subjected to variation, then for locating. What type of locator is used?
   (a) Conical locator
   (b) Cylindrical locator
   (c) Diamond pin locator
95. A workpiece in space to move in any direction can have
   (a) 3 degrees of freedom
   (b) 12 degrees of freedom
   (c) 6 degrees of freedom
96. A process of removing metal by pushing or pulling a cutting tool is called
   (a) Upmilling (b) Forming
   (c) Broaching
97. Ball bearing races are
   (a) Lapped (b) Buffing
   (c) Moving
98. Least Material is removed by
   (a) Grinding (b) Lapping
   (c) Super finishing
99. Which process used for finishing cylindrical holes?
   (a) Lapping (b) Honing
   (c) Polishing
100. While abrasive particles hold in the form of sticks in Honing process
   (a) Al_2O_3 (b) Diamond
   (c) Quartz

ANSWERS

1. (a)	2. (c)	3. (b)	4. (a)	5. (b)	6. (a)	7. (a)	8. (d)
9. (a)	10. (b)	11. (c)	12. (d)	13. (a)	14. (c)	15. (a)	16. (d)
17. (a)	18. (c)	19. (b)	20. (a)	21. (a)	22. (c)	23. (a)	24. (b)
25. (a)	26. (b)	27. (c)	28. (a)	29. (c)	30. (d)	31. (b)	32. (c)
33. (b)	34. (b)	35. (a)	36. (c)	37. (b)	38. (a)	39. (a)	40. (c)
41. (a)	42. (b)	43. (d)	44. (a)	45. (c)	46. (b)	47. (a)	48. (c)
49. (b)	50. (d)	51. (d)	52. (a)	53. (b)	54. (c)	55. (b)	56. (a)
57. (b)	58. (c)	59. (a)	60. (c)	61. (b)	62. (a)	63. (c)	64. (b)
65. (a)	66. (c)	67. (b)	68. (b)	69. (a)	70. (b)	71. (c)	72. (b)
73. (a)	74. (c)	75. (b)	76. (c)	77. (b)	78. (b)	79. (a)	80. (b)
81. (b)	82. (c)	83. (d)	84. (a)	85. (c)	86. (b)	87. (c)	88. (a)
89. (b)	90. (a)	91. (b)	92. (a)	93. (c)	94. (a)	95. (b)	96. (c)
97. (a)	98. (c)	99. (b)	100. (a)				

T

U

V

W